THEORETICAL BIOLOGY OF THE CELL

A Dynamical-Systems Perspective

HIROAKI TAKAGI
Nara Medical University

CHIKARA FURUSAWA
The University of Tokyo

SATOSHI SAWAI
The University of Tokyo

KUNIHIKO KANEKO
University of Copenhagen

CAMBRIDGE
UNIVERSITY PRESS

Shaftesbury Road, Cambridge CB2 8EA, United Kingdom

One Liberty Plaza, 20th Floor, New York, NY 10006, USA

477 Williamstown Road, Port Melbourne, VIC 3207, Australia

314–321, 3rd Floor, Plot 3, Splendor Forum, Jasola District Centre,
New Delhi – 110025, India

103 Penang Road, #05-06/07, Visioncrest Commercial, Singapore 238467

Cambridge University Press is part of Cambridge University Press & Assessment,
a department of the University of Cambridge.

We share the University's mission to contribute to society through the pursuit of
education, learning and research at the highest international levels of excellence.

www.cambridge.org
Information on this title: www.cambridge.org/9781009397841

DOI: 10.1017/9781009397834

When citing this work, please include a reference to the DOI 10.1017/9781009397834

First published 2025

A catalogue record for this publication is available from the British Library

A Cataloging-in-Publication data record for this book is available from the Library of Congress

ISBN 978-1-009-39784-1 Paperback

THEORETICAL BIOLOGY OF THE CELL

To understand life phenomena, we must consider form, structure, organization, motion, and the roles they play in "living" functions. This book explores such elements through mathematical methods. Beginning with an overview of dynamical systems and stochastic processes, the chapters that follow build on experimental advances in quantitative data in cellular processes to demonstrate the applications of these mathematical methods to characterize living organisms. The topics covered include not only cellular motions but also temporal changes in metabolic components, protein levels, membrane potentials, cell types, and multicellular patterns, which are linked to functions such as cellular responses, adaptation, and morphogenesis. This book is intended for undergraduates, graduates, and researchers interested in theory and modeling in biology, in particular cell, developmental, and systems biology. It is also intended for those in the fields of mathematics and physics who are interested in these topics.

HIROAKI TAKAGI is Senior Assistant Professor of Physics at Nara Medical University, and a core member of Q-bioJP (Japanese Society for Quantitative Biology). His expertise lies in biophysics and nonlinear dynamics. He is interested in the functional significance of fluctuations and dynamics at molecular, cellular, and multicellular levels. Through analyses combining quantitative biological data and mathematical models, his research seeks to elucidate their mechanisms.

CHIKARA FURUSAWA is a professor at the University of Tokyo and a team leader at RIKEN (Rikagaku Kenkyūsho [Institute of Physical and Chemical Research]). His research aims to uncover universal characteristics in biological systems by combining theoretical analysis with high-throughput experimental approaches. His primary areas of study include evolutionary dynamics, cell differentiation, and metabolic dynamics, among others.

SATOSHI SAWAI is a professor at the University of Tokyo. He has been fascinated with the slime mold *Dictyostelium* for more than two decades, considering it a hidden gem filled with mathematical ideas and concepts, which are covered in this book.

KUNIHIKO KANEKO has been a professor at the University of Tokyo for 27 years, teaching mathematical biology and nonlinear dynamics, and he is currently at the Niels Bohr Institute, University of Copenhagen. He was also Stanislaw Ulam Fellow at Los Alamos National Laboratory, visiting professor at Osaka University (Frontier Biosciences), University of Lyon, and Freiburg University, and was the Founding Director of the Research Center for Complex Systems Biology and the Universal Biology Institute at the University of Tokyo.

Contents

Preface

Scope of the Book

How can we describe and understand living state? By focusing on dynamical aspects of life, we present basic mathematical methodology for it and apply it mainly to cellular and multicellular phenomena to unveil the logic of life. The methodology covers dynamical systems theory and stochastic processes, whereas the topics include responses, rhythm, excitation, and fluctuations of cells, multicellular pattern formation, cell differentiation and morphogenesis, origin of life, and biological information.

When we try to understand the nature of life phenomena, various aspects come to mind: form, structure, organization, motion, and so forth. Indeed, they lead to the function of "living." In this book, we focus on the dynamical aspect of life phenomena and describe the rudiments of mathematical methods and the "logic of life" revealed by applying them mainly to the cellular scale.

Here, dynamics are not limited to motion of biological elements (e.g., cells or individuals) in real space, as directly observed from the outside. Rather, they include temporal changes in metabolic components, protein levels, or the molecular states in cells, in the electrical potential of neurons, in the number of cell types, in the spatial patterns formed in multicellular organisms, in the population of each species in ecosystems, and so forth. Living systems often include dynamical processes within them. As living systems are not in a static equilibrium state, characterizing dynamics of their state at every scale is essential to our understanding of life.

Of course, intriguing dynamical phenomena are ubiquitous in both chemistry and physics. However, in the case of biological phenomena, we must understand how these are correlated with and generate biological functions, such as cellular responses, adaptation, locomotion, replication, differentiation, memory, information, morphogenesis, and so forth. On the other hand, recent advances in biological experiments enable researchers to gain the data for dynamical processes at cellular and subcellular levels. Measurement techniques for intracellular and intercellular processes have advanced rapidly, and quantitative data on intracellular molecular motion, time-series in the concentration of each intracellular component, cell motility, morphogenesis with dynamical change in multicellular patterns, and changes in cell populations are now available. Now, there is a strong demand to present a "method for understanding" that links these data with the logic of life. This book introduces, on the one hand, mathematical methodologies for dealing with dynamics and, on the other hand, the analysis of life processes, especially cellular states, using these methodologies.

Mathematical methods for analyzing dynamic changes are categorized into deterministic and stochastic approaches. The former is given by theory of dynamical systems. As will be explained in Chapter 1, a dynamical system is represented by a set of state variables, and their time evolution determined by a set of rules, typically given by differential equations. In this case, the rules of time evolution are uniquely determined for each instantaneous state. In contrast, in the latter approach, stochastic processes, the time evolution is not completely determined by the state variables, but it fluctuates within a certain range, and the value to which it changes is probabilistic. Often, due to noise internal or external to a biological system, the state change by the differential equation is valid only on the average, but fluctuates around the mean, where the stochastic approach is relevant. The two standpoints are, of course, not in conflict, and the dynamical systems standpoint can be regarded as a typical (or average) time evolution in the stochastic process, or obtained by ignoring the fluctuations in the latter.

For example, take a problem in which the amount (concentration) of each molecule changes through a reaction. From the dynamical systems standpoint, the concentration of each component is taken as the quantity of state. Here, the rate of reaction depends on the concentration, which gives the time-evolution rule. However, even if the average reaction rate is determined by the concentration of each component, the reaction itself fluctuates around it because it is caused by random collisions between molecules. To take this into account, a stochastic process is necessary. In this case, if the size of the system or the number of molecules is large, there are many reactions occurring at each time within a reasonable time scale. Therefore, although the collisions of each molecule may be random, when averaged, the rate of the reaction is determined only by the concentration, and fluctuations can be neglected. This is the standpoint taken by a dynamical system. The extent to which the dynamical systems approach is sufficient or one needs to adopt a stochastic process to take probability seriously into account depends on the number of molecules in the cell of concern, as well as on our interest. In this book, except for Chapters 5, 6, and 10, the dynamical systems approach is mainly adopted, but it is advisable to be aware of its limitations.

Outline

This book starts with an intuitive introduction to dynamical systems in the context of biological systems in Chapter 1. Then, they are applied to single-cell responses and adaptations in Chapter 2, whereas oscillations and excitability in a cellular state, as observed in experiments, are analyzed in Chapter 3. The classical model of allosteric response by Monod et al. and the classical Hodgkin–Huxley equation for neuronal excitation are also explained. Chapter 4 is devoted to temporal evolution of spatial patterns. The classical theory of pattern formation, known as Turing patterns, is described by adopting dynamical systems with spatial degrees of freedom, in particular reaction–diffusion equations. Chapters 5 and 6 describe the fundamentals of the stochastic approach. Chapter 5 describes the basics of the stochastic approach, and Chapter 6 describes Brownian motion theory and stochastic differential equations for dealing with the time evolution of fluctuations. In particular, the chemical Langevin equation, which has not been treated in standard textbooks on stochastic processes, is also explained in depth, as it is fundamental in dealing with concentration fluctuations in a cell. Chapter 7 describes a dynamical systems approach to cell differentiation in multicellular organisms. In Chapter 8, as an extension of Chapter 4, patterns and motility of cell populations are discussed, focusing on traveling waves, and dynamics of cellular arrangements. Chapter 9 describes a mathematical approach to the origin of life, introducing basic concepts such as error catastrophes and hypercycles. In Chapter 10, problems that deal with biological information are discussed in relationship with probability theory, by also referring to Maxwell's demon.

Depending on the reader's interest, it is not necessary to read the chapters in their current order. Those interested in the phenomenological aspects of cellular response, development, and differentiation may read Chapter 1 first, followed by Chapters 2, 3, 7, and 8, respectively, depending on their interests. Chapter 9 can be read independently after Chapter 1, and Chapter 10 can be read after Chapters 6 and 7. The chapters dealing with individual topics provide many examples, which can be selected according to one's own interests. Of course, it is ideal to read through the entire book in the given order, thereby first gaining the methodology to understand the dynamics of living systems, and then moving on to each topic in succession.

Complementary Books Recommended

Of course, several established textbooks have been published in the field of cell biology, which provide complementary and essential viewpoints to the field. Examples include *Molecular Biology of the Cell* (Alberts et al.), which describes molecular processes that give rise to cellular functions; *Physical Biology of the Cell* (Phillips et al.), focusing on the approaches of statistical physics and

Brownian motion; *Models of Life* (Kim Sneppen), which gives an overview of modeling quantitative biology and principles of life; *Introduction to Systems Biology* (Uri Alon), with a focus on feedback regulation and feed-forward networks; and *Life: An Introduction to Complex Systems Biology* (Kunihiko Kaneko), presenting the logic of life from the dynamics across scales.

Acknowledgments

The content of this book is based on lectures given at the Department of Integrated Sciences, Faculty of Liberal Arts, University of Tokyo, and the Graduate School of Information Science and Technology, Osaka University. In addition, lectures on parts of the contents of this book were also delivered by the authors at Osaka University, University of Tokyo, Nagoya University, Kyushu University, Hiroshima University, Saitama University, Hokkaido University, Kyoto University, Ochanomizu University, Nara Medical University, RIKEN (Rikagaku Kenkyūsho [Institute of Physical and Chemical Research]), and ICTP (International Centre for Theoretical Physics) at Trieste. The feedback from the participants was also a great help in the completion of this book. We would like to thank them again.

We would like to thank the many people who commented on the manuscript. We are grateful to Dr. Jumpei Yamagishi and Dr. Gen Honda (Graduate School of Arts and Sciences, University of Tokyo), and Mr. Riz Noronha (Copenhagen University) for their invaluable suggestions. We also thank Dr. Nobuto Takeuchi (University of Auckland, New Zealand) for valuable comments on Chapter 9, and Dr. Nen Saito (Hiroshima University) for valuable remarks on Chapter 8.

The current volume is based on the Japanese book *Theoretical Biology of the Cell* published by University of Tokyo Press. The main (and sub) contributors for each chapter are as follows: Chapter 1, KK (+ HT); Chapters 2–4, and 8, SS (+ other three); Chapters 5 and 10, HT (+ KK); KK + HT; and Chapters 7 and 9, CF (+KK).

The English version was completed by HT, CF, and KK. We hope that it will be used as a textbook by many people who are interested in biology and want to understand its logic.

1

Introduction to Dynamical Systems for Biology

1.1 Introduction

In this chapter, we introduce the basis of dynamical systems, which is the mathematics for dealing with the dynamics of the states of biological systems. "Dynamical system" is a mathematical term that has as its roots mechanics in physics, and is generalized to deal with a wider range of dynamical processes. As there are already many well-written books on dynamical systems, readers are recommended to refer to them (e.g., [8, 15, 16]). However, some of the standard textbooks may not be easy to read for those in the field of biology. Therefore, in this chapter, we discuss elementary dynamical systems by noting the following points:

- We emphasize the intuitive understanding based on the image of the flow in state space by using pictures, sometimes sacrificing mathematical rigor.
- We often refer to biological examples, such as gene expression and population dynamics. This also helps readers to understand how the concepts of dynamical systems are useful in solving basic problems in cell biology, and will help to connect the basic concept here with biological issues in later chapters.
- We note how the quantities and phenomena in dynamical systems are measured in experimental cell biology.

We start with the concept of state space (also referred to as the phase space). Biological systems (and other systems in the natural and social sciences) have a "state," which describes the system and changes through time. In the case of ecosystems, the state consists of the number of individuals of various species; in the case of multicellular organisms, the number and spatial arrangement of the cells that constitute an organism; and in the case of cells, the composition of chemical components and the arrangement of each molecule, or in the case of proteins, the configuration of amino acids. Dynamical systems are a general mathematical tool used to deal with changes in the internal state of any system.

1.2 Mathematical Representation

How can we discuss the changes and diversification of these "states"? The idea of state space used in the dynamical systems theory is useful. For this purpose, we express the state as a set of variables (e.g., k variables), which is represented by the variables $x_1, x_2, x_3, \ldots, x_k$. Then, the state is represented by a single point in the k-dimensional space. This is the state space.

Example 1.1 For example, the state of a cell is represented by the number (concentration) of molecules (consisting of many chemical components). Consider k chemical components in a cell; then, the state of the cell is expressed by $x_1, x_2, x_3, \ldots, x_k$, that is, the concentration of components $1, 2, 3, \ldots, k$. These components can be, for example, the concentration of calcium, the abundance of proteins, or the amount of various metabolic products.

Example 1.2 Nowadays, "gene expression patterns" are often used to measure the state of cells. The corresponding protein is then produced from each gene. The amount of each mRNA[1] or protein (also called the expression level of a gene), is a major indicator of the cellular state. In this case, the expression level of each gene $x_1, x_2, x_3, \ldots, x_k$ gives the state space.

Of course, there is the question of how many variables are both needed and sufficient to describe a state of concern. This depends on what you want to focus on as a state. Since this question is difficult to answer in general, let us start by expressing the state of an organism (a cell) in terms of an appropriate set of variables.

1.3 In Modeling the Cellular State

In this case, what state variables should be examined? A cell consists of thousands to tens of thousands of proteins, in addition to mRNA and other metabolic components, and a large number of other components. Then, is it a better model if we consider as many variables as are known? First, it is practically impossible to incorporate all of them anyway. Apart from such practical limitations, the idea that it is better to simply consider as many variables as possible is incorrect.

Let us recall thermodynamics. We describe a thermodynamic state (e.g., a gas in a room) in terms of thermodynamic variables such as temperature, pressure, entropy, and volume. If you have several chemical species, it is necessary to

---

[1] The abbreviation mRNA stands for messenger RNA. Messenger RNA is transcribed from a gene in DNA, and then from this mRNA, the corresponding protein is synthesized. Hence, the amount of mRNA can often be used as an indication of the abundances of proteins that are synthesized.

introduce their concentration and chemical potential. At any rate, the number of variables is not so large. In contrast, we now know that the thermodynamic state contains a large number of molecules, so if we want a "complete description" we have to determine a huge number of variables, such as their velocities and positions. However, such complete description is unnecessary to describe the state that we consider. Given the vast amount of data, we are only at a loss. Of course, if we need to understand the motion of a single molecule within the state, we must model the dynamics of the molecule. The description of thermodynamic variables, however, is much more useful for understanding phenomena of our daily scale (macroscopic phenomena). When modeling in this way, it is important to pay attention to the level of the phenomenon we are concerned with and to choose state variables corresponding to it.

In general, biological systems have a hierarchy of scales consisting of molecules, molecular complexes, cells, tissues, individuals, and ecosystems. Although each scale of phenomena can be mutually related, the state in which variables are to be considered will differ depending on which temporal and/or spatial scale phenomena we are focusing on. If we consider the phenomenon of a given scale and incorporate a few related variables, it means that as an approximate description, we do not consider the phenomena at different scales, and assume that our description does not crucially depend on many neglected variables. In other words, we aim to understand many of these variables, in the hope that they will behave "universally" without details of such neglected variables. This assumption of "universality" has been successfully applied in physics, such as in thermodynamics and hydrodynamics. Using the concept of renormalization group, it is also possible to understand that a class of models with different details have the same properties with respect to the behavior of interest [4, 14]. However, it is not clear to what extent the idea of universality is valid in biological systems: There, phenomena at different scales often interfere with each other. These interactions typically need to be taken into consideration, and thus, it is not self-evident that we can be so optimistic. Regardless, we need to start and come back to this issue when necessary.

Naturally, there is a difference in whether the representation is good or not, depending on the choice of variables. However, if one asks which variable is the best in the beginning, it will be difficult to obtain the appropriate answer. Rather, it would be more productive to take a more pragmatic approach by choosing an appropriate set of state variables according to the phenomenon in question, and if it is insufficient, by adding other state variables or eliminating variables if unnecessary.

To sum up, the starting point of the idea of a dynamical system is to choose a state space with a certain number of degrees of freedom in the system, and represent the state as a point in the state space.

1.4 The Dynamical System

In general, the state changes over time. This change is depicted as the trajectory of a point in the state space. Thus, the idea of a dynamical system allows us to consider a change in the motion of a point in a state space. Here, we assume that the set of variables we have taken sufficiently describes the state of the phenomenon we want to consider, and therefore the way the state changes is given by the state variables themselves. In other words, the rule of state change can be expressed as an arrow indicating the direction in which each point in the state space is pointing (Fig. 1.1). The mathematical expression of this is the dynamical system.

According to the formulation of dynamical systems, the state space and the rules of time evolution at each point in the state space are given. Then, given the initial condition, that is, the initial point in the state space, the following change in state variables is derived as change of the locus of the point in the state space. Because the arrow of change is given as a function of each point $(x_1, x_2, \ldots, x_k)$, each state variable deviates from it after a time interval δt by $(f_1(x_1, x_2, \ldots, x_k) \times \delta t, f_2(x_1, x_2, \ldots, x_k) \times \delta t, \ldots, f_k(x_1, x_2, \ldots, x_k) \times \delta t)$. This is described by a differential equation:

$$\frac{\mathrm{d}x_i}{\mathrm{d}t} = f_i(x_1, x_2, \ldots, x_k) \quad (i = 1, 2, \ldots, k), \tag{1.1}$$

where $(f_1(x_1, x_2, \ldots, x_k), f_2(x_1, x_2, \ldots, x_k), \ldots, f_k(x_1, x_2, \ldots, x_k))$ are a vector, as indicated by the arrows in the figure. In this case, the flow in the k-dimensional space is completely determined by the k variables, and subsequent changes in the state are completely given by the arrows in this space. Accordingly, it is also called a deterministic equation because the future state is determined once the present state is given. In contrast, factors other than the variables that we are dealing with may also affect the time evolution to some extent. This effect is treated as a fluctuation (or noise) in the time-evolution change, as discussed in Chapters 5 and 6.

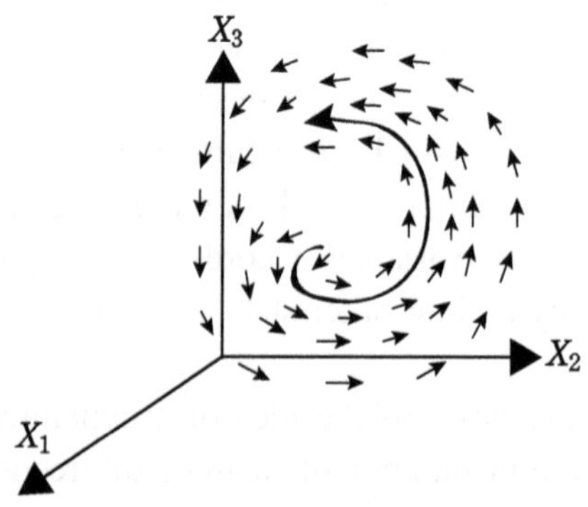

Figure 1.1 Schematic representation of a dynamical system.

Here, it is more important to understand how the state flows along the arrows in the k-dimensional state space than to solve the equations analytically.

Example 1.3 Consider a system with chemical components whose concentrations change by chemical reactions. For example, consider a system with three components X_1, X_2, X_3, whose concentrations are given by x_1, x_2, x_3, respectively. Assume that X_1 is synthesized from a certain component A, X_2 is synthesized from another component B, and X_3 is formed by the reaction of components X_1 and X_2 (the synthesis rate is r), and the components are decomposed at the rate of r_d. Assuming that the reaction proceeds unidirectionally, the rate of increase in component 3 is proportional to the product of concentrations x_1 and x_2; thus, the dynamical system of the concentrations is given by

$$\frac{dx_1}{dt} = a - rx_1x_2, \quad \frac{dx_2}{dt} = b - rx_1x_2, \quad \frac{dx_3}{dt} = rx_1x_2 - r_dx_3. \tag{1.2}$$

Of course, the number of events of each reaction is not exactly given by the preceding rate equation, because the reaction originates from a collision of molecules that move randomly in space. On average, it is given by the preceding rate, and if the number of molecules of concern is large, the fluctuation around the average is negligible (please refer to Chapters 5 and 6 for the formulation to take into account the fluctuation).

By using the dynamical systems representation of concentration changes due to chemical reactions, one can consider applications of dynamical systems to intracellular processes. For instance, consider a pair of chemical reactions consisting of appropriate enzymatic reactions and consider a dynamical system in which the number of chemical components is used as a state variable to represent the change. As a simple example, let us consider the reaction, *Substrate + Enzyme $\rightarrow$ Product + Enzyme*. Further, let us assume that the product is degraded with the rate k_d and the substrate is supplied at the rate k_s. Then the dynamical system for the change in the concentrations of the product x_p and the substrate x_s is given by $\frac{dx_p}{dt} = rx_ex_s - k_dx_p$, $\frac{dx_s}{dt} = k_s - rx_ex_s$, where x_e is the concentration of the enzyme and r is the rate of the reaction.

Exercise 1.4.1 Draw the arrows in the state space (x_s, x_e) and discuss the behavior of the preceding dynamical system of two variables.

Toggle switch: An example of mutually repressive gene expression

Next, we discuss the dynamical systems of concentration of two proteins that mutually regulate. As a specific example, let us consider the mutual regulation of expression by two proteins. This example appears a few times later throughout the text.

Consider the case in which two genes repress each other. Specifically, protein 1 expressed by gene 1 suppresses the expression of gene 2, whereas protein 2 expressed by gene 2 suppresses the expression of gene 1. Let mRNA1 be synthesized from gene 1, protein 1 (whose concentration is p_1) from gene 1, and protein 2 from gene 2 via mRNA2 (whose concentration is p_2). The state space is given by the two protein concentrations (p_1, p_2). The reason why only proteins are chosen without considering the concentration of mRNAs is that changes in mRNA concentrations have a faster time scale for their synthesis and degradation, which are assumed to be proportional to the concentration of proteins. (The elimination of such fast variables is discussed in Section 1.12.) As will be shown in the following discussion, this system has two stable states, either with the expression of protein 1 ("on") and the suppression of 2 ("off") or vice versa. This type of bistable switch with an on/off knob is called a toggle switch herein.

Proteins synthesized from mRNAs degradate at certain rates. For simplicity, let us assume that both rate constants are the same, that is, set to 1 by taking the unit of time appropriately. The rate of the expression of each gene, that is, the synthesis rate, is reduced by the partner protein, which is protein 2 for protein 1 and protein 1 for protein 2. Specifically, we assume that the synthesis rate of each protein is expressed as $\alpha/(1 + p_2^2)$, $\alpha/(1 + p_1^2)$. For example, if two protein molecules form a dimer, which attaches to the promoter of the gene and inhibits the production of mRNA, this form is expected, as the repression follows the second order of the concentrations of the corresponding proteins. (This is just an example, and there are many other possible cases.) Then, in this case, the dynamical system of the time evolution of p_1, p_2 is given by

$$\frac{dp_1}{dt} = \frac{\alpha}{1 + p_2^2} - p_1, \tag{1.3}$$

$$\frac{dp_2}{dt} = \frac{\alpha}{1 + p_1^2} - p_2. \tag{1.4}$$

These are the instances in which proteins inhibit each other in their synthesis, but when they promote each other, an increasing function for p_2 (say $\alpha p_2^2/(1 + p_2^2)$) is used instead of a decreasing function such as $\alpha/(1 + p_2^2)$.

1.5 Deterministic Dynamical Systems vs. Stochastic Process

As mentioned in the preface, there are two basic methods for analyzing dynamic changes: deterministic and stochastic approaches. The former is given by the theory of dynamical systems, as introduced previously. As mentioned, in this case, the rules of time evolution are uniquely determined for each instantaneous state. In contrast, the time evolution in stochastic processes is not completely determined by the state variable, but it fluctuates within a certain range, and the value to which it changes is probabilistic. Often, due to noise internal or external to a biological system, the state change is given on average, but fluctuates around it. The stochastic approach is then relevant, as will be discussed in Chapters 5 and 6. The two standpoints are, of course, not in conflict, and the dynamical systems standpoint can be regarded as a typical (or average) time evolution in the stochastic process, or obtained by ignoring the fluctuations in the latter.

For example, consider a problem in which the amount (concentration) of each molecule changes through a reaction, as mentioned in the previous section. From the dynamical systems standpoint, the concentration of each component is taken as the quantity of state. Here, the rate of reaction depends on the concentration, which gives the time-evolution rule. However, even if the average reaction rate is determined by the concentration of each component, the reaction itself fluctuates around it because it is caused by random collisions between molecules. To take this into account, a stochastic process is necessary. In this case, if the size of the system or the number of molecules is large, there are many reactions occurring at each time within a reasonable time scale. Although the collisions of each molecule may be random, when averaged, the rate of the reaction is determined only by the concentration, and fluctuations can be neglected. This is the standpoint taken by a dynamical system. When the dynamical systems approach breaks down, one needs to adopt a stochastic process to take seriously into account that probability depends on the number of molecules in the cell of concern, as well as on our interest (see Chapter 5).

1.6 Nullcline and Fixed Point

Now, given Eq. (1.1) for a dynamical system, it is necessary to determine whether each state variable increases or decreases when drawing an arrow flow from it. This is given by the sign of $f_i(x_1, x_2, \ldots, x_k)$. Therefore, it is important to determine the space (plane) given by $f_i(x_1, x_2, \ldots, x_k) = 0$ to draw the arrows of increase and decrease for each variable. The space that satisfies this condition is called the nullcline.

nullcline: subspace that satisfies the change in each variable is zero, that is, $f_i(x_1, x_2, \ldots, x_k) = 0$.

Example 1.4 Nullcline of Toggle Switch For the model of the genes of mutual repression in Section 1.4, let us draw the nullclines of p_1 and p_2 in the dynamic system of the protein. From Eqs. (1.3) and (1.4), respectively, we obtain $p_1 = \alpha/(1 + p_2^2)$ and $p_2 = \alpha/(1 + p_1^2)$. How these two nullclines intersect differs depending on whether $\alpha < 2$ or $\alpha > 2$, as shown in Fig. 1.2(a) and (b), respectively. Using this, the arrows (flows) of the dynamic system are shown in Fig. 1.2.

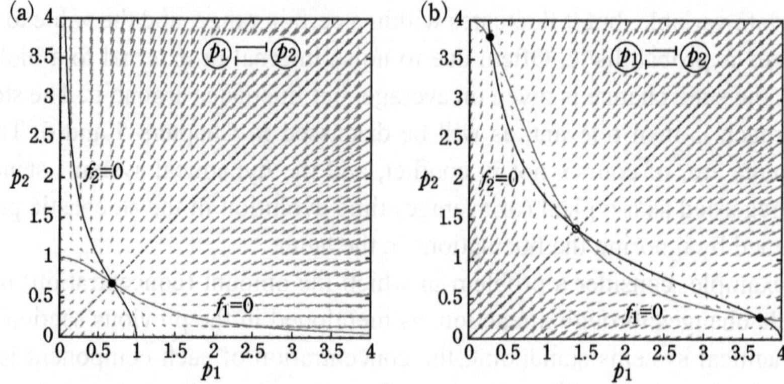

Figure 1.2 Nullcline and flow of a toggle-switch system. (a) $\alpha < 2$, (b) $\alpha > 2$. For reference the dotted line is drawn at $p_1 = p_2$.

Next, consider the case in which the nullcline conditions are simultaneously satisfied. This gives the point at which the change in all x_i ($i = 1, 2, \ldots, k$) with time vanishes simultaneously. That is, if we start from this point $(x_1, x_2, \ldots, x_k)$, the values of these variables will not change thereafter. The state was constant over time. Such a point is called a fixed point.

fixed point: the point $(x_1^*, x_2^*, \ldots, x_k^*)$, in which the temporal changes of the variables are zero according to the dynamical system, that is, $f_i(x_1^*, x_2^*, \ldots, x_k^*) = 0$ for all $i = 1, 2, \ldots k$.

Example 1.5 Fixed Point of Toggle Switch Let us obtain a fixed point for a model of a toggle switch. In the case of (a) $\alpha < 2$, it is given by the solution of the cubic equation $(1 + p_1^2)p_1 = \alpha$, where $p_1 = p_2$. In the case of (b) of $\alpha > 2$, two more fixed points appear in addition to this fixed point, as shown in Fig. 1.2. The two are located in mutually symmetric positions, (p_1^*, p_2^*) and (p_2^*, p_1^*).

1.7 Linear Stability of Fixed Points

A fixed point is a state in which variables, if left exactly as they are, will stay there. Then, if we start from the point where the values are shifted slightly, would the state go back to that fixed point? This can be judged by whether or not the arrow around the fixed point is pointing in the direction of the fixed point. If the arrows are going out of the fixed point in some directions, the fixed point is unstable, and if they move toward the fixed point from all directions, the fixed point is stable. The behavior near the fixed point can be determined by making a Taylor expansion of the original Eq. (1.1) around the fixed point and taking the lowest order of the deviation from the fixed point to obtain a linear equation. Now let the fixed point be $(x_1^*, x_2^*, \ldots, x_k^*)$. It satisfies $f_j(x_1^*, \ldots, x_k^*) = 0$ $(j = 1, 2, \ldots, k)$. Let $x_i(t) = x_i^* + \delta x_i(t)$ and substitute it for f_j and perform a Taylor expansion around x_i^*, leaving only the first order of $\delta x_i(t)$. Considering that the lowest order disappears under the condition $f_j(x_i^*) = 0$, then

$$\frac{\mathrm{d}\delta x_i(t)}{\mathrm{d}t} = \sum_j J_{ij}\delta x_j(t). \tag{1.5}$$

Here, $J_{ij} = \frac{\partial f_i}{\partial x_j}$ is the Jacobi matrix around the fixed point. In a vectoral form, it is written as

$$\frac{\mathrm{d}\delta \boldsymbol{x}(t)}{\mathrm{d}t} = \boldsymbol{J}\delta \boldsymbol{x}(t). \tag{1.6}$$

Because Eq. (1.6) is linear, it can be solved by diagonalizing the matrix $\boldsymbol{J} = \{J_{ij}\}$. Let the transformation matrix for diagonalization be $\boldsymbol{T}$ and $\boldsymbol{\Lambda} = \boldsymbol{TJT}^{-1}$ be a diagonal matrix, and let the eigenvalues of the matrix be $\lambda_1, \lambda_2, \ldots, \lambda_k$. (As long as the eigenvalues are not degenerate, the matrix can be diagonalized as shown earlier. For the degenerate case, the Jordan standard form can be used, but the degenerate case is special, so we will not go into it. In this case, it is easier to assume that λ_j is not equal and to take the limit of the difference to zero.)

Now, we transform the vector $\delta \boldsymbol{x}$ by $\boldsymbol{T}$ such that $\boldsymbol{v} = \boldsymbol{T}\delta \boldsymbol{x}$. Then, we multiply $\boldsymbol{T}$ from the left of Eq. (1.6). By inserting $\boldsymbol{T}^{-1}\boldsymbol{T}$ between $\boldsymbol{J}$ and $\delta \boldsymbol{x}(t)$, we obtain

$$\frac{\mathrm{d}\boldsymbol{v}(t)}{\mathrm{d}t} = \boldsymbol{\Lambda}\boldsymbol{v}(t). \tag{1.7}$$

Hence, we have $\mathrm{d}v_i(t)/\mathrm{d}t = \lambda_i v_i$, which is solved as $v_i(t) = v_i(0)\exp(\lambda_i t)$. Thus, we have

$$\delta x_i(t) = \sum_j \exp(\lambda_j t)T_{ij}^{-1}v_j(0) = \sum_{j,m} \exp(\lambda_j t)T_{ij}^{-1}T_{jm}\delta x_m(0). \tag{1.8}$$

In summary, the time evolution of $\delta \boldsymbol{x}(t)$ is given by the linear superposition of the terms with $\exp(\lambda_j t)$. Here, the original equation involves only real variables (and parameters), and the eigenvalues λ_j can be complex in general. In this case, the complex conjugate λ_j^* is also an eigenvalue. If we write $\lambda_j = a + i\omega$ (i is the pure imaginary number $[i^2 = -1]$, a is the real part of λ_j, and ω is real), then because $\exp(at)\exp(\pm i\omega t)$, the preceding solution is given by the superposition of $\exp(at)\cos(\omega t)$ and $\exp(at)\sin(\omega t)$.

Now, we intend to investigate whether a slight deviation from x^* is reduced and the state falls onto the original fixed point x^*. If the real part a of the eigenvalue λ_j is positive for some j, then the deviation increases exponentially. In other words, the deviation is amplified (at least) in one direction, and thus, the fixed point is unstable. Hence, for the fixed point to have (linear) stability, the real part of all the eigenvalues of $\boldsymbol{J}$ must be negative. (If it is exactly 0, it is a boundary between

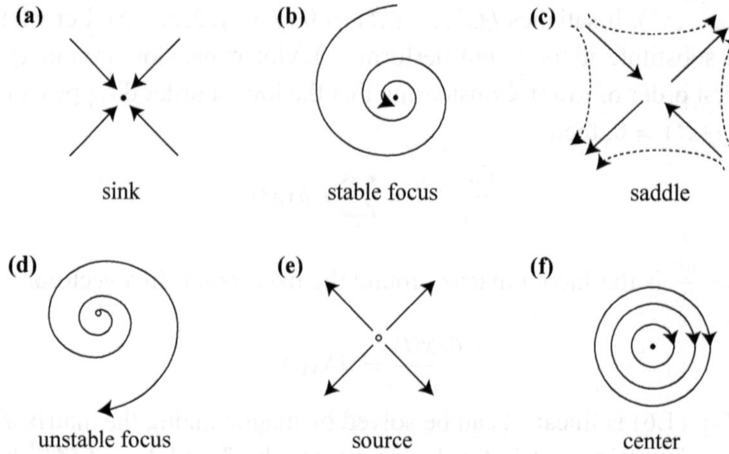

Figure 1.3 Classification of fixed points.

stability and instability, so we need to check it further. An example of this will be discussed in Section 1.11.)

The behavior around a fixed point can be checked by the sign of the real part of the eigenvalues and whether or not there is an imaginary component. We now discuss the two-variable case as an example (see Fig. 1.3).

(a) Both eigenvalues are real and negative (sink): directed to the fixed point without rotation.

(b) A pair of complex-conjugate eigenvalues with a negative real part (stable focus). By writing the real part as $a < 0$ and the imaginary part as ω, the dynamics around the fixed point are written by the superposition of $\exp(at)\cos(\omega t)$ and $\exp(at)\sin(\omega t)$, so that the orbit is attracted to a fixed point with rotation at an angular velocity of ω.

(c) The eigenvalues are real with one positive and one negative (saddle): Let the eigenvalues be $\lambda_1 > 0 > \lambda_2$. Then $v_2(t)$ decays with time, so that the orbit is attracted to a fixed point from the direction of this eigenvector. On the other hand, as $\lambda_1 > 0$, the orbit exits from a fixed point along the eigenvector v_1.

(d) A pair of complex conjugate eigenvalues with a positive real part (unstable focus): In contrast to case (b), the trajectory exits from the vicinity of the fixed point with rotation, because $a > 0$.

(e) Both eigenvalues are real and positive (source): The arrows of case (a) are reversed, so that the orbit exits from the vicinity of the fixed point in any direction.

(f) The border between cases (b) and (d), where the real part is 0, and the imaginary part is nonzero (center), and the orbit rotates around the fixed point.

Example 1.6 Stability of Fixed Points in Toggle Switch Let us examine whether each of the fixed points of the toggle switch is stable or not. As expected from Fig. 1.2, in the case of $\alpha < 2$ in (a), the only fixed point is stable; in the case of $\alpha > 2$ in (b), the fixed point $p_1 = p_2$ is unstable and the two fixed points $p_1 \neq p_2$ are stable.

The preceding example is in the two-variable case, while combinatorial discussion is possible even in the multivariable case by considering whether the real part of the eigenvalues is positive or negative, and the imaginary part is zero or not, respectively.

1.8 Temporal Evolution of State Variables and Attractors

As we have seen so far, as long as the solution does not diverge to infinity (which is a reasonable postulate for a model for natural phenomena), we would expect it to approach an asymptotic state after sufficient time. The region in the state space where the trajectory passes after a sufficiently long (infinite, mathematically speaking) time span is called an attractor. In other words, the orbit always passes through the attractor region (any neighborhood of the region) after any period of time. After some time, the state asymptotically falls into this attractor and stays within it. In contrast, starting from any initial state, there are regions where the orbit passes through but to which it never returns. These are called transients (transient states).

The stability of a state implies that when the state deviates slightly, it returns to its original state. Usually, attractors have such stability because they are attracted from the initial conditions around them. After falling into an attractor, if one perturbs the state, the orbit returns to it.[2]

attractor: set in the state space to which many initial points evolve by the dynamical system

Now, the simplest attractor is the stable fixed point. As we have already seen, starting from a suitable initial state, the trajectory is pulled toward its fixed point and stays there. If we use the expression "falling down," it is easy to imagine that there is a quantity characterizing the height of landscape (potential) in the state space, and the trajectory is falling as it reduces it. One can imagine a contour line drawn in the state space, and that a trajectory is drawn along a direction perpendicular to the contour line. For example, one can imagine a particle falling over the landscape with energy differences; under a condition with high friction, the

---

[2] There is an exceptional case in which the attractor does not have such asymptotic stability: the orbits are attracted from a variety of initial conditions (i.e., with a finite measure), but it will jump out if one perturbs to a special direction (see Section 1.10). As this is a special case, we will not consider it for now.

ball falls onto the bottom of the landscape, that is, the location of the minimum potential energy (see Example 1.9). In dynamical systems, such a potential exists in some cases, but it is not always so. For example, in case (ii) in Section 1.7, that is, the case with rotation to fall down to a fixed-point attractor, such a contour cannot be drawn and the potential does not exist.

1.9 Limit-Cycle

So far, we have understood the fixed-point attractor in which the orbit falls down to a stable fixed point. Are there other types of attractors? For example, consider the case of (iv) in Section 1.7, where the state goes out from the fixed point with rotation. As it is, the variables will diverge infinitely. Such divergence could be avoided in natural/life phenomena. The reason for this unnatural behavior is that in Section 1.7, δx_i is assumed to be small for use in the analysis within a linear range. Then, as the value increases, it cannot be used. Usually, there is a higher-order term than the first order, which suppresses divergence (Fig. 1.4). If the suppression by the nonlinear term and the growth of the deviation by the linear term are balanced, a continuous rotation state is expected to exist. In this case, an oscillatory state in which the variables change periodically over time would eventually be realized.

Example 1.7   As an example, let us consider an ordinary differential equation of variables x and y:

$$\frac{dx}{dt} = x - \omega y - (x^2 + y^2)x, \tag{1.9}$$

$$\frac{dy}{dt} = y + \omega x - (x^2 + y^2)y. \tag{1.10}$$

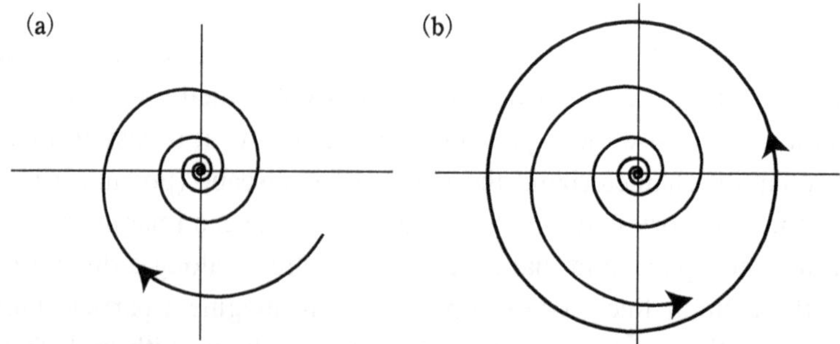

Figure 1.4 (a) Rotating attraction to a stable fixed point (stable focus). (b) Rotating exit from an unstable fixed point (unstable focus). With the aid of a nonlinear term, divergence to infinity is suppressed.

One can see that $x = y = 0$ is a fixed point. With straightforward calculations, the eigenvalues of the Jacobi matrix around the point are given by $1 \pm i\omega$. As their real part is positive, we have the situation of (iv) in Section 1.7, and the orbit rotates from the point. Here, the third term in the equation is included to avoid divergence of the orbit to infinity. Now, we introduce the polar coordinates $r(t), \theta(t)$, given by $x(t) = r(t)\cos\theta(t)$, $y(t) = r(t)\sin\theta(t)$ $(r > 0)$. With the transformation to these polar coordinates, the preceding equation is written as

$$\frac{\mathrm{d}r}{\mathrm{d}t} = r - r^3, \quad \frac{\mathrm{d}\theta}{\mathrm{d}t} = \omega. \tag{1.11}$$

If the r^3 term is ignored while retaining only the linear terms, as in Section 1.7, r increases exponentially, while θ rotates at a constant angular velocity ω, such that the orbits of r and θ show the exit from the fixed point while rotating. On the other hand, expression (1.11) leaves the term r^3, so that one can see that the increase in r is suppressed. From the equation $\mathrm{d}r/\mathrm{d}t$, $r = 0$ or 1 becomes a fixed point of the equation for $\mathrm{d}r/\mathrm{d}t$. As r increases near $r = 0$ and decreases from $r > 1$, r approaches 1 with time. Here, because θ increases at a constant velocity ω, the orbit eventually approaches the state of turning a circle of $r = 1$ at a constant angular velocity ω. From the original variables x and y, the solution is given by $(x, y) = (\cos(\omega t + \theta_0), \sin(\omega t + \theta_0))$, where θ_0 is given by the initial condition. Hence, the orbit is attracted to the periodic solution moving with the period $T = 2\pi/\omega$.

Such periodic solutions are represented by closed orbits returning to the original state. The closed orbit representing the periodic solution is called the limit-cycle.

limit-cycle: periodic orbit that is invariant by the temporal evolution of dynamical system

It is a stable limit-cycle when the orbit approaches a periodic solution over time. In the preceding example, the limit-cycle was a circle, but in general, the limit-cycle orbit showed a distorted trajectory and could have a more complex shape (Fig. 1.6). This stable limit-cycle is an attractor because it is asymptotically approached by the time evolution and remains there over time. On the other hand, it is an unstable limit-cycle when slight perturbation from it leads to the deviation from the periodic solution. (In other words, if one reverses the time, the limit-cycle is reached as time goes to $-\infty$.) Because only the stable limit-cycle should be considered as an attractor, in the present book we often do not explicitly write *stable* and abbreviate it as a limit-cycle.

Poincaré–Bendixson Theorem

Let us consider a two-dimensional dynamical system in which the orbit is neither attracted to a fixed point nor diverges to infinity. In other words, the orbit is bounded within a finite region in a two-dimensional plane. By drawing orbit curves in a

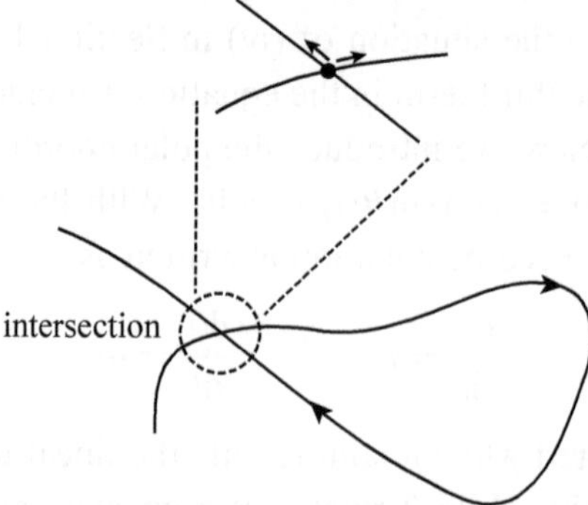

Figure 1.5 Schematic representation to show that the orbit in a dynamical system cannot cross itself.

plane, one can expect the orbit is attracted to a limit-cycle. This "expectation" in a picture is indeed proven mathematically as a theorem.

Poincaré–Bendixson Theorem: If the trajectory of a two-dimensional (2 variables) dynamical system is confined within a finite range (i.e., if it does not diverge), its attractor is a fixed point or limit-cycle.

Poincaré–Bendixson Theorem (meaning): In a two-dimensional dynamical system, if a trajectory enters and does not leave a closed and bounded region which contains no fixed-point attractors, then a periodic orbit is approached with the time evolution of dynamical system.

This can sometimes be used to establish the existence of a (stable) periodic orbit for a planar vector field. Intuitively, this theorem is easy to understand. As long as the orbit remains in the finite range, if we consider the possibility other than a fixed point and periodic orbit, the orbit inevitably intersects with itself somewhere. However, the orbit of a dynamical system cannot intersect with itself, because it moves along a given vector at each point. If the orbit intersected with a finite velocity, the orbit, when returned to the point next time, would go to a different direction. Heading in a different direction, however, is impossible because the direction is given by the vector on the point (Fig. 1.5). For the mathematical proof of the theorem see mathematics textbooks, for example, [8].

In other words, attractors for a two-dimensional dynamical system are either fixed points or limit-cycles. Note that this statement is true only for two-dimensional dynamical systems. If the degrees of freedom are more than two, there is a different type of attractor. One of the most remarkable ones, in this case, is a chaos attractor, as will be discussed in Section 1.16.

Now we discuss a few examples of limit-cycle attractors in a chemical reaction model, and gene expression dynamics.

Example 1.8 BZ Reaction and Brusselators In a chemical reaction system in an open (nonequilibrium) system, oscillation of the component concentration was found by Belousov and established by Zhabotinsky, and termed the BZ reaction. They adopted the oxidation reaction of citric acid using the catalyst as a part of the research on the citric acid cycle, which is the metabolic pathway of the organism, and discovered the phenomenon in which redox occurs repeatedly and the color of the solution oscillates over time. This reaction is complicated and involves dozens of intermediate steps. A simplified abstract example of such an oscillatory reaction was devised by Brussel's group, known as Brusselators [13]. They adopted two components, X and Y, with reactions with components A and B, which are externally supplied. This abstract reaction is written as

$$2X + Y \rightarrow 3X, \quad B + X \rightarrow Y + D, \quad A \rightarrow X, \quad X \rightarrow E \tag{1.12}$$

where the concentrations of A and B are assumed to be sustained as constant values. Then, by setting the concentrations of the X, Y as X, Y, their temporal change is written as

$$\frac{dX}{dt} = A - (B + 1)X + X^2 Y, \tag{1.13}$$

$$\frac{dY}{dt} = BX - X^2 Y. \tag{1.14}$$

(For simplicity, the reaction rate constant was set to unity.) The fixed point of this equation is given by $(X^*, Y^*) = (A, B/A)$. Then, the dynamics of the deviations from the point are given using the Jacobi matrix

$$\begin{pmatrix} B - 1 & A^2 \\ -B & -A^2 \end{pmatrix}.$$

By computing the eigenvalues of the matrix, we show that the fixed point is a stable focus if $B < 1 + A^2$, whereas for $B > 1 + A^2$, it is an unstable focus. In the latter case, a limit-cycle attractor exists, as shown in Fig. 1.6.

Regulation Network of Two Genes with Positive and Negative Feedback In place of the toggle switch, a mutually repressing gene control network, consider a case in which the gene X activates itself and the gene Y, whereas the gene Y represses X, as shown in Fig. 1.7 [6]. Using x, y as the concentration of X, Y, and following the example of a toggle switch, the suppression can be written as $1/\left(1 + \left(\frac{y(t)}{K_y}\right)^n\right)$, and the activation as $\left(\frac{x(t)}{K_x}\right)^n / \left(1 + \left(\frac{x(t)}{K_x}\right)^n\right)$. It would be possible to model gene regulation in this form with K_x and K_y as thresholds for activation

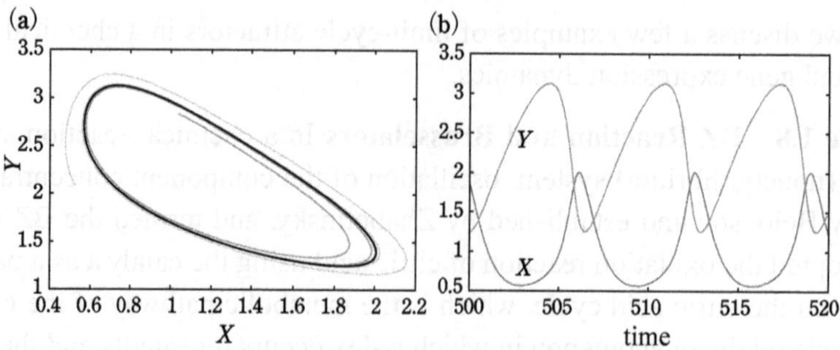

Figure 1.6 Attractor of Brusselator (a) and the timeseries of x and y (b).

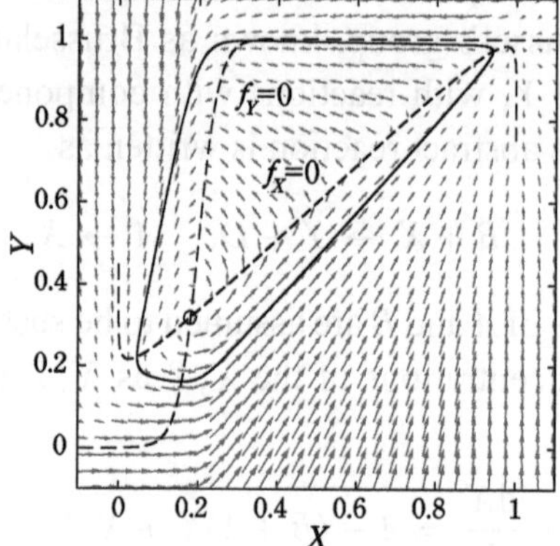

Figure 1.7 The limit-cycle attractor of the model (1.15), with $J_{11} = J_{21} = 1$, $J_{12} = -1$, $J_{22} = 0$, $\theta_1 = -0.1$, $\theta_2 = 0.2$, $\gamma_1 = \gamma_2 = 1$, $\beta = 8$. The dotted lines represent the nullclines $\frac{dx_1}{dt} \equiv f_X = 0$, and $\frac{dx_2}{dt} \equiv f_Y = 0$.

and inhibition. The former decreases with y and asymptotically approaches zero at $y \gg K_y$, whereas the latter increases with x and asymptotically approaches 1 at $x \gg K_x$. Instead of this form, we can use another form of modeling that exhibits similar behavior, the threshold dynamics, which are often used in theoretical physics.

Consider genes x_i with $i = 1, \ldots, k$, that represents the expression level by x_i. Here, x_i is 0 if the gene is not expressed, and 1 if it is fully expressed (In other words, it can be considered that the protein concentration p_i has a maximum (saturated) concentration $p_{\max}$ and $x_i = p_i / p_{\max}$.). In a simple form, the effect of the gene (protein) j on i is written as $J_{ij} x_j$. Here, if j activates i, $J_{ij} > 0$; if it is suppressed, $J_{ij} < 0$; and $J_{ij} = 0$ when j does not affect i. In addition, the effect of other genes is assumed to be simply superposed by addition, whereas the effect of many genes as a combination is not considered. That is, the input to gene i is given by $\sum_j J_{ij} x_j$. In response to the input, protein x_i was synthesized. Let us assume that the synthesis rate starts to increase around a certain threshold, θ_i, for the input. As the input increases, the synthesis rate saturates and approaches a constant value. (This corresponds to the Hill type described in Section 2.1.2.) Accordingly, $F\left(\sum_j J_{ij}x_j - \theta_i\right)$ is adopted for the synthesis rate of the gene expression x_i, where $F(x)$ is a monotonically increasing function, which rises sharply from 0 to 1 at $x \sim 0$. As $x < 0$ decreases, $F(x) \to 0$, while $x > 0$ increases, $F(x) \to 1$. Such $F(x)$ is called a sigmoid function; for example, the form $1/(\exp(-\beta x) + 1)$ is often adopted, where $\beta > 1$ shows the steepness of the increase at the threshold. This corresponds to exponent 2 of p_1^2 or p_2^2 in expressions (1.3) and (1.4). If we replace this 2 with a number n, known as the Hill coefficient (see Section 2.1.2), this n gives an index of sharp increase at the threshold, and corresponds to β. In both forms, as β or n increases to infinity, the function approaches a step function. We further introduce the decay term proportional to $-x_i$, which represents the decomposition. The time scale for the synthesis and decomposition is denoted by γ_i. Then the time evolution of x_i is given by

$$\frac{\mathrm{d}x_i(t)}{\mathrm{d}t} = \gamma_i\left(F\left(\sum_j J_{ij}x_j - \theta_i\right) - x_i\right). \tag{1.15}$$

This form can represent various networks with mutual activation or suppression by taking positive, negative, and 0 values for the elements of matrix J_{ij}. Equation (1.15) represents the dynamical systems for such cases, and thus is often used as a simplified model of gene expression dynamics. On the other hand, if x_i represents the activity (firing rate) of a neuron, it is often adopted as a simple model equation for neural networks, where positive J_{ij} represents the activation of neuron j by i, and a negative J_{ij} represents the inhibition connection. Thus, this type of threshold network model has been extensively studied in theoretical neuroscience.

Now let us return to the original gene expression problem and consider the case of two genes. (Let the concentration of the gene X be x_1 and the concentration of Y be x_2.) In the toggle switch, the genes are mutually suppressed, and thus $J_{12} < 0, J_{21} < 0, J_{11} = J_{22} = 0$. On the other hand, when the gene X activates itself and the gene Y, and the gene Y represses the gene X, $J_{11} > 0$, $J_{21} > 0$, $J_{12} < 0$, $J_{22} = 0$ is used.

In this case, if the parameter $\theta_{1,2}$ is chosen appropriately, the flow of the dynamical system is as shown in Fig. 1.7 (when β is large). There is a fixed point at the intersection of two nullclines, and the eigenvalues of the Jacobi matrix around the fixed point are complex numbers with a positive real part. Then, the flow exits from the fixed point while rotating, and the nullcline does not cross at the upper right. In addition, X and Y are reduced by decomposition as they increase, so that they remain in a finite range. Examining this flow diagram, the orbit remains within a finite range in a two-dimensional space, and is not attracted to a fixed point. Then according to the Poincaré–Bendixson theorem, the attractor will be a limit-cycle. In fact, the concentrations of X and Y, in this case, oscillate periodically.

Exercise 1.9.1 Set $J_{12}=J_{21}=-1$, $J_{11}=J_{22}=0$, $\theta_1=\theta_2=0.5$, $\gamma_1=\gamma_2=1$, $\beta=8$ in Eq. (1.15), draw the flow in a two-dimensional state space, and confirm the behavior of the toggle switch (Eqs. (1.3) and (1.4)).

Exercise 1.9.2 Set $J_{11}=J_{12}=1$, $J_{21}=-1$, $J_{22}=0$, $\gamma_1=\gamma_2=1$, $\beta=8$ in Eq. (1.15), draw the flow in a two-dimensional state space, and show that the system can have a limit-cycle attractor by choosing the values (θ_1,θ_2), appropriately.

1.10 Basin of Attraction for a Multiattractor System

In general, a dynamical system can have multiple attractors. In this case, depending on the initial condition, the state is attracted to one of the attractors. Then, the set of initial conditions that are attracted to one attractor is termed as the basin.

basin (of attraction): set (region) of initial conditions that are attracted to each attractor by the evolution of dynamical systems

Example 1.9 Consider the motion of a ball rolling down (under large friction) in a landscape with several valleys. A particle of mass m is in motion under the potential U, such that the force is given by the negative gradient (derivative) of the potential. Suppose that the friction is proportional to the speed. Then, the dynamical systems for this motion of the ball are given by

$$\frac{\mathrm{d}x}{\mathrm{d}t}=v,\quad m\frac{\mathrm{d}v}{\mathrm{d}t}=-\gamma v-\frac{\mathrm{d}U}{\mathrm{d}x} \tag{1.16}$$

according to Newtonian mechanics. Assume that the friction is so large that the acceleration is assumed to be zero. Accordingly, $\mathrm{d}v/\mathrm{d}t=0$. Then, we get $v\sim-(1/\gamma)\frac{\mathrm{d}U}{\mathrm{d}x}$. Hence,

$$\frac{\mathrm{d}x}{\mathrm{d}t}=-c\frac{\mathrm{d}U}{\mathrm{d}x}, \tag{1.17}$$

where $c=1/\gamma$. (For the elimination of variable v, see also the adiabatic elimination method of Section 1.12.) As shown in Fig. 1.8, the potential has a few hills and valleys (extrema). Then, the fixed point of this dynamical system is given by $\mathrm{d}U/\mathrm{d}x=0$, that is, the top of the hills and the bottom of the valleys. As is easily confirmed, the tops are unstable and the bottoms are stable fixed-point attractors. In this case, each basin is given a valley region separated by the top of the hills.

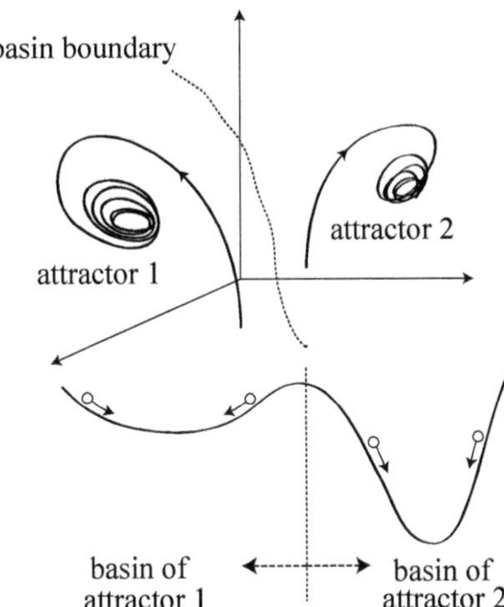

Figure 1.8 Multiple attractors.

Example 1.10 **Toggle Switch** Now, consider the model in Eqs. (1.3) and (1.4) (Fig. 1.2). In this case, there exists a unique attractor if $\alpha < 2$, whereas for $\alpha > 2$, there are two fixed-point attractors with $p_1 > p_2$ and $p_2 > p_1$. In this case, the basin for each attractor is given by $p_1 > p_2$ and $p_2 > p_1$, respectively, and the basin boundary (watershed) is given by the diagonal line $p_1 = p_2$.

The preceding is a basin for the case of multiple fixed-point attractors. Basins are not limited to this case but can be applied to any (complex) attractor. In general, basin boundaries can be more complex.[3] In addition, if the attractor touches with the boundary of the basin, it may be pulled in from one side sufficiently, but on the other hand, it may be pulled in by another attractor with only a slight shift. It is suggested by Milnor to include such cases in the attractor, although in this case, there is no stability in the sense that slightly shifted orbits return to the original.[4]

1.11 Lotka–Volterra Equation

There is a classic study of a model for biological oscillation by Lotka and Volterra, in which the population of two species oscillates over time. This example is simple

---

[3] For instance, they may be fractal: a tiny change in the initial conditions can make the orbits fall into different attractors, and the boundary involve finer and finer structures.

[4] Some attractors, albeit rather unusual, attract a finite domain of initial conditions, but by slight perturbation from the attractor, the state is kicked out of it [11, 12]. However, such a case is rather special and appears only at an exceptional parameter, as for dynamical systems with few degrees of freedom. However, for dynamical systems with many degrees of freedom, such exceptional attractors can appear more often [10].

and significant, even though the oscillation is not a limit-cycle attractor and is a bit special. In this section, we discuss this model.

Consider a system of predator Y (say, a fox) and prey X (e.g., a rabbit). We denote the population of prey X as x and that of the predator as y. Assume that X grows proportionally to its number (set the proportional coefficient to a) without the existence of prey, and decreases proportionally to the product of x and y, as X is eaten by Y when they are encountered (set the proportional coefficient to b). On the other hand, Y increases by eating X, and hence proportionally to the product of x and y, whereas it dies in proportion to its population (let each proportional coefficient be c and d). The evolution equation of this model is called the Lotka–Volterra equation, which is given by

$$\frac{dx}{dt} = ax - bxy, \qquad \frac{dy}{dt} = cxy - dy. \tag{1.18}$$

The fixed points of this model are given by setting the preceding equations to zero, leading to (i) $x^* = y^* = 0$, (ii) $x^* = d/c$, and $y^* = a/b$. Here, we are not interested in the former case, as no population exists. Therefore, we focus on case (ii). First, we investigate the linear stability around this fixed point. By linearizing the equation around it, we obtain the Jacobi matrix:

$$\begin{pmatrix} 0 & -bd/c \\ ac/b & 0 \end{pmatrix}.$$

The eigenvalues of the matrix are $\lambda = \pm i\omega$, where $\omega = \sqrt{ad}$. Hence, the real part is zero, and the fixed point lies at the border between stability and instability and is a center. The imaginary part is $\pm\omega$, x, and y in the state space is expected to rotate around the fixed point, with an angular velocity of ω.

Then, does this equation have a limit-cycle attractor? Because the eigenvalues are purely imaginary, the orbit neither approaches nor departs from the fixed point. Depending on the initial condition, the orbit rotates around the fixed point with a different amplitude (this corresponds to (f) in Fig. 1.3). Indeed, if we draw the orbits from different initial conditions, they rotate on different closed curves (see Fig. 1.9).

Undeniably, this behavior is due to a rather exceptional property of the Lotka–Volterra equation. Equation (1.18) is written as follows: $d(\log(x))/dt = a - by$, $d(\log(y))/dt = cx - d$ and hence we obtain $d(d\log(x) + a\log(y))/dt = acx - bdy$. Noting that $c\,dx/dt + b\,dy/dt = acx - bdy$ from the original equation, it follows that $d(d\log(x) + a\log(y) - (cx + by))/dt = 0$. In other words,

$$d\log(x) + a\log(y) - (cx + by) = \text{constant}. \tag{1.19}$$

Thus the preceding quantity takes a constant value independent of time. Such a quantity is called a conserved quantity.

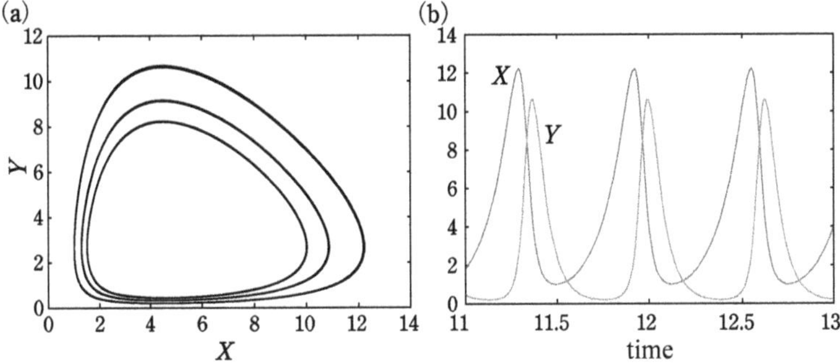

Figure 1.9 The orbit (a) and time-series (b) of the Lotka–Volterra equation.

Once this conserved quantity is given by an initial condition, it never moves to a state that has a different quantity value. The curve for a given conserved quantity provides a closed circuit that rotates around the fixed point. Hence, depending on the initial condition, oscillations of different amplitudes are given. The orbit does not approach a single curve, as in the case of a limit-cycle.

Such dynamical systems with a conserved quantity (quantities) are rather exceptional. In fact, a slight change in Eq. (1.18) (for instance, replacing bxy with $bxy/(1 + \delta y)$) destroys the existence of a conserved quantity. With such a slight change, the revised model has a limit-cycle attractor. Thus, a slight change in the system alters the qualitative behavior of the Lotka–Volterra equation. In mathematics, there is a concept of *structural stability*, which means that the qualitative (topological) behavior of a model is stable against small changes in the equations (e.g., a slight change in a parameter, the addition of a tiny term). In this sense, the Lotka–Volterra equation is not structurally stable, and although the equation is a pioneering model for oscillatory population dynamics, caution is needed when using it as a model for a biological system.

Structural stability is a mathematically rigorous concept, whereas it is not clear whether it can be adopted as a guideline for models of life and natural phenomena, because it postulates stability against any possible perturbations and may be too strict. However, biological systems have a certain degree of robustness to environmental changes and genetic variations. Thus, some modification of the concept of structural stability will be important in the criterion for choosing a model.

1.12 Elimination of Variables

In general, biological dynamical systems involve many variables. Multivariate dynamical systems are, of course, not easy to deal with. Therefore, theoretical methods are often devised to reduce the degrees of freedom (number of variables) without losing the essence of the original dynamical systems. One of the standard methods is the adiabatic elimination of fast variables. Suppose there is a dynamical system of (x, y), where x changes slowly and y changes rapidly. Consider

$$\frac{dx}{dt} = f(x, y), \qquad \frac{dy}{dt} = \gamma g(x, y). \qquad (1.20)$$

Assuming that $\gamma \gg 1$, x changes more slowly than y. Then, while x changes, y is assumed to reach the stationary state. In considering the time evolution of x, y is assumed to reach an attractor for dynamical systems with a given fixed x. Now, consider the case of a fixed-point attractor (When y is a single degree of freedom, the time evolution of y leads to a fixed point for given x, but the method here can be applied to a case with many variables.) Hence, we set $\frac{dy}{dt} = 0$. In other words, we solve $g(x, y^*) = 0$ and represent y^* as a function of x. Now, we write this solution as $y^* = G(x)$, and substitute it into the evolution of x to obtain

$$\frac{dx}{dt} = f(x, G(x)). \qquad (1.21)$$

Thus, the time evolution of x without y is obtained, which is called the adiabatic elimination of a fast variable (e.g., [7]).

Example 1.11   The gene expression dynamics of the protein concentration for the toggle switch discussed so far was obtained by eliminating the concentration of mRNAs. Here, mRNA1 (whose concentration is m_1) is transcribed from the corresponding gene. Protein P1 (whose concentration is p_1) is synthesized from mRNA1 at a given rate. The process of synthesizing mRNA2 (with concentration m_2) from gene 2 and then to protein P2 (with concentration p_2) progresses in the same way. The degradation rate of mRNAs is assumed to be γ, whereas the rates of synthesis and degradation of proteins were set at 1. (The rates could be different between 1 and 2; however, for simplicity, we assume they are equal.) As already mentioned, the rate of mRNA synthesis decreases with the concentration of the other protein, and the rates for mRNA1 and mRNA2 are assumed to be given by $S/(1 + p_2^2)$ and $S/(1 + p_1^2)$, respectively. All dynamical systems are then in the state space (m_1, p_1, m_2, p_2), and their time evolution is given by

$$\frac{dm_1}{dt} = \gamma\left(\frac{\alpha}{1 + p_2^2} - m_1\right), \qquad \frac{dp_1}{dt} = m_1 - p_1, \qquad (1.22)$$

$$\frac{dm_2}{dt} = \gamma\left(\frac{\alpha}{1 + p_1^2} - m_2\right), \qquad \frac{dp_2}{dt} = m_2 - p_2, \qquad (1.23)$$

where we set $\alpha = S/\gamma$.

The changes in mRNA concentration are much faster than those in proteins; in other words, γ and S are much larger than 1, the rates of protein synthesis and degradation. Then, for given protein concentrations, mRNA concentrations were balanced with a much faster time scale. Hence, we can eliminate mRNA concentrations. Assuming that (m_1, m_2) reached a steady state (a fixed point) for given protein concentrations, we obtain $dm_1/dt = \gamma(\alpha/(1 + p_2^2) - m_1) = 0$, $dm_2/dt = \gamma(\alpha/(1 + p_1^2) - m_2) = 0$. Solving these, we find $m_1 = \alpha/(1 + p_2^2)$ and $m_2 = \alpha/(1 + p_1^2)$. By using these, we obtain the dynamical system for p_1, p_2 as

$$\frac{dp_1}{dt} = \frac{\alpha}{1 + p_2^2} - p_1, \quad \frac{dp_2}{dt} = \frac{\alpha}{1 + p_1^2} - p_2. \tag{1.24}$$

These are the Eqs. (1.3) and (1.4) used in Section 1.4.

Example 1.12   Assume that X and Y make a complex C and decompose; that is, there are reactions $X + Y \leftrightarrow C$, and from the complex C, the product P is synthesized in proportion to the amount of C. In this case $\frac{dC}{dt} = k_+ XY - k_- C$. If the reactions $X + Y \leftrightarrow C$ are faster than $C \leftrightarrow P$, we eliminate C adiabatically so that $C = \frac{k_+}{k_-} XY$. Then, the synthesis rate of P is proportional to $\frac{k_+}{k_-} XY$. The elimination of the complex concentration is often adopted in reactions involving enzymes, as described in Section 2.2.

1.13 Dynamical Systems Approach in Cell Biology Experiments

In the twentieth century, the study of dynamic changes in biological systems was mainly based on macroscopic measurements, such as the changes in population numbers mentioned in Section 1.11. The dynamical systems approach to cell biology has been discussed by Goodwin [5], Rosen [15], and others theoretically since the 1960s, whereas direct experimental measurements were not possible at that time. This situation has changed significantly, with major technological advances ever since the 1990s. Now it has been possible to measure the dynamics of the cellular state using several variables.

In this respect, the method of fluorescent proteins has had a significant impact. Originally, Shimomura identified a protein called green-fluorescent-protein (GFP, green fluorescent protein) that emits jellyfish fluorescence; subsequently, a large number of proteins have been synthesized to improve fluorescence in a cell, together with changes in fluorescence wavelength and increases in fluorescence intensity. The expression level of the gene can be measured as the intensity of fluorescence by placing the gene of this fluorescent protein downstream of a specific promoter, or by placing the gene of this fluorescent protein into a fusion protein with a protein of interest. In addition, the development of a method for continuous measurement of a single living cell under a microscope has made it possible to follow the "variable" of intracellular protein concentration. The experiments to measure the variables x_1, x_2, and ... listed in the examples thus far have been the standard methods used.

The preceding discussion concerns the measurement of the intracellular dynamical system, but the introduction of a dynamical system into a cell has also become possible. Genetic engineering has advanced the technology of introducing genes into cells by plasmids (cyclic DNA that is replicated and inherited outside the genome) or into the genome, enabling the incorporation of a simple gene network so that cells exhibit specific dynamics. This allows for the engineering study of designing cells with specific properties, which is known as "synthetic biology." On the other hand, we set up simple conditions and measured the behavior of cells under these conditions to see what kind of logic of life can be drawn from them. This was proposed earlier, as "constructive biology" [11]. In both constitutive and synthetic biology, it is standard practice to introduce novel gene networks into cells and measure protein abundances.

Column I: Constructive vs. Synthetic Biology

Kaneko and Yomo proposed constructive biology as the standpoint to create the basic properties of biological systems in as simple a setting as possible, in order to uncover universal logic and laws in biological systems. Here, rather than creating an elaborately tuned functional system, we aim to extract the properties that biological systems should have even before they are optimized through evolution. On the other hand, synthetic biology was proposed and has been widespread [3]. Both share a common goal of creating some property in a cell, but synthetic biology is more oriented toward an engineering side to design a specific functional property. Constructive biology, on the other hand, stands at a heuristic position in which, by setting up simple conditions, one tries to extract the basic logic of life. One of the pioneering studies of synthetic biology is shown in Fig. 1.11, in which a gene circuit was designed and embedded in *Escherichia coli* so that the amount of protein oscillates over time, as shown in [3]. This led to the design of a cell with an oscillatory dynamics similar to that of a biological clock. In addition, by introducing the gene circuit of the toggle switch, the cells were designed to have bistable expression. Contrastingly, in constructive biology, the logic of cellular adaptation has been extracted through the discovery that, despite embedding this bistable genetic circuit, only one state adapted to the environment is chosen [9].

Example 1.13   Construction of a Toggle Switch Experiments were carried out to embed mutually repressive gene networks, or toggle switches, into cells, which we have discussed several times (Fig. 1.10). In particular, by introducing genes for fluorescent proteins that emit different wavelengths of light (e.g., green and red) downstream of each promoter, we can confirm that a bistable system with two attractors is actually constructed, with one expressed in green and the other in red.

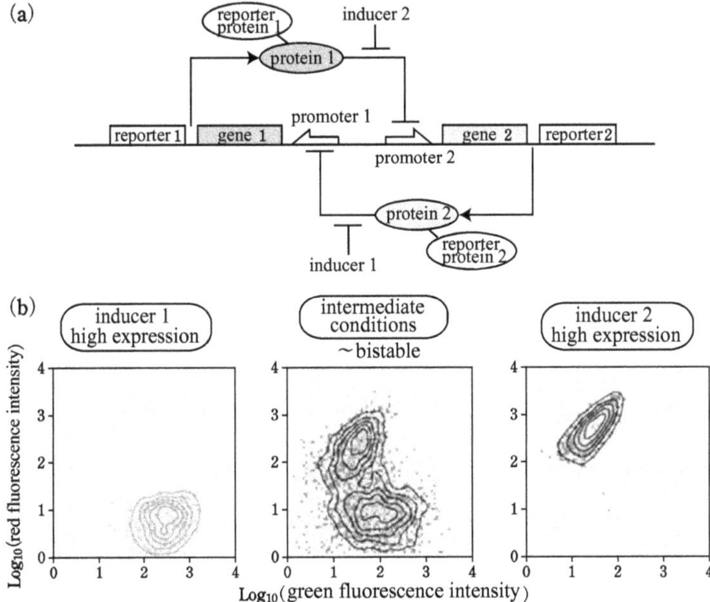

Figure 1.10 (a) Schematic diagram of the toggle switch. Protein 1 binds to the promoter region of gene 2 and represses its expression. Similarly, protein 2 represses the expression of gene 1. Each gene expression can be measured at the single-cell level using fluorescent proteins at different wavelengths. The strength of each expression or repression can be controlled by adding a molecule called an inducer. For example, the addition of more inducer 1 weakened the repression of gene 1 expression, resulting in an increase in gene 1 expression. (b) Experimental results: The horizontal and vertical axes show the expression of genes 1 and 2, respectively, and each point corresponds to the state of a single cell. (The density distribution is overwritten because the points are too densely concentrated.) Bistability was achieved under intermediate conditions (by courtesy of Dr. Saburo Tsuru (The University of Tokyo)).

Example 1.14   Repressilator: Constitution of a Limit-Cycle Elowitz and Leibler designed a gene circuit and embedded it in *E. coli* so that the expression level of a protein oscillates over time. This was a pioneering study in synthetic biology. Specifically, they introduced a gene network, as shown in Fig. 1.11. There are three genes, 1, 2, and 3, where expression of 1 suppresses 2, expression of 2 suppresses 3, and expression of 3 suppresses 1.

By writing the abundances of each mRNA as m_i ($i = 1, 2, 3$), and those of the corresponding protein as p_i, the temporal evolution is given by $\mathrm{d}m_i/\mathrm{d}t = \gamma(\alpha/(1 + p_{i-1}^2) - m_i)$, $\mathrm{d}p_i/\mathrm{d}t = m_i - p_i$, where we used the notation $p_0 = p_3$. By solving the evolution of dynamical systems, it is shown that the attractor is a limit-cycle for a certain range of parameter values. Indeed, Elowitz and Leibler numerically obtained the limit-cycle attractor from the preceding equations, as shown in the figure.

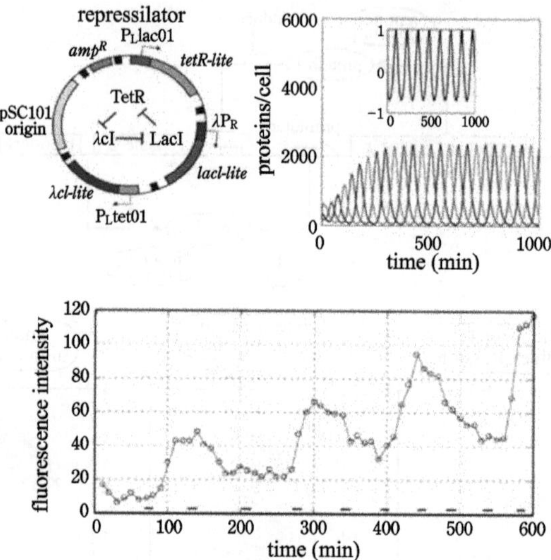

Figure 1.11 Expression oscillation in the repressilator. Left top: Plasmid introduced in *E. Coli*, where three genes (TetR, LacI, λcI) that mutually inhibit their expression are coded. Right top: Time series of the expression levels of the three genes, obtained from numerical simulations of the model. Bottom: The time series of the expression of the level of TetR obtained from the experiment using the fluorescent protein (reproduced from [3] with the courtesy of Dr. Elowitz and under the permission of Nature).

In contrast, in the experiment, p_1 was measured from the amount of fluorescence using a green fluorescent protein, and the temporal oscillation was confirmed (Fig. 1.11). The oscillation itself is observed as designed, although it does not perfectly match the behavior of the preceding equations, since cells grow and the concentrations are diluted according to the volume increase, and the intracellular environment is not constant.

However, in reality, there are many cases where the embedded gene circuit does not behave as designed. Various gene circuits are synthesized by incorporating activation, repression, and their combination. In this sense, one might expect that if we design a circuit diagram, one can perform a kind of engineering much the same as the design of electronic circuits that behave accordingly. This expectation has led to the establishment of a field of synthetic biology. However, things are not so easy. First, in a cell, interactions with various other components are not always negligible. Next, a dilution term of protein concentration is introduced as a result of cell volume growth, the rate of which depends on the expression of some genes. In general, the rate of each reaction varies depending on whether the cell state is adapted, and the concentrations of amino acids and ATP used for protein synthesis also vary. In this sense, the intracellular behavior of gene expression does not always follow the engineering design. On the contrary, in constructive biology, one end extracts the logic of cellular adaptation from the deviation from the design.

Multidimensional Measurement

As we have already mentioned, there are many protein species in a cell, and the dynamics of a cell cannot be tracked only by changes in the amount of a single protein. To obtain multidimensional information on changes in protein abundances, it is necessary to introduce fluorescent proteins of different colors into each gene. However, it is not easy to introduce many genes at the same time, and there is also the problem of spectral overlap between fluorescent proteins. In addition, the fluorescent proteins themselves are not harmless to the cells. Hence, it is difficult to express a variety of fluorescent proteins simultaneously. (Currently, there is an example to embed seven different fluorescent proteins, but usually only two or three can be stably measured.)

Instead, transcriptome analysis is widely adopted to examine the multidimensional state of cells, by using concentrations of several thousand $\sim$ tens of thousands of mRNAs that are expressed from all genes in the genome at once. As of 2020, a technique called RNA-Seq, which uses massively parallel sequencers, has been mainstream. This is a measurement method for quantifying the number of mRNAs corresponding to each gene by extracting mRNAs from cells and decoding their nucleotide sequences. In a typical RNA-Seq analysis, mRNA is extracted from a large number of cells (e.g., 10^6 or more), and the amount obtained gives the average of expression levels over the population of cells. In contrast, single-cell RNA-Seq has been developed to quantify the expression level of mRNA extracted from a single cell by using microfluidic techniques, such as aligning the cells in a flow channel and letting them lyse in a microdomain. Using this method, the multidimensional cellular state over a large number of cells (e.g., 10^6) as the expression levels of mRNA species can be obtained.[5] In addition, if some gene expression for cells that are examined falls to a limit-cycle attractor, the corresponding attractor structure should be embedded in a cloud around a one-dimensional closed curve in the expression space.

In addition to the measurement of mRNA expression, proteomic analysis, which uses mass spectrometry to comprehensively quantify the number of proteins present in a cell, and metabolome analysis, which uses mass spectrometry to simultaneously quantify the amount of a large number of metabolites (e.g., several hundred types), are also used as multidimensional measurements of cellular states.

---

[5] For instance, in [2], single-cell RNA-Seq was carried out in over 2 million cells at different stages of development, and multidimensional dynamics were analyzed.

Furthermore, several methods using massively parallel sequencers have recently been developed, including the Hi-C method to analyze the structure of the genome, ChIP-Seq to measure the amount of proteins that are bound to the genome, and techniques to measure the quantities of epigenetics (changes that do not involve genome sequence changes, such as DNA modification and histone modification). Currently, with the preceding methods, only an instant cellular state, not the time course, is measured, because the measurement involves destroying the cells, so that they cannot be tracked further. Hence, these methods to measure multidimensional quantities are complementary to fluorescence imaging technology, which can capture the dynamics of a few components. We have not yet reached the point of measuring the time-series of multidimensional cellular state of a single cell as a trajectory, but as measurement and informatics techniques in this field are progressing rapidly, the analysis of the trajectory in a high-dimensional state space will be possible in the near future.

Measurement of the Distribution of States over Cells

Up to this point, we had in mind the measurement of the internal state of a single cell, for example, protein abundances. In many cases, it is necessary to obtain the distribution of such states (e.g., protein abundances) across cells from an ensemble of cells, instead of one by one, and to investigate how the distribution changes depending on the condition. For this purpose, a flow cytometer is effective. Using this equipment, one can obtain the distribution of cell states, such as cell size, density, and intensity of fluorescence of proteins. By putting and flowing cells in a droplet one by one, and with the use of laser injection, one measures the forward and bilateral scattering of a droplet containing a cell. The distribution of points in a state space can be obtained simultaneously over a large number of cells. For example, from a population of cells with embedded genes to code green and red fluorescent proteins, one can measure the expression levels of the two proteins in the toggle switch with a large number of cells to confirm that they are indeed bistable (see Fig. 1.10(b)).

Furthermore, by using a device called a cell sorter, cells are classified according to certain conditions, based on the flow cytometer. For example, it is possible to obtain only cells that exhibit a certain range of fluorescence. In this way, it will be possible to perform measurements from a population of cells whose state is within a certain range. As mentioned earlier, it is often necessary to arrange (nearly) uniform initial cell states, for which a cell sorter can be used. For example, it is possible to align the phases of the oscillating cell states or to select only one of the bistable states. Of course, since the state of a cell has many variables, it is important to note that even if some of the variables are aligned in the cell sorter, it is possible that differences remain due to "hidden variables" that were not measured.

1.14 Biological Significance of Attractors

As we have seen so far, it is hard for any dynamical systems model to fully represent the system in concern of interest. Considering that molecules react with other molecules when they collide with each other stochastically, the equations of the dynamical system describe an average behavior, and fluctuations exist around them. As we will see in Chapter 6, one can add a noise term to the equations of dynamical systems to include the stochasticity around the average behavior. Therefore, it is important to determine whether the behavior of the dynamical system is not significantly altered by this noise. From the standpoint of the cell, the question is whether the state of the cell can be stable even if the number of molecules is not so large and may cause deviation from the average value described by the dynamical system.

In this regard, attractors are an important concept. If the state is represented as an attractor, then it is stable to return to its original state even if it is slightly perturbed from it. As long as the state remains within the basin of the attractor, the state after perturbation will be attracted to the state. In this sense, attractors are essential in considering the stability of biological states.[6]

For example, the number of proteins in each cell is not identical, but varies slightly, even among the same type of cells. However, if the variation is not too large and the cells are regarded as belonging to the same type, these states are considered to belong to the same attractor even though they are perturbed by noise. Therefore, it would be natural to consider different cell types as different attractors. Even if there are some molecular fluctuations or external disturbances that cause the state to deviate from the attractor, the state will remain around its original attractor. Cells belonging to different attractors remain close to their attractors unless they are disturbed enough to exceed the basin of the attractor. In other words, if many cells are represented in the state space, they are expected to be represented as some "clouds" around the attractors.

For example, cells in the epidermis are slightly different from each other, but belong to the same attractor: cells in the myocardium also slightly differ from each other, but they belong to the same attractor, and so forth. Each of the clouds around attractors may be regarded as a cell type.

When there are many attractors in a single dynamical system, each stable state can be regarded as a different cell type. Thus, we can represent the existence of many kinds of stable cell types. The cell differentiation under this viewpoint will be discussed in Chapter 7.

---

[6] For the exceptional attractors without such asymptotic stability, see Section 1.10.

1.15 Bifurcation

In general, a dynamical system has several control parameters that determine its time evolution, in addition to variables that change with time. These parameters are given in terms of conditions, for example, specifying the environmental conditions of an organism. External concentrations of chemical components (e.g., nutrients) outside the cell, temperature, and rates of reaction are such examples. As we change these parameters, the flow of the dynamical system changes, as does the attractor. In the case of living organisms, it can be said that their state changes in response to the changes in environmental conditions. Naturally, if one changes the parameters continuously, the attractor usually changes continuously as well. The value of a fixed point, the position of the limit-cycle, amplitude, period, and so on, changes continuously with the parameters. In contrast, changing the parameters may result in qualitative changes in the attractor. In this case, the original attractor disappears and is replaced by another attractor at a certain parameter: The flow of the dynamical system around the attractor changes, for example, the trajectory that was previously attracted to a point turns out of it. Such qualitative changes are called bifurcations. If the number of parameters associated with the bifurcation is one, it is called a co-dimension 1 bifurcation. If the corresponding number is two, it is called a co-dimension 2 bifurcation. The classification of possible bifurcations has been thoroughly investigated in mathematics. Here, we explain a few examples that often appear in biological systems, by focusing on the case of co-dimension 1 bifurcation. First, consider the simplest case in which the variables of the relevant dynamical system are just one.

(i) Exchange of stability (transcritical bifurcation)

Consider the case in which there are two fixed points A and B. Now, assume the following situation: For the parameter $a < a_c$, A is stable and B is unstable, and at $a = a_c$, the two fixed points collide, and for $a > a_c$, A is unstable and B is stable.

Example 1.15   Consider

$$\frac{\mathrm{d}x}{\mathrm{d}t} = ax - x^2. \tag{1.25}$$

The fixed points are $x = 0$ and $x = a$. As can be verified straightforwardly, for $a < 0$, $x = 0$ is stable and $x = a\,(< 0)$ is unstable. At $a = 0$, both the fixed points have the same value. For $a > 0$, $x = 0$ is unstable, and $x = a$ is stable. The plot of the fixed-point attractor as a function of a is shown in Fig. 1.12(a). As shown, the value is constant for $a \leq 0$, and for $a > 0$, it increases linearly with a.

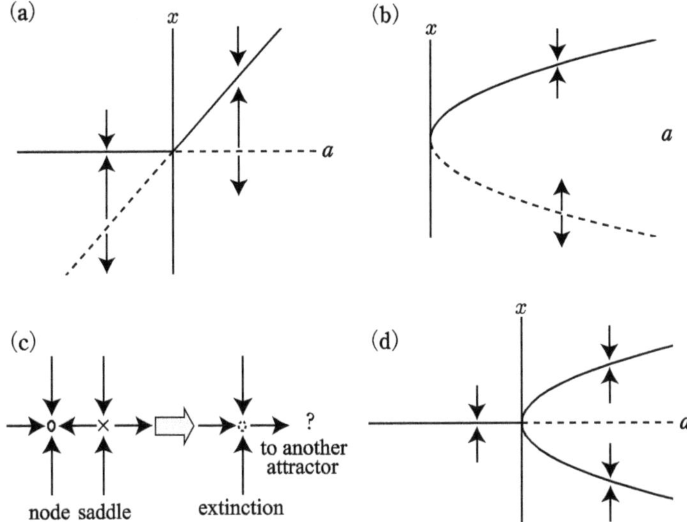

Figure 1.12 Schematic representation of the bifurcations. (a) Exchange of stability, (b) saddle-node bifurcation, (c) collision of saddle and node, and (d) pitchfork bifurcation. The dotted line represents unstable fixed points, and the solid line represents the stable fixed point (attractor).

Example 1.16   Consider the case where protein X regulates the expression of the gene that encodes it. That is, mRNA synthesis from a gene is given by $f(x)$, a function of the protein concentration x. Here, again, we adiabatically eliminate the mRNA concentration, and set the degradation rate of the protein (or dilution) as 1. Then, the change in x is given by

$$\frac{\mathrm{d}x}{\mathrm{d}t} = f(x) - x. \tag{1.26}$$

Here, $f(x)$ is proportional to x if X simply increases its own synthesis. However, if the concentration of the protein increases too much, usually the synthesis cannot make it in time, and it is saturated. In the limit where x is large, $f(x)$ approaches a constant value. Let this constant value be a and let the concentration at which the saturation effect begins to appear be x_c. Then, a simple form for $f(x)$ will be $f(x) = ax/(1 + x/x_c)$ (also see the Michaelis–Menten form in Section 2.2.1). In this case, the fixed points in Eq. (1.26) are $x = 0$ and $x = x_c(a - 1)$. A simple calculation shows that for $a < 1$, $x = 0$ is stable and $x = x_c(a - 1)$ is unstable (in fact, it is negative and therefore not meaningful in the current problem). At $a = 1$, they both take the same value, and for $a > 1$, $x = 0$ is unstable, and $x = x_c(a - 1)$ is stable. Hence, $a = 1$ is the point at which bifurcation occurs. For this cell, by increasing the parameter a, the protein X is expressed whose concentration increases proportionally to $(a - 1)$ if a exceeds 1 (Fig. 1.12(a)).

(ii) Saddle-node bifurcation

By changing the parameter, a stable fixed point (attractor) and an unstable fixed point collide and a pair annihilates (or if we change the parameter from the opposite direction, they pair-create).

Example 1.17

$$\frac{\mathrm{d}x}{\mathrm{d}t} = a - x^2. \tag{1.27}$$

In this case, a stable fixed point $x = \sqrt{a}$ and an unstable fixed point $x = -\sqrt{a}$ exist for $a > 0$. They take the same value at $a = 0$ and disappear for $a < 0$, where x is negative. With this equation, only x goes to $-\infty$, and if there is an additional term to suppress this divergence, another attractor will be generated at $x < 0$. Hence, the nature of attractors changes at $a = 0$, which is the bifurcation point (see Fig. 1.12(b)).

Example 1.18　By taking Example 1.16, assume that the synthesis of X is activated by the dimer of protein X, so that the synthesis rate is proportional to x^2, but for $x \gg 1$, it is saturated to approach as constant value. For simplicity, we consider the case with $f(x) = ax^2/(1 + x^2)$. In this case, the fixed point $x = 0$ is always stable for all $a > 0$, whereas another pair of fixed points $x = (a \pm \sqrt{a^2 - 4})/2$ exists for $a > 2$. In this pair of fixed points, the larger one (i.e., $(a + \cdots)$) is stable, and the smaller one $(a - \cdots)$ is unstable (see Fig. 1.12(b)). They take the same value at $a = 2$. Hence, by decreasing a from $a > 2$, the stable and unstable fixed points collide at $a = 2$ and pair-annihilate, where only attractor $x = 0$ remains. In other words, for $a > 2$, a high expression state of x and $x = 0$ coexist, but for $a < 2$, the former disappears, and only the null expression state remains. Here at $a = 2$, the high-expression state takes the value $x = 1 > 0$. Hence, by decreasing a, there is a discontinuous transition in the expression to 0.

(iii) Pitchfork bifurcation

By changing a parameter value, a stable fixed point becomes unstable, and at this parameter value, a pair of stable fixed points appears.

Example 1.19

$$\frac{\mathrm{d}x}{\mathrm{d}t} = ax - x^3. \tag{1.28}$$

The fixed points are $x = 0$ for all a and $\pm\sqrt{a}$ for $a > 0$. From the linear stability analysis, $x = 0$ is stable for $a < 0$, and for $a > 0$, $x = \pm\sqrt{a}$ is stable (see

Fig. 1.12(d)). When a is increased from a negative value, the fixed point $x = 0$ loses its stability at $a = 0$ and becomes an unstable fixed point; instead, a new pair of stable fixed points $x = \pm\sqrt{a}$ is created. $x = 0$ is still a fixed point but is unstable, so that the orbit moves away from its vicinity. In this case, increasing the parameter a from a negative value to the bifurcation point $a = 0$, the attractor $x = 0$ loses its stability, which is replaced by the two fixed-point attractors $x = \pm\sqrt{a}$. As a increases further, x changes continuously in the form of $x = \pm\sqrt{a}$.

Example 1.20 The toggle switch we have discussed a few times shows this pitchfork bifurcation at $\alpha = 2$, with the increase in α. At $\alpha < 2$, only the stable fixed point with $p_1 = p_2$ exists, whereas for $\alpha > 2$, the fixed point is unstable and is replaced by a pair of stable fixed points with $p_1 > p_2$ and $p_2 > p_1$ (see Fig. 1.2).

So far, we have mainly presented a case with one variable. The last example has two variables, but the bifurcation occurs only from one direction. The flow is always attracted to a one-dimensional manifold; thus, it is essentially a one-variable case. As long as the flow from all other directions except the one-dimensional manifold remains stable while the stability is changed only in one direction, the preceding three examples can be used in the same way for more than one variable. In particular, when the stable and unstable fixed points collide in case (ii), the unstable fixed point is a saddle because the flow from the other directions is stable, whereas the stable fixed point before the bifurcation is a node because it is attracted from any direction. Hence, the saddle and node collide in case (ii); thus, it is called a saddle-node branch (see Fig. 1.12(c)).

(iv) Hopf bifurcation

Thus far, we have discussed a bifurcation between fixed-point attractors, but for dynamical systems with more than one variable, a limit-cycle appears. In this case, in correspondence with cases (i) and (ii), the limit-cycle collides with an unstable fixed point or an unstable periodic solution. However, the case inherent to the limit-cycle is seen when a fixed-point attractor loses its stability with a change in a parameter value and is replaced by a limit-cycle attractor. Consider the case in which the eigenvalues of the Jacobi matrix around a fixed point are complex. For the parameter $a < a_c$, the real part is negative, so that the fixed point is stable. Then, for $a > a_c$, assume that the real part turns positive (the imaginary part is nonzero, without much change). Then, the fixed point becomes unstable, and the orbit going out of it is often attracted to a limit-cycle, as discussed in Section 1.9.

This is called the Poincaré–Andronov–Hopf bifurcation (in this volume, we simply call it Hopf bifurcation as usually adopted).

Example 1.21 Let us include a bifurcation parameter in Eqs. (1.9)–(1.10). Specifically, we considered the model

$$\frac{dx}{dt} = a(x - \omega y) - (x^2 + y^2)x, \qquad \frac{dy}{dt} = a(y + \omega x) - (x^2 + y^2)y. \qquad (1.29)$$

Using the transformation adopted in the derivation of Eq. (1.11), we obtain $dr/dt = ar - r^3$, $d\theta/dt = a\omega$. Here, $x = y = 0$ ($r = 0$) is a stable fixed point for $a < 0$, whereas for $a > 0$, the fixed point becomes unstable, and the attractor is replaced by a limit-cycle with $r = \sqrt{a}$. Hence, at $a = 0$, Hopf bifurcation occurs.

Example 1.22 Consider the Brusselator model given by Eqs. (1.13) and (1.14), respectively. For $B < 1 + A^2$, $(X, Y) = (A, B/A)$ is a stable fixed point and at $B = 1 + A^2$ bifurcation occurs, and for $B > 1 + A^2$, the attractor is a limit-cycle.

Example 1.23 By using a complex variable $z = x + iy$, and considering

$$\frac{dz}{dt} = a\exp(i\theta)z - |z|^2 z, \qquad (1.30)$$

this equation can be used as a normal form of Hopf bifurcation, to which systems with Hopf bifurcation are universally reduced near the bifurcation point. As can be seen, there is a limit-cycle attractor for $a \geq 0$.

(v) Global bifurcation

The bifurcation discussed so far concerns a local change, in which the stability of a fixed point changes to show a bifurcation, whereas there is a case that the stability of attractor is changed globally in the case of limit-cycle: the limit-cycle collides with the fixed point and disappears globally. For example, in the case of Fig. 1.13, changing the parameters results in a change in flow, as shown in the figure, so that the limit-cycle is replaced by a fixed-point attractor. Viewed from the side of the fixed point, this change can be regarded as a saddle-node bifurcation; thus, it is called a saddle-node on an invariant curve (SNIC) bifurcation. As a result, the periodic solution disappears and is replaced by a fixed-point attractor. This is an example of a global bifurcation.

Recall, an example of the activation-inhibition gene network, discussed in Eq. (1.15). There is no fixed point on the upper right side of Fig. 1.7. When θ_1 in the model equation (Eq. (1.15)) is decreased, the nullclines collide in the upper right, resulting in two fixed points. One is a stable node and the other is a saddle. With the creation of fixed points, the limit-cycle disappears. This bifurcation leads to a change from an oscillatory to a fixed expression state.

There can also be other complex bifurcations. In particular, when there are two types of control parameters, there can be a combination of the bifurcations

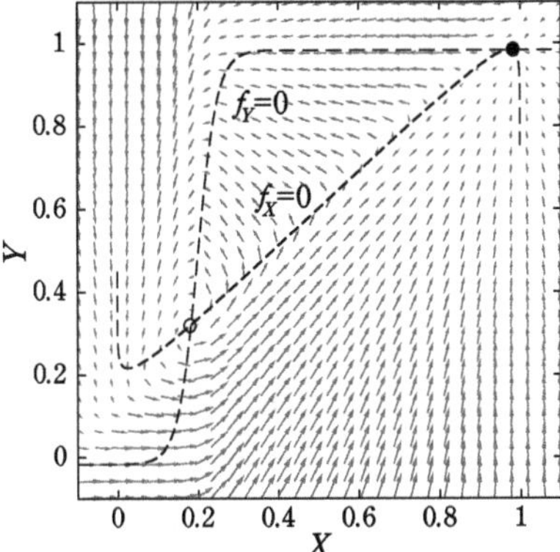

Figure 1.13 Consider the gene expression model of Eq. (1.15) and change the parameter θ_1. At $\theta_1 \sim -0.117$, the limit-cycle collides with a fixed point. For θ_1 below this value, the limit-cycle attractor disappears and is replaced by a fixed-point attractor. Here, (X, Y) denotes (x_1, x_2), and the dotted curve denotes the nullcline for X and Y.

described previously, which is sometimes important. Still, they are a bit advanced, and we will not go further in depth here.

Bifurcation is important in cell biology because changing conditions such as the environment often result in qualitative changes in the state of the cell. When the concentration of signal molecules is changed, the cellular state changes; it may go to a different stage of the cell cycle, such as G0 and G1; depending on the external signal concentration, the morphology of cells is changed; cells start to move or assemble in response to external signals and start to differentiate.

Such qualitative changes in the cellular state can be captured from the perspective of bifurcation. Furthermore, the history dependence of the cellular state may also be understood by bifurcation. If we can express the influence of other cells as a control parameter, the change in cellular state upon interaction with other cells can be treated also as bifurcation (see Section 7.7). If the same bifurcation occurs in different models, the mechanisms behind them can be regarded as qualitatively equivalent. This will allow us to understand the various behaviors of biological systems coherently in terms of dynamical systems.

1.16 Chaos

Attractors are not limited to fixed points and limit-cycles when the number of variables (dimension) is greater than two. In the two-dimensional case, due to the Poincaré–Bendixson theorem, any attractor that is not a fixed point is a limit-cycle, that is, an attractor with time-periodic behavior. If the space is beyond two dimensions, nonperiodic attractors can exist because the state space is not divided into the inside and the outside by a one-dimensional curve, and the Poincaré–Bendixson theorem is not applicable.

First, there is an attractor called the quasi-periodic state (torus), in which the attractor is not a simple periodic solution but is given by a superposition of limit-cycles of two different (incommensurate) frequencies. Geometrically, this attractor is located on a two-dimensional surface, for example, on the surface of a doughnut.

Another type of attractor which shows more complex behaviors is the so-called chaotic attractor. The time-series of the chaos attractor shows an irregular, aperiodic oscillation (see Fig. 1.14a), in which the state does not completely return to the original value. However, the trajectory is restricted within a certain area in the state space, and the orbit is attracted toward the attractor (see Fig. 1.14b). Even though the state value of an attractor does not return to the earlier value exactly, it will eventually return to its neighborhood, as the orbit is bounded within a finite region.

There are several well-explored examples of dynamical systems that show such chaotic behavior. Here we give one of the simplest equations that exhibits a chaotic attractor, the Rössler equation, given by

$$\frac{\mathrm{d}x}{\mathrm{d}t} = -y - z, \quad \frac{\mathrm{d}y}{\mathrm{d}t} = x + ay, \quad \frac{\mathrm{d}y}{\mathrm{d}t} = b + xz - cz. \tag{1.31}$$

In Fig. 1.14, we give an example of the time series of x for this equation, whereas the attractor in the space (x, y, z) is given as the orbits for time interval $550 < t < 900$. Here the motion of (x, y, z) is not periodic, and the state does not return to exactly the same value, even though it can come back close to the previous state.

Let us consider a slight perturbation to the initial value in a chaotic attractor, and compare the two orbits. In this case, the difference between the two orbits, which is initially tiny, increases exponentially with time t. Infinitesimal difference in the orbits increases as $exp(\lambda t)$, where λ, the rate of increase, is called the Lyapunov exponent. The positivity of λ gives a criterion of chaos.

It should be noted that the orbit of chaotic attractor is confined within a finite region, and after a slight perturbation in the initial condition, the perturbed orbit, even though the difference from the original is amplified, remains on the same attractor. Hence, the increase in the difference between the two orbits is bounded. However, amplification of the slight initial difference always exists on the attractor (sensitive dependence on the initial condition). Hence, when we are concerned with the prediction of a single orbit from a given initial condition, there is unpredictability. Any small difference in the initial condition is amplified, to make the prediction of future value at a given time is harder. The prediction of the state value after a sufficient time later is not possible, as long as the precision of the initial condition is finite.

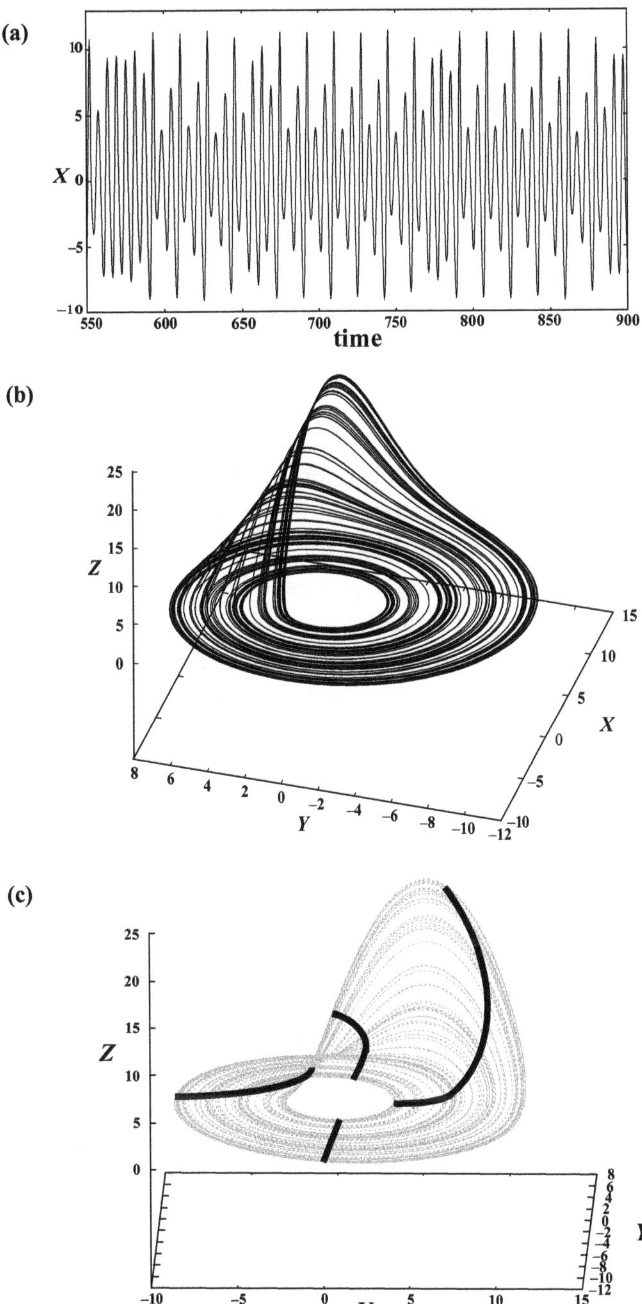

Figure 1.14 Time series, attractor, and stretch-folding process of the Rössler equation: (a) the time series of x, (b) the orbit in the state space (x, y, z), and (c) stretch-folding process of the dynamics.

In contrast to this orbital instability, the orbits from different initial conditions ultimately fall on the same attractor. Hence, statistical prediction over an ensemble of orbits is possible. For example, in the Rössler equation, the probability that the value of x is larger than 0, over a long period of time, is predictable, even though whether $x(t)$ is larger than 0 after sufficiently large time t is unpredictable, as long as the initial value $(x(0), y(0), z(0))$ is measured under a finite precision. In other words, even though chaos is a consequence of a "deterministic" dynamical system in the sense that the later time evolution is determined by a dynamical system, for the prediction of the state values, one needs a stochastic description. Below, we list up some characteristics of chaos.

- Stretching and folding: As the initial difference is expanded, the flow in the state space is diverted, whereas the orbits return to the original region. To satisfy these two postulates, the orbits are stretched in the state space and folded to remain within a finite region (see Fig. 1.14c for this process for the Rössler attractor). This flow structure is similar to the process of kneading bread or pie dough.
- Fractal attractor: The set of chaotic attractors is not a simple object, such as points (zero-dimension, fixed point), curves (one-dimension, limit-cycle), or surfaces (two-dimensional torus). The chaos in a three-variable system consists of infinitely many surfaces. When we coarse-grain and see, the chaotic attractor appears to be located on a two-dimensional surface, but if we look finely, there are more surfaces arbitrarily close to each other. If we expand and see, we can find more surfaces. By expanding the fine structure, a similar structure is observed. The attractor involves infinitely fine, self-similar structures. Such a geometric object with a self-similar fine structure is called a *fractal*, which is represented by a noninteger dimension (fractal dimension).
- Infinitely many frequencies are needed to decompose: chaos appears as an attractor of differential equations with few degrees of freedom, but when one decomposes into oscillators, infinitely many frequencies are required. This is in strong contrast with periodic or quasiperiodic attractors. In the case of a limit-cycle, the state is represented by a combination of Fourier modes $\sin(2n\pi ft)$ with $n = \pm 1, \pm 2, \pm 3, \ldots$, where the basic frequency f is given by $f = 1/T_p$ with a period T_p. In the case of quasiperiodic attractor on a two-dimensional torus, it is represented by the combination of Fourier modes with two basic frequencies. To represent the time series of chaotic attractor, continuously many frequencies are necessary: continuous spectra are needed to represent the chaotic motion by Fourier modes.
- Diversity of orbits: Consider the measurement of a chaotic orbit with a finite resolution (i.e., with a finite number of symbols, say $0, 0.1, 0.2, \ldots$). Then, the trajectory can be represented as a sequence of symbols. For example, let us divide

the state space into two parts, say $x > 0$ or $x \leq 0$, and then examine if the state lies in which part per given time span. For instance, we term the state right (R) if $x > 0$ and left (L) otherwise. Then the time series of a trajectory is represented by a symbol sequence as RLRLRRLLRLLL .... The diversity of this sequence of symbols can be judged by the degree to which different symbol sequences (RLL ...) appear when a certain length of subsequences is taken. In the case of periodic solutions, even if the diversity of subsequences starts to increase with the length, the increase in the diversity of symbol sequence terminates. For example, if a period-2 cycle is repeated (RLRLRLRL ...), increasing the number of subsequences does not increase the diversity; for example, LR and RL for two-character subsequences, or LRL and RLR for three-character subsequences; the variety of symbol sequences remains at two, and no more increase occurs.

On the other hand, in the case of chaos, as the length of the subsequences increases, novel symbol sequences continue to appear. As will be discussed in Chapter 10, the rate at which novel symbol sequences appear (including its probability) can be computed as the amount of information, or Shannon entropy. Using this entropy, the diversity of chaotic trajectories can be characterized as having a positive amount of information per character, as the length of the subsequence goes to infinity. In this sense, a chaotic attractor can be regarded to generate a sequence with positive information amount.

Chaos is often present in differential equations with more than two variables, so that they can be present in chemical reactions or in the dynamics of gene expression. Indeed, there are some examples of biological phenomena that appear to show chaos, particularly in neuroscience. In the present volume, however, such examples are not presented, and therefore, we do not discuss chaos further here. If one is interested in this topic, please refer to some introductory textbooks on chaos (e.g., [1, 8, 16]).

References

[1] P. Bergé, Y. Pomeau, and C. Vidal. *Order within Chaos*. New York, John Wiley and Sons, 1984.

[2] J. Cao *et al.* The single-cell transcriptional landscape of mammalian organogenesis. *Nature*, **566**(7745): 496–502, 2019.

[3] M. B. Elowitz and S. Leibler. A synthetic oscillatory network of transcriptional regulators. *Nature*, **403**(6767): 335–338, 2000.

[4] N. Goldenfeld. *Lectures on Phase Transitions and the Renormalization Group*. Boston, Addison-Wesley, 1992.

[5] B. Goodwin. *Temporal Organization in Cells*: *A Dynamic Theory of Cellular Control Processes*. London, Academic Press, 1963.

[6] Y. Goto and K. Kancko. Minimal model for stem-cell differentiation. *Physical Review E*, **88**(3): 032718, 2013.

[7] H. Haken. *Synergetics*. Berlin, Springer, 1979.

[8] M. W. Hirsch, S. Smale, and R. L. Devaney. *Differential Equations, Dynamical Systems, and an Introduction to Chaos*. Cambridge, MA, Academic Press, 2003.

[9] K. Kaneko. *Life: An Introduction to Complex Systems Biology*. New York, Springer, 2010.

[10] K. Kaneko. Dominance of Milnor attractors and noise-induced selection in a multi-attractor system. *Physical Review Letters*, **78**(14): 2736, 1997.

[11] K. Kaneko and I. Tsuda. *Complex Systems: Chaos and Beyond – A Constructive Approach with Applications in Life Sciences*. Berlin, Springer, 2000.

[12] J. Milnor. On the concept of attractor. In B. R. Hunt, T-Y. Li, J. A. Kennedy, and H. E. Nusse (eds.), *The Theory of Chaotic Attractors* (pp. 243–264). New York, Springer, 1985.

[13] G. Nicolis and I. Prigogine. *Self-Organization in Nonequilibrium Systems*. New York, Wiley, 1977.

[14] Y. Oono. *The Nonlinear World: Conceptual Analysis and Phenomenology*. Tokyo, Springer, 2012.

[15] R. Rosen. *Dynamical System Theory in Biology*. New York, John Wiley and Sons, 1970.

[16] S. H. Strogatz. *Nonlinear Dynamics and Chaos: With Applications to Physics, Biology, Chemistry, and Engineering*. Boulder, Westview Press, 2014.

2

Input–Output Relationships of Cell Systems

2.1 Input–Output Functions of Association–Dissociation Reactions

Cells regulate their proliferation, differentiation, and motility in response to the external environment. In many cases, these responses are realized through a combination of binding, dissociation, and catalytic reactions that involve intermolecular interactions with very high specificity. This chapter discusses the cellular responses due to such chemical reactions from a mathematical perspective.

2.1.1 Association and Dissociation Equations

Consider the binding and dissociation between a molecule X and a molecule Y in a sufficiently dilute solution. If XY is assumed to be the complex of X and Y, then

$$X + Y \underset{k_{\text{off}}}{\overset{k_{\text{on}}}{\rightleftharpoons}} XY. \tag{2.1}$$

Let us denote the concentration of X as $[X]$ and so on. Then, the reaction rate equation is represented as follows:

$$\begin{aligned}
\frac{\mathrm{d}[XY]}{\mathrm{d}t} &= k_{\text{on}}[X][Y] - k_{\text{off}}[XY] \\
&= k_{\text{on}}[X](Y_{\text{total}} - [XY]) - k_{\text{off}}[XY],
\end{aligned} \tag{2.2}$$

where $Y_{\text{total}} = [Y] + [XY]$, k_{on} is a binding rate constant, and k_{off} is a dissociation rate constant. The dimensions of k_{on} and k_{off} are "1/(concentration· time)" and "1/time," respectively.

In the steady-state condition, by using the dissociation constant

$$K_d = \frac{k_{\text{off}}}{k_{\text{on}}} = \frac{[X][Y]}{[XY]}, \tag{2.3}$$

we obtain

$$\frac{[XY]}{Y_{\text{total}}} = \frac{[X]}{K_d + [X]}. \tag{2.4}$$

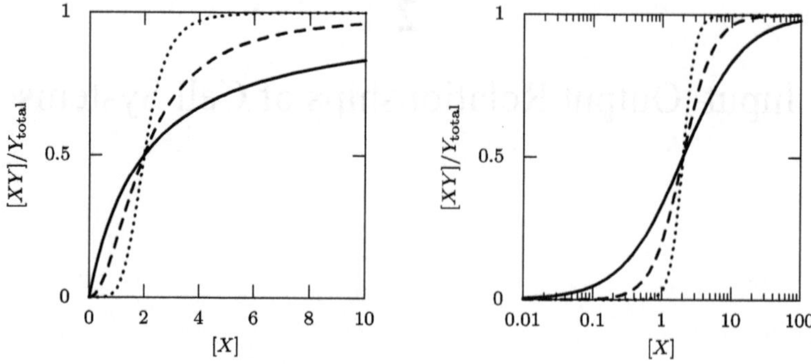

Figure 2.1 Hill equation (Eq. (2.8), $K_d = 2$). (left) normal plot, (right) semilog plot, $n = 1$ (solid line), $n = 2$ (dashed line), $n = 5$ (dotted line).

The solid line in Fig. 2.1 represents the Eq. (2.4). Such an S-shaped curve is often called a sigmoidal curve. The value of $[XY]/Y_{\text{total}}$ goes to zero for small values and asymptotes to 1 for large values. By substituting $[X] = K_d$, the value takes 0.5. Thus, at the concentration of K_d, half of all Y molecules are bound to X. The smaller K_d is, the higher the affinity between the molecules. For example, the binding of avidin to biotin (vitamin B7) is one of the strongest known bonds, with K_d below the picomolar concentration (pM).

Exercise 2.1.1 Show that Eq. (2.4) can be transformed into $(1 + \tanh(x/2))/2$ by using $x = \log([X]/K_d)$.

2.1.2   Hill Equation

In the association or dissociation reactions inside the cells and organisms, there are some cases in which multiple molecules of the same species are bound to a target molecule. Let $X_n Y$ denote the state in which n molecules of X are bound to a single Y molecule simultaneously. Furthermore, we assume that n molecules of X and Y simultaneously associate and dissociate, that is,

$$nX + Y \underset{k_{\text{off}}}{\overset{k_{\text{on}}}{\rightleftharpoons}} X_n Y. \tag{2.5}$$

As in the previous section, the reaction rate equation of the preceding process is represented by the following equation:

$$\frac{\mathrm{d}[X_n Y]}{\mathrm{d}t} = k_{\text{on}}[X]^n[Y] - k_{\text{off}}[X_n Y]$$

$$= k_{\text{on}}[X]^n(Y_{\text{total}} - [X_n Y]) - k_{\text{off}}[X_n Y]. \tag{2.6}$$

In the steady state, by using

$$(K_d)^n = \frac{k_{\text{off}}}{k_{\text{on}}} = \frac{[X]^n[Y]}{[X_n Y]}, \tag{2.7}$$

we obtain

$$\frac{[X_nY]}{Y_{\text{total}}} = \frac{[X]^n}{(K_d)^n + [X]^n},$$ (2.8)

which is called the Hill equation [12]. The n is the Hill coefficient, which corresponds to the steepness of the sigmoid curve around K_d (Fig. 2.1).

2.1.3 Adair Equation

In the derivation of the Hill equation, multiple molecules are assumed to associate or dissociate simultaneously. As an alternative model, let us consider that each process in a chemical reaction occurs sequentially, that is,

$$X + Y \underset{k_{\text{off},1}}{\overset{k_{\text{on},1}}{\rightleftharpoons}} XY,$$

$$X + XY \underset{k_{\text{off},2}}{\overset{k_{\text{on},2}}{\rightleftharpoons}} X_2Y,$$

$$\cdots$$

$$X + X_{n-1}Y \underset{k_{\text{off},n}}{\overset{k_{\text{on},n}}{\rightleftharpoons}} X_nY.$$ (2.9)

The reaction rate equation is

$$\frac{\mathrm{d}[X_nY]}{\mathrm{d}t} = k_{\text{on},n}[X][X_{n-1}Y] - k_{\text{off},n}[X_nY].$$ (2.10)

In the steady state, by using the inverse of the dissociation constant (association constant) $K_{a,n} = k_{\text{on},n}/k_{\text{off},n}$, we obtain

$$[X_nY] = K_{a,n}[X_{n-1}Y][X]$$ (2.11)

$$= \left\{ \prod_{k=1}^{n} K_{a,k} \right\}[Y][X]^n.$$ (2.12)

Considering each Y has n binding sites, the percentage of molecule X bound to Y (i.e., the occupancy of the binding sites) can be represented by

$$\frac{\sum_{j=1}^{n} j[X_jY]}{n \cdot Y_{\text{total}}} = \frac{\sum_{j=1}^{n} j[X_jY]}{n \sum_{j=0}^{n} [X_jY]}$$ (2.13)

$$= \frac{\sum_{j=1}^{n} j\left\{ \prod_{k=1}^{j} K_{a,k} \right\}[X]^j}{n \sum_{j=0}^{n} \left\{ \prod_{k=1}^{j} K_{a,k} \right\}[X]^j},$$ (2.14)

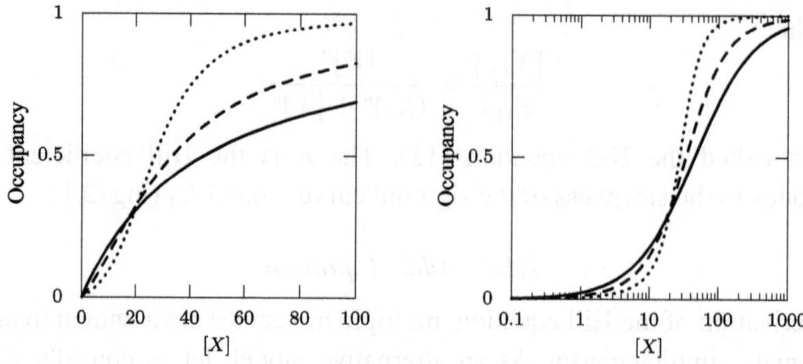

Figure 2.2 Adair equation ($K_1 = 1/45, K_2 = 1/25, K_3 = 1/250, K_4 = 1/1.5$). (left) the normal plot, (right) semilog plot. $n = 1$ (solid line), $n = 2$ (dashed line), $n = 4$ (dotted line).

which is called the Adair equation [1]. Here we use $[X_0 Y] \equiv [Y]$ for notational convenience. For instance, in the case of $n = 4$,

$$(right\ side) = \frac{K_1[X] + 2K_1K_2[X]^2 + 3K_1K_2K_3[X]^3 + 4K_1K_2K_3K_4[X]^4}{4(1 + K_1[X] + K_1K_2[X]^2 + K_1K_2K_3[X]^3 + K_1K_2K_3K_4[X]^4)},$$

(2.15)

where we use K_j as the abbreviation of $K_{a,j}$. The plot of this equation is shown in Fig. 2.2. The slope of the dissociation curve becomes steeper as the number of reaction steps n is increased. As can be seen from the equation and Fig. 2.2, the concentration of X at half the maximum value decreases as n increases.

2.1.4 MWC Model

Oxygen in the blood and its carrier, the hemoglobin protein, are some of the earliest systems studied as examples of binding and dissociation. It is known that the dissociation curve of the oxygen molecule and hemoglobin has a steep rise, as discussed in the previous section, but it is not well described by Hill's equation or Adair's equation.

The MWC model proposed by Monod, Wyman, and Changeux provides a more detailed picture of binding and dissociation [22]. In this model, as is shown in Fig. 2.3, hemoglobin is a tetramer in which each subunit transits between the T (tense) and R (relaxed) states. Let T_i and R_i denote a state in which i ($i = 0, 1, \ldots, 4$) oxygen atoms are bound to the molecule. Here, we focus only on the number of oxygen molecules bound, and neglect which subunit is binding to the oxygen.

The number ratio between hemoglobin and the oxygen molecules binding to hemoglobin is given by

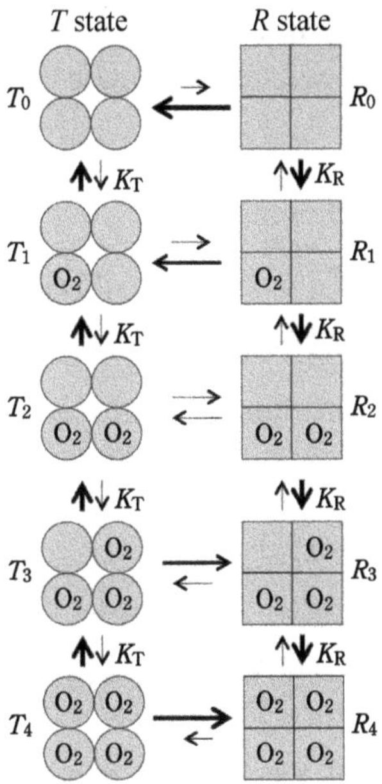

Figure 2.3 Schematic illustration of MWC model.

$$\frac{\sum_{i=0}^{4}\left\{\left(i \times [R_i]\right) + \left(i \times [T_i]\right)\right\}}{\sum_{i=0}^{4}\left\{[R_i] + [T_i]\right\}}. \tag{2.16}$$

The equilibrium constant between two states is $L_i = [T_i]/[R_i]$. Assuming that the association constants of T state (K_T) and R state (K_R) to oxygen molecule do not depend on the number of binding oxygen molecules (i), they are represented as

$$[T_i] = K_T[O_2][T_{i-1}], \tag{2.17}$$

$$[R_i] = K_R[O_2][R_{i-1}]. \tag{2.18}$$

At the equilibrium state, the frequency of direct transitions between T_1 and R_1 state equals one of the indirect transitions, in which the T_1 state switches to the R_1 state via the T_0 and R_0 states (which is called "detailed balance").

$$\begin{aligned}
L_1 &= \frac{[T_1]}{[R_1]} \\
&= \frac{[T_1]}{[T_0]} \times \frac{[T_0]}{[R_0]} \times \frac{[R_0]}{[R_1]} \\
&= \frac{K_T L_0}{K_R} \\
&= L_0 c.
\end{aligned}$$

Note that $K_T = cK_R$ and $c \equiv \frac{K_T}{K_R}$ is a constant. Since the same relationship holds in the subsequent cases of state $i = 2, \ldots$, we obtain

$$L_i = L_0 c^i. \tag{2.19}$$

Then, we consider the case with n-mers:

$$[T_i] = C_n^i (K_T[O_2])^i [T_0], \tag{2.20}$$

$$[R_i] = C_n^i (K_R[O_2])^i [R_0], \tag{2.21}$$

where C_n^i represents the combination and the binomial theorem and its derivatives are used:

$$\sum_{i=0}^{n} C_n^i x^i = (1 + x)^n,$$

$$\sum_{i=0}^{n} i C_n^i x^i = nx \sum_{i=0}^{n-1} C_{n-1}^i x^i = nx(1 + x)^{n-1}.$$

As in Eq. (2.16), the ratio of the binding number of oxygen molecules Y is given by

$$Y = \frac{\displaystyle\sum_{i=0}^{n} \left\{ i C_n^i (K_R[O_2])^i [R_0] + i C_n^i (cK_R[O_2])^i [R_0]L_0 \right\}}{\displaystyle\sum_{i=0}^{n} \left\{ C_n^i (K_R[O_2])^i [R_0] + C_n^i (cK_R[O_2])^i [R_0]L_0 \right\}}$$

$$= nK_R[O_2] \frac{(1 + K_R[O_2])^{n-1} + L_0 c(1 + cK_R[O_2])^{n-1}}{(1 + K_R[O_2])^n + L_0(1 + cK_R[O_2])^n}.$$

For simplicity, we use $\tilde{Y} = Y/n$ and $x = K_R[O_2]$, and obtain

$$\tilde{Y} = \frac{x(1 + x)^{n-1} + L_0 cx(1 + cx)^{n-1}}{(1 + x)^n + L_0(1 + cx)^n}. \tag{2.22}$$

In the case of $c = 1$ or $L_0 \to 0$, this equation becomes $\tilde{Y} = x/(1 + x)$, while in the case of $L_0 \to \infty$, it is $\tilde{Y} = cx/(1 + cx)$. This indicates that the occupancy $\tilde{Y}$ does not increase sharply with respect to oxygen concentration if there is no distinction between the T and R states or if the oxygen is only bound to the T state.

For $\tilde{Y}$ to have a sharply rising Hill function-like shape, the slope around $x = 0$ must be small. That is,

$$\left. \frac{d\tilde{Y}}{dx} \right|_{x=0} = \frac{L_0 c + 1}{L_0 + 1} \tag{2.23}$$

needs to be small, leading $L_0 \gg 1$ and $c \ll 1$. This means that the hemoglobin is mostly in the T state at low oxygen concentrations, but at high oxygen concentrations, the hemoglobin tends to the R state. Since $c \ll 1$ means $K_R \gg K_T$, once the

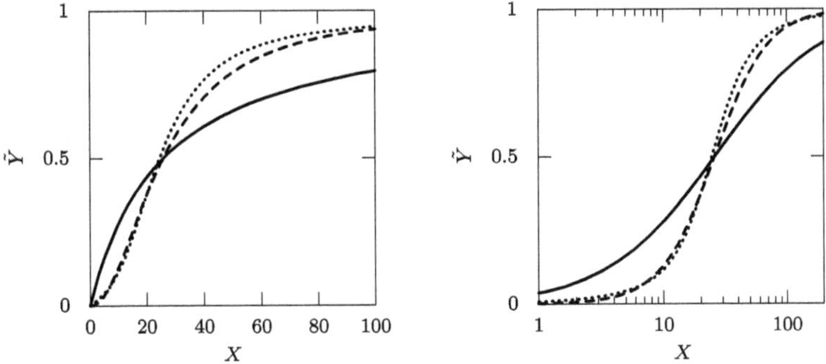

Figure 2.4 Comparison of the results of Hill's function and the MWC model. (left) the normal plot, (right) semilog plot. Hill function ($K_d = 25.85$, Hill coefficient $n = 1$ (solid line), $n = 2$ (dashed line)), and MWC ($n = 4$, $K_R = 0.24$, $c = 0.03$, $L_0 = 1{,}200$ (dotted line)).

equilibrium is tilted to the R state, oxygen molecules tend to bind rapidly. For the fraction occupied by the oxygen molecules to rise rapidly by this mechanism, the following condition is necessary:

$$\left.\frac{d^2 \tilde{Y}}{dx^2}\right|_{x=0} = \frac{2}{(L_0 + 1)^2}\left\{nL_0(1 - c)^2 - (L_0 c^2 + 1)(L_0 + 1)\right\} > 0, \tag{2.24}$$

which yields

$$n > \frac{(L_0 c^2 + 1)(L_0 + 1)}{L_0(1 - c)^2} \tag{2.25}$$

$$\Rightarrow n > (L_0 c^2 + 1), \tag{2.26}$$

by recalling that $L_0 \gg 1$ and $c \ll 1$. The curve for these parameters is shown in Fig. 2.4.

When fitting the hemoglobin data with the Hill function, the Hill coefficient takes a noninteger value between $n = 2$ and $n = 3$. Such a noninteger value of the coefficient is difficult to explain using the original binding and dissociation model. In contrast, the MWC model can account for such an intermediate curve. Although the MWC model was originally used in the binding of oxygen molecules to hemoglobin, it has been applied to various regulatory enzymes and receptors [6], including the ultrasensitivity of the cellular response to bacterial attractants [2, 26]. This mechanism is often called allosteric control because there are multiple binding sites.

Exercise 2.1.2 Show that $\tilde{Y} = \frac{x(1+x)^{n-1}}{(1+x)^n + L_0}$ holds in the extreme case where the substrate does not bind the T state at all.

Example 2.1   KNF Model The MWC model assumes that the molecule of interest transitions between two states, R and T, even in the absence of the ligand. On the other hand, ligand binding may cause a change in the state of the subunit, making it easier for ligand binding to occur in other subunits. This is known as the sequential model or the KNF (Koshland, Némethy, and Filmer) model [18].

2.2   Input–Output Functions of Enzyme Reactions

2.2.1   Michaelis–Menten Equation

Many biochemical reactions in living organisms and cells are catalyzed by enzymes. The rate of enzymatic reactions increases as substrate concentration increases, and the concentration dependency often exhibits a nonlinear curve, as shown in the previous section. Here, we discuss the nonlinear dependency of reaction rates on substrate concentration. Let the substrate be X, the product be X^*, the enzyme be E, and the complex of X and E be XE, and the enzymatic reaction is given by

$$X + E \underset{k_{-1}}{\overset{k_1}{\rightleftharpoons}} XE \overset{k_2}{\longrightarrow} X^* + E. \tag{2.27}$$

The reaction rate equations are

$$\frac{d[XE]}{dt} = k_1[X][E] - (k_{-1} + k_2)[XE], \tag{2.28}$$

$$\frac{d[X^*]}{dt} = k_2[XE]. \tag{2.29}$$

We make the assumption that the total concentration of enzymes, denoted by E_T, remains constant, that is, $E_T = [E]+[XE]$. Additionally, we presume that the generation of XE occurs rapidly enough on the relevant time scale, and its concentration remains constant. Based on these assumptions, we obtain

$$[XE] = E_T \frac{[X]}{K_M + [X]}, \tag{2.30}$$

with

$$K_M = \frac{k_{-1} + k_2}{k_1}. \tag{2.31}$$

Here, we used adiabatic elimination (see Section 1.12). K_M is called Michaelis–Menten constant. Note that Eq. (2.30) has the same form as the sigmoid curve obtained for the binding and dissociation reactions (Eq. (2.4)). However, instead of the dissociation constant K_d, the Michaelis–Menten coefficient K_M represents the concentration of the substrate at which the value of the sigmoid curve is half of the maximum.

To sum up, Eq. (2.29) is written as

$$\frac{d[X^*]}{dt} = V_{\max}\frac{[X]}{K_M + [X]},\tag{2.32}$$

and $V_{\max} = k_2 E_T$. The value of the right-hand side increases as the substrate X increases and approaches the maximum value $V_{\max}$. If the concentration of the substrate X is sufficiently larger than K_M, then

$$\frac{d[X^*]}{dt} \approx V_{\max}\tag{2.33}$$

is satisfied. This is called a zero-order reaction because the rate of production depends only on the total amount of the enzyme and not on the substrate concentration. In contrast, if the substrate X is sufficiently small, then it can be approximated as a first-order reaction as follows:

$$\frac{d[X^*]}{dt} \approx V_{\max}\frac{[X]}{K_M}.\tag{2.34}$$

2.2.2 Inhibitory Reaction

In cellular dynamics, there are many components that inhibit reactions among other components. The function of such inhibitory factors plays a pivotal role in the system-level behavior of biochemical reactions and signaling systems. The function of an inhibitory reaction takes different forms depending on the specific reaction mode, and there is often insufficient information about which one should be adopted. Depending on the problem to be solved by a coarse-grained mathematical model, it is often sufficient to have a reduced representation of the nonlinear function that does not depend on the details of the reaction; for example, the Hill function is often sufficient without the details of reaction sequences as the MWC model. For inhibitory reactions, the following two functional forms are often used. One is inhibition by competitive binding, where the value of K_M is effectively increased by the presence of the inhibitor. The other is noncompetitive allosteric inhibition, in which the inhibitory effect is due to complex formation that does not produce any product.

First, let us discuss competitive inhibition. Suppose that X^* is produced from a substrate X catalyzed by an enzyme E. Similar to the derivation of the Michaelis–Menten model, let XE be the complex of substrate and enzyme. The inhibitor I binds at the same site on the enzyme E where the substrate binds, forming the complex IE, but X^* is not produced from IE. That is,

$$X + E \underset{k_{-1}}{\overset{k_1}{\rightleftharpoons}} XE \overset{k_2}{\longrightarrow} X^* + E,\tag{2.35}$$

$$I + E \underset{k_{-3}}{\overset{k_3}{\rightleftharpoons}} IE. \tag{2.36}$$

The reaction rate equations are given by

$$\frac{\mathrm{d}[XE]}{\mathrm{d}t} = k_1[X][E] - (k_{-1} + k_2)[XE], \tag{2.37}$$

$$\frac{\mathrm{d}[X^*]}{\mathrm{d}t} = k_2[XE], \tag{2.38}$$

$$\frac{\mathrm{d}[IE]}{\mathrm{d}t} = k_3[I][E] - k_{-3}[IE]. \tag{2.39}$$

Assume that there are enough inhibitors I compared to the enzyme E, so that the concentrations of $[IE]$ and $[XE]$ are in a steady state, and that the overall concentration of the enzyme E, that is, E_T, is constant. Then, we obtain

$$[XE] = \frac{E_\mathrm{T}[X]}{K_\mathrm{M}(1 + \frac{[I]}{K_I}) + [X]}, \tag{2.40}$$

where $[E] = E_\mathrm{T} - [XE] - [IE]$ and $K_I = k_{-3}/k_3$. Then, we obtain

$$\frac{\mathrm{d}[X^*]}{\mathrm{d}t} = V_{\max} \frac{[X]}{K_\mathrm{M}(1 + \frac{[I]}{K_I}) + [X]}, \tag{2.41}$$

in which $V_{\max} = k_2 E_\mathrm{T}$. Comparing the equation with Eq. (2.32), it is clear that an increase in I shifts the response region of the sigmoid function, which effectively suppresses the productive reaction.

Next, let us consider the reaction rate equation with an inhibition by noncompetitive binding. Unlike in the case of competitive binding, the binding site of X and the binding site of I are assumed to exist independently of the enzyme E. Assume that the binding of the substrate to the enzyme and the binding of the inhibitor to the enzyme do not affect each other. Then,

$$X + E \underset{k_{-1}}{\overset{k_1}{\rightleftharpoons}} XE \overset{k_2}{\longrightarrow} X^* + E, \tag{2.42}$$

$$I + E \underset{k_{-3}}{\overset{k_3}{\rightleftharpoons}} EI, \tag{2.43}$$

$$I + XE \underset{k_{-3}}{\overset{k_3}{\rightleftharpoons}} XEI, \tag{2.44}$$

$$X + EI \underset{k_{-1}}{\overset{k_1}{\rightleftharpoons}} XEI. \tag{2.45}$$

From the same calculation in the previous paragraph,

$$[XE] = \frac{E_\mathrm{T}[X]}{K_\mathrm{M} + [X]} \frac{K_\mathrm{I}}{K_\mathrm{I} + [I]} \tag{2.46}$$

is obtained, where $E_T = [E] + [XE] + [EI] + [XEI]$ and $K_I = k_{-3}/k_3$, respectively. Then, the production rate of X^* is given by

$$\frac{\mathrm{d}[X^*]}{\mathrm{d}t} = V_{\max} \frac{[X]}{K_M + [X]} \frac{K_I}{K_I + [I]}. \tag{2.47}$$

Note that the left part of the right-hand side of the equation is a sigmoid function whose value increases as X increases. In contrast, the right part of the right-hand side of the equation is a sigmoid function whose value decreases as I increases, and the total reaction is governed by the product of the two.

Exercise 2.2.1 Verify that Eq. (2.41) and Eq. (2.47) hold.

Exercise 2.2.2 Using Cartesian coordinates (x, y, z), which take X and I on the x-axis and y-axis, and the production rate of X^* on the z-axis, plot Eq. (2.41) and Eq. (2.47), respectively. Examine how the behavior changes depending on the values of K_M and K_I.

2.3 Input–Output Relationship of Network Modules

2.3.1 Push–Pull Reaction and the Zero-Order Ultrasensitivity

Many biochemical reactions that characterize signal transduction in cells involve proteins with the catalytic activity of phosphorylation and dephosphorylation. Regarding the role of these phosphorylation and dephosphorylation reactions, it has been proposed that their simple regulation can result in the emergence of switch-like behavior [11]. Let us discuss the mechanism of this switch-like behavior in signal transduction machinery. The phosphorylated protein X is denoted by X^*. The phosphorylation and dephosphorylation reactions are catalyzed by kinase E_1 and phosphatase E_2, respectively.

$$X + E_1 \underset{k_{-1}}{\overset{k_1}{\rightleftharpoons}} XE_1 \overset{k_2}{\longrightarrow} X^* + E_1,$$

$$X^* + E_2 \underset{k_{-3}}{\overset{k_3}{\rightleftharpoons}} X^* E_2 \overset{k_4}{\longrightarrow} X + E_2$$

The reaction rate equation is given as follows:

$$\frac{\mathrm{d}[X]}{\mathrm{d}t} = -k_1[X][E_1] + k_{-1}[XE_1] + k_4[X^*E_2], \tag{2.48}$$

$$\frac{\mathrm{d}[XE_1]}{\mathrm{d}t} = k_1[X][E_1] - (k_{-1} + k_2)[XE_1], \tag{2.49}$$

$$\frac{\mathrm{d}[X^*]}{\mathrm{d}t} = -k_3[X^*][E_2] + k_{-3}[X^*E_2] + k_2[XE_1], \tag{2.50}$$

$$\frac{d[X^*E_2]}{dt} = k_3[X^*][E_2] - (k_{-3} + k_4)[X^*E_2]. \tag{2.51}$$

Assuming that the concentrations of the complexes are in a steady state, we obtain

$$[XE_1] = \frac{E_{1T}[X]}{K_{M1} + [X]}, \tag{2.52}$$

$$[X^*E_2] = \frac{E_{2T}[X^*]}{K_{M2} + [X^*]}, \tag{2.53}$$

where $K_{M1} = (k_{-1} + k_2)/k_1$, $K_{M2} = (k_{-3} + k_4)/k_3$, and the total concentration of each enzyme E_{1T}, E_{2T} are kept constants. Considering a steady state of X and taking the sum of Eq. (2.48) and Eq. (2.49), we obtain

$$k_4[X^*E_2] = k_2[XE_1]. \tag{2.54}$$

Then, from Eq. (2.52) and Eq. (2.54), we obtain

$$[X] = \frac{K_{M1}[XE_1]}{E_{1T} - [XE_1]} = \frac{K_{M1}\frac{k_4}{k_2}[X^*E_2]}{E_{1T} - \frac{k_4}{k_2}[X^*E_2]}.$$

Considering the case where there is a much larger amount of substrate than enzyme, we obtain

$$[X^*] = X_T - [X] - [XE_1] - [X^*E_2]$$
$$\cong X_T - [X]$$
$$= X_T - \frac{K_{M1}\frac{k_4}{k_2}[X^*E_2]}{E_{1T} - \frac{k_4}{k_2}[X^*E_2]}. \tag{2.55}$$

Then, from Eq. (2.53) and Eq. (2.55), the ratio of phosphorylated protein Y^* is given as follows:

$$Y^* = \frac{[X^*]}{X_T} = 1 - \frac{\frac{K_{M1}}{X_T}\frac{k_4}{k_2}[X^*E_2]}{E_{1T} - \frac{k_4}{k_2}[X^*E_2]}$$
$$= 1 - \frac{\frac{K_{M1}}{X_T}k_4\frac{E_{2T}[X^*]}{K_{M2}+[X^*]}}{k_2E_{1T} - k_4\frac{E_{2T}[X^*]}{K_{M2}+[X^*]}}. \tag{2.56}$$

By introducing the notation $K_1 = K_{M1}/X_T$, $K_2 = K_{M2}/X_T$, $V_1 = k_2E_{1T}$, and $V_2 = k_4E_{2T}$, the equation is simplified to a quadratic equation of Y^*, as follows:

$$(V_1 - V_2)Y^{*2} - (V_1 - V_2 - K_2V_1 - K_1V_2)Y^* - V_1K_2 = 0. \tag{2.57}$$

As expected from the symmetry of the original reaction rate equation, this solution strongly depends on the value of the parameters characterizing the forward

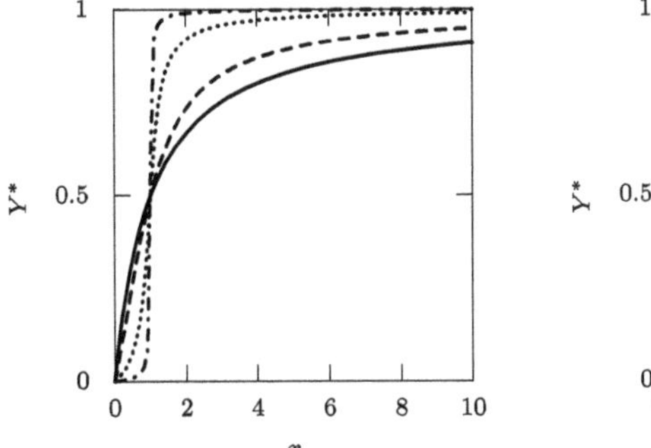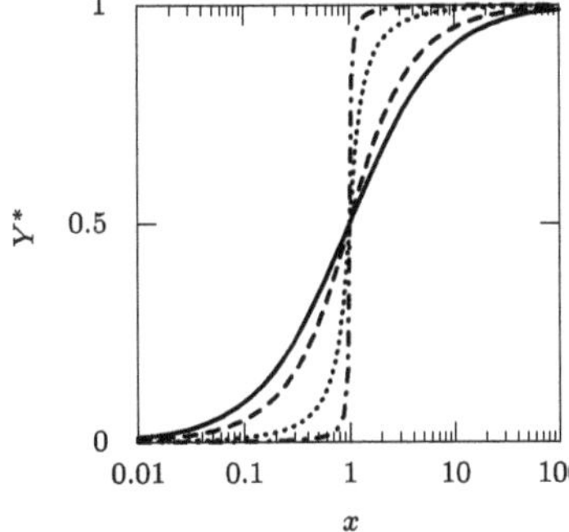

Figure 2.5 The zero-order ultrasensitivity. (left) the normal plot, (right) semilog plot. Hill function ($K_d = 25.85$, $n = 1$ (solid line)), push-pull reaction ($\kappa = 1$). $K_2 = 1$ (dashed line), $K_2 = 0.1$ (dotted line), $K_2 = 0.01$ (dot-dash line). It can be seen that the rise is sharper as the Michaelis–Menten coefficient is smaller.

and reverse directions. Here, by setting $x = V_1/V_2$ and $\kappa = K_1/K_2$, and from the condition $Y^* > 0$, we obtain

$$Y^* = \frac{((x-1) - K_2(x+\kappa)) + \sqrt{((x-1) - K_2(x+\kappa))^2 + 4K_2 x(x-1)}}{2(x-1)}. \qquad (2.58)$$

This function [11] represents the strength of the response, which depends on the parameter x, and shows a sharp rise as shown in Fig. 2.5. This sharp response to x is called zero-order ultrasensitivity. This property is based on the assumption that the amount of substrates is much larger than the amount of enzymes. In the situation of near-zero-order reactions for a substrate, the faster of the forward (push; $X \to X^*$) or reverse (pull; $X^* \to X$) reactions will eventually convert all the substrate to product. Therefore, the curve shows which direction of the reaction is dominant according to the ratio of $V_{\max}$ in the forward and reverse directions, that is, V_1/V_2. In the signal transduction pathway, the upstream reaction often modulates the downstream kinase activity. The preceding mechanism suggests that in such cases, continuous changes in the upstream factors can induce switch-like state transitions downstream.

2.3.2 *The Effect of Multiple Phosphorylation Modifications*

In intracellular signaling systems, phosphorylation modifications occur at multiple sites on the same protein. This can lead to input–output relationships with strong nonlinearity [7]. For example, let us assume that modification reactions proceed step by step, as follows:

$$X \underset{\gamma_Y}{\overset{V_X}{\rightleftharpoons}} Y \underset{\gamma_Z}{\overset{V_Y}{\rightleftharpoons}} Z \underset{\gamma_W}{\overset{V_Z}{\rightleftharpoons}} W.$$

Here, the concentrations of X, X^*, X^{**}, and X^{***} are denoted as X, Y, Z, and W, respectively (the number of $*$ represents phosphorylation modification). The V_X, V_Y, and V_Z, are the maximum reaction rates of each modification reaction, and γ_Y, γ_Z, and γ_W are the degradation rate constants of the protein, and m and n are the Hill coefficients. The Michaelis–Menten coefficients are denoted as K_X and K_Y, respectively. Then, these reaction processes are represented as follows:

$$\frac{dY}{dt} = V_X \frac{X^m}{(K_X)^m + X^m} - \gamma_Y Y \tag{2.59}$$

$$\frac{dZ}{dt} = V_Y \frac{Y^n}{(K_Y)^n + Y^n} - \gamma_Z Z \tag{2.60}$$

Let $\bar{X}$, $\bar{Y}$, $\bar{Z}$ be the concentration in the steady state; then we obtain

$$\frac{\bar{Y}}{Y_T} = \frac{(\bar{X})^m}{(K_X)^m + (\bar{X})^m}, \tag{2.61}$$

$$\frac{\bar{Z}}{Z_T} = \frac{(\bar{Y})^n}{(K_Y)^n + (\bar{Y})^n}, \tag{2.62}$$

where $Y_T = V_X/\gamma_Y$, $Z_T = V_Y/\gamma_Z$. In this case, $\bar{Z}$ can be represented by $\bar{X}$ as follows.

$$\begin{aligned}
\frac{\bar{Z}}{Z_T} &= \frac{(Y_T)^n \left(\frac{(\bar{X})^m}{(K_X)^m + (\bar{X})^m}\right)^n}{(K_Y)^n + (Y_T)^n \left(\frac{(\bar{X})^m}{(K_X)^m + (\bar{X})^m}\right)^n} \\
&= \frac{(Y_T)^n (\bar{X})^{mn}}{(K_Y)^n ((K_X)^m + (\bar{X})^m)^n + (Y_T)^n (\bar{X})^{mn}} \\
&= \frac{(\bar{X})^{mn}}{(\tilde{K}_Y)^n ((K_X)^m + (\bar{X})^m)^n + (\bar{X})^{mn}}, \tag{2.63}
\end{aligned}$$

where $\tilde{K}_Y = K_Y/Y_T$. The plot of this relationship is shown in Fig. 2.6 (left). Compared with $\bar{Y}$, the rise of $\bar{Z}$ is shifted to the lower concentration side of $\bar{X}$, and its steepness is also increased.

For simplicity, let us consider the case of $n = 1$. In this case, $\bar{Z}$ is written as

$$\frac{\bar{Z}}{Z_T} = \frac{1}{(\tilde{K}_Y + 1)} \frac{(\bar{X})^m}{\frac{\tilde{K}_Y (K_X)^m}{(\tilde{K}_Y + 1)} + (\bar{X})^m}. \tag{2.64}$$

Note that when $\tilde{K}_Y = K_Y/Y_T < 1$ is expected, and the Michaelis–Menten coefficient decreases effectively from K_X to $K_X/(1 + 1/\tilde{K}_Y)^{1/m}$ by the reaction cascade

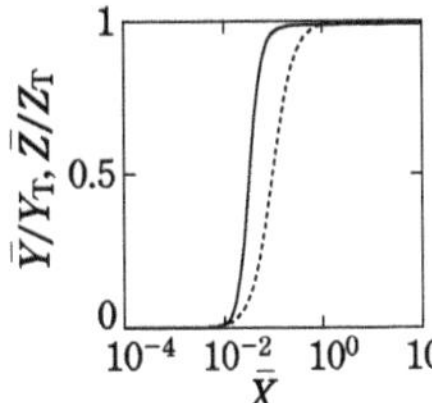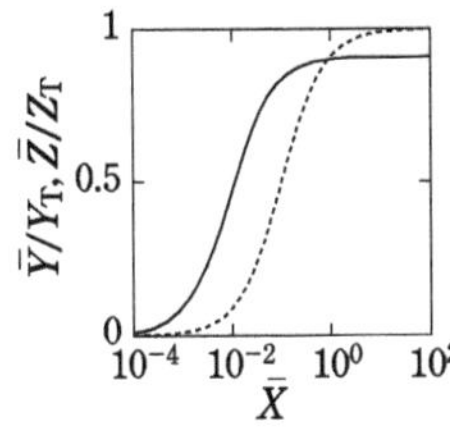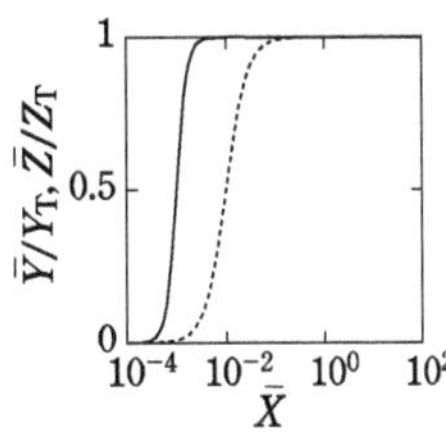

Figure 2.6 Effects of cascading signal transmission on input–output relationships ($\bar{Y}$: dashed line, $\bar{Z}$: solid line). (left) Eq. (2.63), $K_X = 0.1, \tilde{K}_Y = 0.1, m = 2, n = 2$, (center) Eq. (2.64), $K_X = 0.1, \tilde{K}_Y = 0.1, m = 1, n = 1$, (right) Eq. (2.65), $K_X = 0.01, \tilde{K}_Y = 0.01, m = 2, n = 2$.

(Fig. 2.6 (center)). This means that the downstream of the signal transduction cascade is more likely to respond to the changes in the low-concentration region of the upstream signal.

In the case of $m \neq 1, n \neq 1$, the equation is difficult to solve analytically, but if we assume that X moves only in the low-concentration region, then $K_X \gg \bar{X}$ and Eq. (2.63) can be approximated as

$$\frac{\bar{Z}}{Z_T} \simeq \frac{(\bar{X})^{mn}}{(\tilde{K}_Y^{1/m} K_X)^{mn} + (\bar{X})^{mn}}. \tag{2.65}$$

Since $\tilde{K}_Y < 1$ leads to $\tilde{K}_Y^{1/m} < 1$, Eq. (2.65) means that the effective Michaelis–Menten coefficient decreases from K_X to $\tilde{K}_Y^{1/m} K_X$, and the Hill coefficient is enhanced to $m \times n$ (Fig. 2.6 (right)).

Example 2.2 Ultrasensitivity by Receptor Clustering The cascade reaction discussed in this section enables a sensitive response to weak signals in a well-mixed solution. On the other hand, in a cell, some receptors on the cell membrane are spatially localized, and such spatial structure also affects the sensitivity. For example, many G protein-coupled receptors are thought to form dimers. The receptors involved in the chemotaxis of *E. coli* are also known to be localized in the anterior part of the cell. Statistical mechanics models suggest that the interaction between receptors may cause cooperative state transitions with spatial spreading and that such a mechanism may give rise to ultrasensitivity [23].

2.3.3 Switch with Positive Feedback

Let us return to the push–pull type reaction (Section 2.3.1). In the previous section, we looked at the ultrasensitivity that appears in the zero-order reaction, but let us now consider a situation that can be approximated by the first-order reaction [8].

The enzyme in the push direction reaction is denoted as S, and the enzyme in the pull direction reaction is denoted as I. Then, we obtain

$$\frac{dx^*}{dt} = k_+ S(X_T - X^*) - k_- I X^*, \tag{2.66}$$

where k_+ and k_- are the binding rate constant and the dissociation rate constant, respectively. In the steady state, the following relationship holds:

$$\frac{X^*}{X_T} = \frac{(S/I)}{k_-/k_+ + (S/I)}. \tag{2.67}$$

As can be seen, the input–output relationship from the first-order reaction is expressed as a sigmoid function with a Hill coefficient of 1, and there is no ultrasensitivity as in the zero-order reaction. The first and second terms on the right-hand side of Eq. (2.66) are the increasing and decreasing parts of $\dot{X}^*(= dx^*/dt)$, respectively. Describing them separately on the phase plane of $(X^*, |\dot{X}^*|)$ is useful to visualize the fixed points of the system and its stability.[1]

Since it is a first-order reaction, two lines with slopes determined by S and I represent the reaction rates in the push and pull directions. The X^* at the intersection is the fixed point of this dynamics (upper left of Fig. 2.7). If we consider the left side of the fixed point, the reaction rate in the push direction (represented by the solid line) is greater than the reaction rate in the pull direction (represented by the dashed line). As a result, X^* moves in the positive direction as time goes on. On the right side of the fixed point, the reaction rate in the push direction is smaller than the reaction rate in the pull direction. Hence, X^* changes with time in the negative direction. Thus, the fixed point is stable.

Next, consider a system with positive feedback [8]:

$$\frac{dx^*}{dt} = k_+ S(X_T - X^*) + k_2 \frac{(X^*)^n}{K^n + (X^*)^n}(X_T - X^*) - k_- I X^*. \tag{2.68}$$

The second term on the right-hand side represents positive feedback, meaning that an increase in X^* leads to more X^*. Here, let us consider the case where $S = 0$ for simplicity. On the phase plane of $(X^*, |\dot{X}^*|)$, the curve representing the reaction in the positive direction has an inflection point, which results in a new fixed point at $X^* = 0$. If the Hill coefficient equals 1, the fixed point at $X^* = 0$ is unstable, as seen in the upper center panel of Fig. 2.7. As the Hill coefficient increases, the curve intersects the line representing the reaction in the negative direction three times, resulting in the fixed point at $X^* = 0$ to be stable (upper right panel of Fig. 2.7). Thus, strong nonlinear feedback can lead to bistability.

---

[1] While the nullcline analysis in Chapter 1 dealt with the relationships on the phase plane (space) of multiple variables, here we discuss the stability of fixed points in a one-variable system.

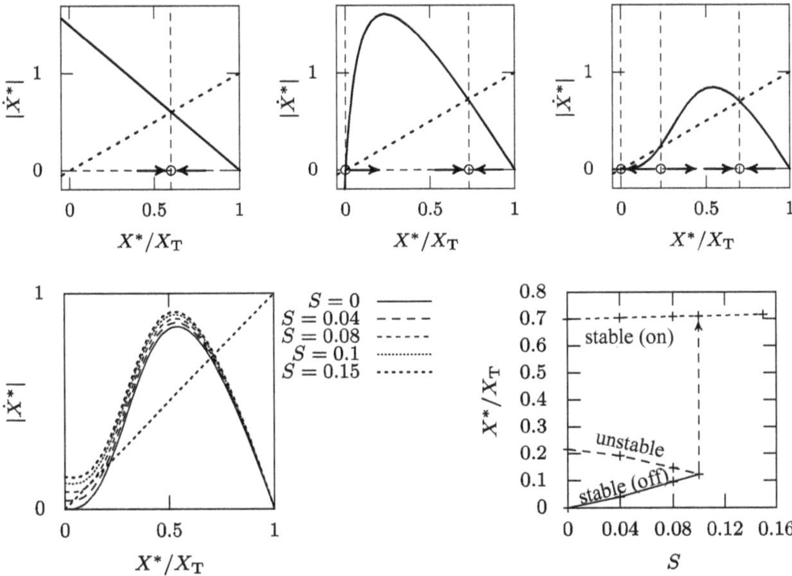

Figure 2.7 Phase planes of push-pull reactions with or without feedback [8]. (upper left) without feedback, (upper center) with feedback ($S = 0, n = 1$), (upper right) with feedback ($S = 0, n = 3$). The bifurcation when the input S is changed ($n = 3$). (lower left) The change of the position and number of fixed points on the phase plane. (lower right) Annihilation of stable and unstable fixed points and discontinuous transition to the on-state branch.

Let us examine how bistability is affected by changes in the input S (Fig. 2.7, lower panels). If we set the system's initial condition around $X^* = 0$, the system remains at the fixed point around $X^* = 0$ as long as S is small. However, when S is increased beyond a certain threshold, the stable and unstable fixed points collide and disappear (Fig. 2.7, lower right; solid and dashed lines), which is a saddle-node bifurcation (see Section 1.15). As a result, the system abruptly transitions to the other stable fixed point (Fig. 2.7, lower right; vertical arrows). Conversely, if we choose $X^*/X_T = 0.7$ as the initial condition and decrease S, the system remains at this fixed point (Fig. 2.7, lower right; dashed line). Therefore, positive feedback exhibits an irreversible switch-like behavior (toggle switch; see Section 1.4), whereby the system cannot return to the previous state once it has transitioned. This is one of the mechanisms that underlie hysteresis and memory.

2.3.4 *Feed-Forward Loop: Unimodal Type of Input–Output Relation*

Let us now consider the case where the input S regulates both the intermediate molecule I and the output R, while I regulates R, as illustrated in Fig. 2.8 (upper left). Such a network, due to the reaction's unidirectional nature from S to R, is referred

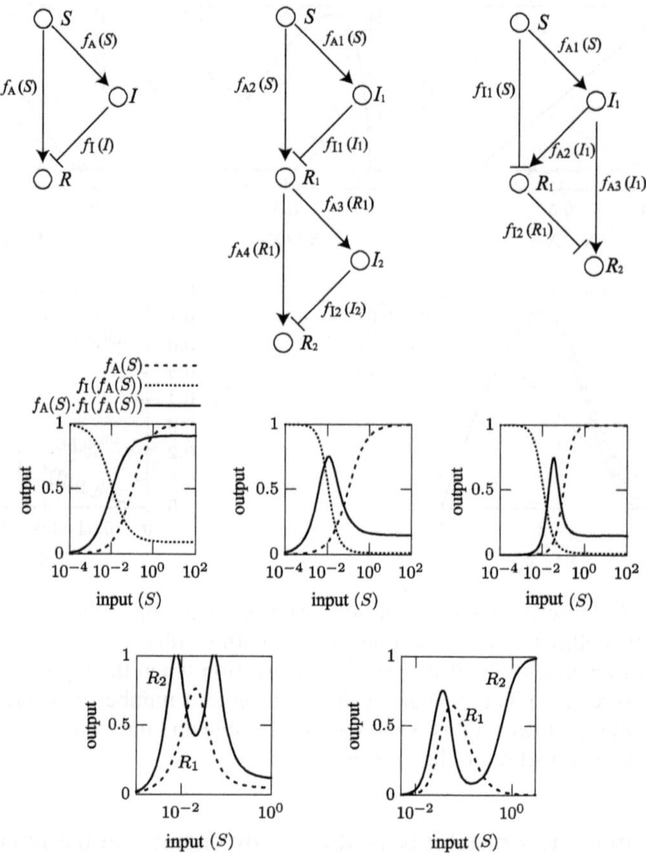

Figure 2.8 Steady-state input–output relationship using feed-forward loop [15]. (top) Schematic diagram of the loop (middle) Input–output relationship of feed-forward loop $f_A(S) \cdot f_I(f_A(S))$ (solid line), $f_A(S)$ (dashed line), $f_I(f_A(S))$ (dotted line). $K_d = 0.1$. (left) $m = 1, n = 1$, (center) $m = 1, n = 2$, (right) $m = 2, n = 2$. (bottom) Input–output relationship in hierarchical feed-forward loops (left) loop in the upper center $R_1(S) = f_{A2}(S) \cdot f_{I1}(f_{A1}(S))$ (dashed line), $R_2 = f_{A4}(R_1(S)) \cdot f_{I2}(f_{A3}(R_1(S)))$ (solid line). (right) loop in the upper right $R_1(S) = f_{I1}(S) \cdot f_{A2}(f_{A1}(S))$ (dashed line), $R_2 = f_{I2}(R_1(S)) \cdot f_{A3}(f_{A1}(S))$ (solid line). The output is properly normalized.

to as a feed-forward loop.[2] These circuits are commonly observed in biochemical and transcriptional networks within cells. In this section, we will discuss the characteristics of the input–output relationship in the steady state of the feed-forward loop.

As discussed in Section 2.1.2, let us use the following functions $f_A(S)$ and $f_I(S)$ to represent the activation and inhibition reactions in the feed-forward loop, respectively (Fig. 2.8, upper left):

---

[2] It is not a loop, but just a feed-forward (path). Still we use this term as it is commonly adopted.

$$f_A(S) = V \frac{S^m}{(K_d)^m + S^m}, \tag{2.69}$$

$$f_I(S) = V \frac{(K_d)^n}{(K_d)^n + S^n}. \tag{2.70}$$

Let us denote the time constant of the degradation reaction as γ^{-1}. Using this constant, we can describe the dynamics of the concentrations of I and R as follows:

$$\frac{dI}{dt} = -\gamma(I - f_A(S)),$$

$$\frac{dR}{dt} = -\gamma(R - g(S, I)).$$

Suppose that the effect of I on R in the feed-forward loop is inhibitory. The form of $g(S, I)$ will depend on the specific reaction mode of the inhibitory interaction. In this study, we consider the case of noncompetitive inhibition, as discussed previously. This means that the two reactions act by multiplication, as shown in Eq. (2.47) [15]. In the steady state, from $I = f_A(S)$, we obtain

$$\begin{aligned} g(S, I) &= f_A(S) f_I(I) \\ &= f_A(S) f_I(f_A(S)). \end{aligned}$$

The bifurcated pathways in the feed-forward loop have opposing effects on the output, with one being positive and the other negative. For this reason, this type of loop is sometimes referred to as an incoherent feed-forward loop [21]. The steady-state R is a function of S, which is determined as

$$R(S) = f_A(S) f_I(f_A(S)). \tag{2.71}$$

What input–output relationship would this form of the function generate? The functions $f_A(X)$ and $f_I(X)$ are monotonically increasing and monotonically decreasing functions that take values in $[0, V]$, respectively. If something interesting were to happen at intermediate values of X, the values of K_d would have to be close to the intermediate value. As shown in the middle left of Fig. 2.8, If we look at the nested function $f_I(f_A(X))$, it is similar to $f_I(X)$ in the form (even though it asymptotes to a finite value in $X \to \infty$). When the value of the Hill coefficient is 1, $R(S)$ is a monotonically increasing function, and the suppression by $f_I(f_A(S))$ only has the effect of shifting the curve to a smaller value of X. When the Hill coefficient of the inhibitory response exceeds 1, the inhibition exerts a stronger effect prior to activation, resulting in a single peak at the intermediate substrate concentration S (2.8, middle center). The steeper the Hill coefficient is, the steeper the peak becomes (Fig. 2.8, middle right). Hence, it is apparent that a feed-forward loop with strong nonlinearity can establish an input–output relationship that elicits responses at intermediate values of input S.

Next, let us consider a reaction network consisted of multiple feed-forward loops. Fig. 2.8 (top center, right) shows a network with two loops. In the first-stage loop, S to R_1 outputs a unimodal response as described earlier. This principle of input-output conversion is repeated in the next loop so that a hump can be formed for each slope of the bell-shaped input R_1 (Fig. 2.8, bottom left and right). This hierarchical arrangement of the feed-forward loop shows that the circuit can be configured to output responses for multiple concentration regions.

2.4   Input–Output Relationship in Adaptation and Chemotaxis

2.4.1   *Feed-Forward Loop: Adaptive Response and Fold-Change Detection*

Here, we discuss the transient response characteristics of the feed-forward loop. To enhance the equation's symmetry, we assume that the input S increases both the activator A and the inhibitor I, and the output switches between the inactive state (R_{inactive}) and the active state (R), depending on A and I. For simplicity, we approximate all reactions as first-order processes: [19].

$$\frac{dA}{dt} = k_A S - \gamma_A A,$$

$$\frac{dI}{dt} = k_I S - \gamma_I I, \tag{2.72}$$

$$\frac{dR}{dt} = k_R A (R_T - R) - \gamma_R I R.$$

Here, we assume that $R_{\text{inactive}} + R = R_T = $ constant, and consider the case where the product reaction rates are $k_A > k_I$. In this case, as the input S increases, A initially rises before I (Fig. 2.9). Consequently, for a step input, the output R undergoes a transient rise before returning to its original state. This phenomenon is known as adaptive response [17, 20] (Fig. 2.9). Notably, R returns to its initial value despite the persistent input S.

In the steady state of Eq. (2.72),

$$R = \frac{k_R R_T (A/I)}{k_R (A/I) + \gamma_R} \tag{2.73}$$

is stable. Since $A/I = k_A \gamma_I / k_I \gamma_A$, it follows that R is completely independent of the input S. This phenomenon is known as perfect adaptation (exact adaptation). A negative response is observed when the input S decreases, highlighting the transient nature of the system's response to changes in the input over time.

In cellular and individual signal processing, adaptation to input stimuli is often observed, where the system state returns to its original state after the stimulus,

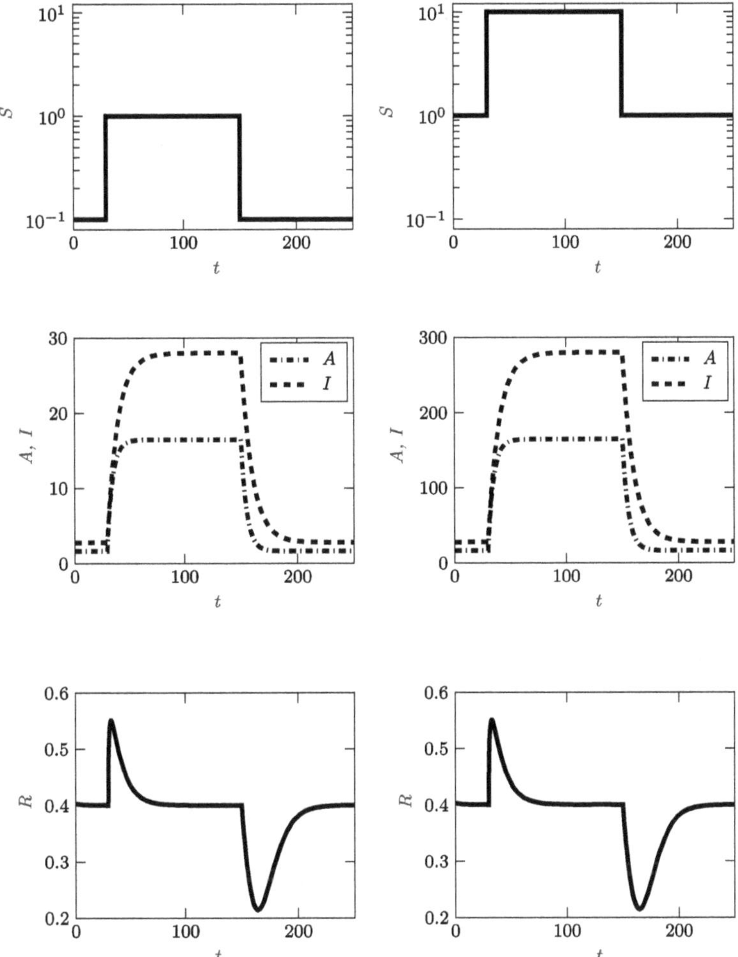

Figure 2.9 The response of the feed-forward loop to a step input. (left) $S = 0.1, 1$, (right) The response to the $S = 1.0$, 10-step stimulus.

as in the case of R in Eq. (2.72), while still responding to changes in the input stimulus. In slime molds and neutrophils' chemotaxis, the reaction that forms the cell's front side promoted by attraction stimulation is adaptive and thought to be based on the feed-forward loop Eq. (2.72). In some systems, including the one described by Eq. (2.72), the strength of the response R depends only on how many times S increases or decreases. For example, the response triggered by changing S from 0.1 to 1 is the same as changing S from 1 to 10 (Fig. 2.9). This type of response is called a fold-change detection response [10, 24], which is analogous to the Weber–Fechner law in psychology.

2.4.2 *Adaptive Response by Negative Feedback*

The adaptive response and fold-change detection, as described previously, are also generated by circuits of the feedback type. Here, we discuss the role of feedback regulation in the adaptive response using a simple model of signal transduction.

For the conversion from input S to output R, we consider the following simplified model [16, 27]:

$$\frac{dR}{dt} = k_R \frac{S^m}{(K_R I)^m + S^m} - \gamma_R R,$$
$$\frac{dI}{dt} = k_I R I - \gamma_I I. \tag{2.74}$$

Here, the inhibitory factor I increases as the output signal R increases, resulting in the inhibition of R itself. This network structure is commonly referred to as a feedback loop. In the previous section, we discuss that in the transient response of a feed-forward loop, the time delay between the slow rise of the inhibitory factor and the activating factor plays a critical role. In contrast, in a feedback loop, the system intrinsically produces a time delay.

In the steady state, we can derive the fixed point of R as $R^* = \gamma_I / k_I$ from the latter equation of Eq. (2.74). Therefore, the steady-state value of R is independent of the input S, reflecting the characteristics of perfect adaptation. Fig. 2.10 displays the response to a step input. The key difference from the adaptive response in a feed-forward loop is that the waveform of the response exhibits a transient drop below the steady-state value (as seen in the middle portion of Fig. 2.10, around $t = 50$). Furthermore, damping oscillations can be observed, depending on the system parameters (as shown in the right panel of Fig. 2.10).

Adaptive responses are essential for detecting temporal changes in stimulus concentration. This type of response is widely observed in biological systems, such as the transient elevation of calcium ions and second messengers, such as cAMP in response to hormones. Adaptive responses are also universally observed in unicellular organisms, including bacteria and amoebae, where they play a role in chemotaxis and stress response. There are two types of circuits that can generate adaptive responses: feed-forward and feedback. Feedback circuits are characterized by vortex damping and damping vibration [13]. It is known that some adaptive systems with feedback circuits exhibit the characteristics of fold-change detection, similar to feed-forward circuits.

2.4.3 *Chemotaxis of* E. coli: *Asakura–Honda Model*

The chemotaxis of *E. coli* is one of the best examples of cellular sensory information processing [5]. When the flagella of *E. coli* rotate counterclockwise, they form

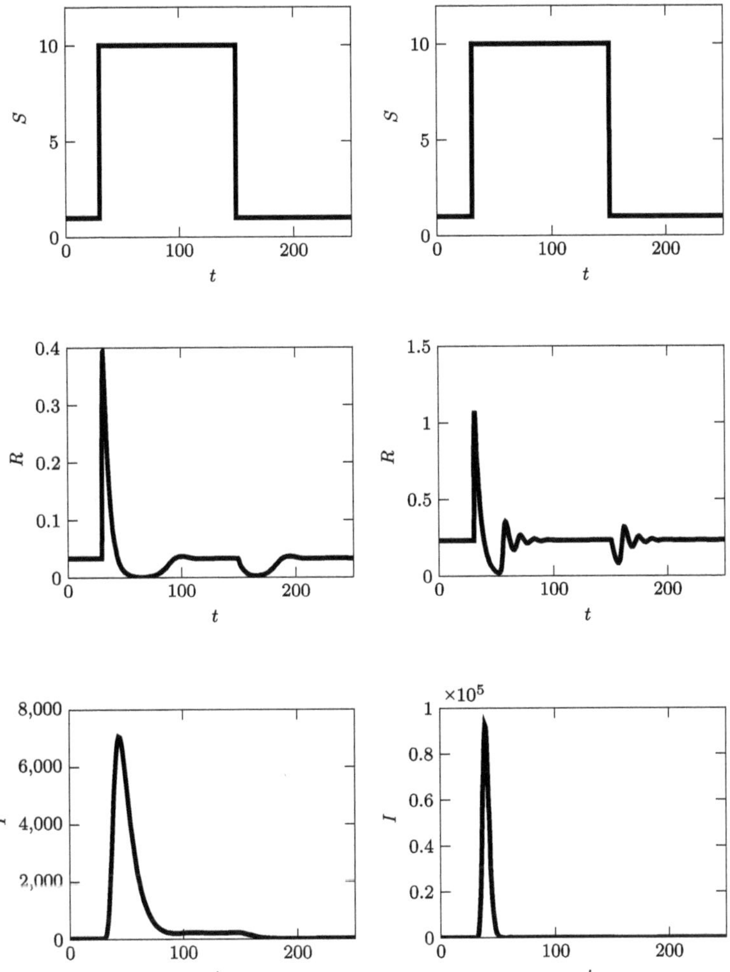

Figure 2.10 The response of the feedback loop to the step input $S = 0.1, 1$. (left) $\gamma_I = 0.1$, (right) $\gamma_I = 0.7$.

a bundle that propels the cell forward. In contrast, when they rotate clockwise, the bundle unravels, causing the cell to reorient randomly (see Fig. 5.10). This process is known as tumbling. If the concentration of a molecule, such as a specific amino acid that acts as a chemotactic attractant, increases over time, the tumbling frequency decreases. Conversely, if the concentration decreases over time, the frequency of tumbling increases. In this way, *E. coli* exhibits positive chemotaxis by performing a biased random walk toward higher concentrations of attractant molecules. In contrast, for repellent molecules, cells move toward lower concentrations (negative chemotaxis) due to the opposite regulation of tumbling frequency, where tumbling frequency increases with increasing concentration. The regulation

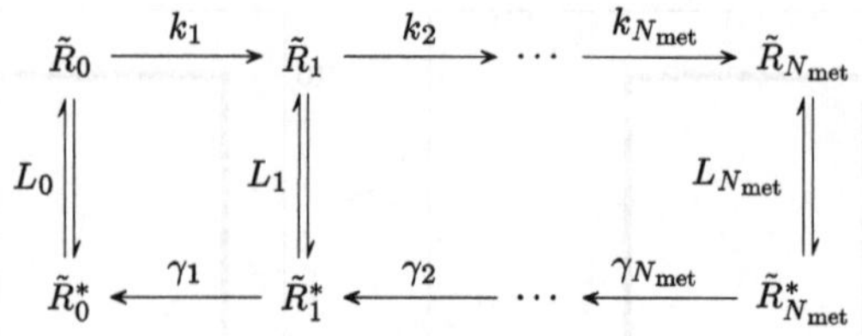

$$\tilde{R}_0 \xrightarrow{\ k_1\ } \tilde{R}_1 \xrightarrow{\ k_2\ } \cdots \xrightarrow{\ k_{N_{\mathrm{met}}}\ } \tilde{R}_{N_{\mathrm{met}}}$$

$$L_0 \quad\quad L_1 \quad\quad\quad\quad L_{N_{\mathrm{met}}}$$

$$\tilde{R}_0^* \xleftarrow{\ \gamma_1\ } \tilde{R}_1^* \xleftarrow{\ \gamma_2\ } \cdots \xleftarrow{\ \gamma_{N_{\mathrm{met}}}\ } \tilde{R}_{N_{\mathrm{met}}}^*$$

Figure 2.11 The reaction scheme of the Asakura–Honda model. $\tilde{R}_i = R_i + R_i S_a$, $\tilde{R}_i^* = R_i^* + R_i^* S_b$ ($i = 0, 1, \ldots, N_{\mathrm{met}}$).

of tumbling frequency is a typical example of an adaptive response in which the methylation of transmembrane receptor proteins plays a major role.

Let us look at a model that considers the adaptive response of *E. coli* chemotaxis from the viewpoint of multiple methylations of receptors (Asakura–Honda two-state model [3]). This model was proposed long before the molecular details for chemotaxis became clear. Although the variables do not necessarily correspond to the molecular reality, it has provided important guidance for understanding the logic of adaptation [4, 25].

The receptor is assumed to be in either R or R^* states, with *E. coli* moving straight ahead in the R state and tumbling in the R^* state. The states in which $i = 0, 1, \ldots, N_{\mathrm{met}}$ methyl groups are bound to the receptor are denoted as R_i and R_i^*, respectively. We assume that methylation only occurs in the R state and demethylation only occurs in the R^* state (Fig. 2.11). We further assume that the receptor in the R state can only receive one methyl group at a time, while the receptor in the R^* state can only release one methyl group at a time. The equilibrium constant in the absence of ligand is denoted as $L_i^0 = R_i/R_i^*$, and we assume the following relationship:

$$L_0^0 > L_1^0 > \cdots > L_{N_{\mathrm{met}}}^0. \tag{2.75}$$

The equilibrium is biased toward the R^* state as the number of methyl groups increases in the absence of extracellular ligand concentration. Additionally, we assume that the attractant S_a binds only to the R state and the repellent S_b binds only to the R^* state, and both binding constants are independent of the number of methyl groups.

Under these assumptions, if the coupling constants are K_a and K_b, the following equation holds between the receptor in the R state and the concentration of the attractant molecule:

$$S_a/K_a = R_0 S_a/R_0 = R_1 S_a/R_1 = \cdots = R_{N_{\mathrm{met}}} S_a/R_{N_{\mathrm{met}}}. \tag{2.76}$$

Similarly, the following equation holds between the receptor in the R^* state and the concentration of the repellent molecule:

$$S_b/K_b = R_0^* S_b/R_0^* = R_1^* S_b/R_1^* = \cdots = R_{N_{\mathrm{met}}}^* S_b/R_{N_{\mathrm{met}}}^*. \tag{2.77}$$

It is also assumed that methylation and demethylation occur in the same way, regardless of whether the receptor is alone or in a ligand–receptor complex. Here, we use the notations $\tilde{R}i = Ri + R_i S_a$ and $\tilde{R}_i^* = R^*i + R^*iS_b$. Additionally, we define the normalized ligand concentrations as $\alpha = S_a/K_a$ and $\beta = S_b/K_b$.

The equilibrium between the states of the receptor at each i is determined by the binding of the attractant and repellent molecules, as follows:

$$L_i = \frac{\tilde{R}_i}{\tilde{R}_i^*} = L_i^0 \frac{(1+\alpha)}{(1+\beta)}. \tag{2.78}$$

This relationship implies that an increase in the number of attractant molecules leads to an increase in the R-state, while an increase in the number of repellent molecules leads to an increase in the R^*-state. Conversely, an increase in the R-state promotes methylation reaction (and an increase in the R^*-state promotes demethylation reaction). From Eq. (2.78) and Eq. (2.75), we obtain

$$L_0 > L_1 > \cdots > L_{N_{\mathrm{met}}}. \tag{2.79}$$

This relationship indicates that the methylation of the receptor shifts its equilibrium toward the R^* state, irrespective of the ligand concentration.

From the preceding discussion, it is evident that if the transition between R_i and R_i^* states occurs rapidly enough to attain equilibrium, and the methylation/demethylation reaction is comparatively slower, then the transitions between receptor states operate in the opposite direction with a time lag. As S_a increases, R_i^* transiently decreases (i.e., the tumbling frequency decreases) at each i (Eq. (2.78)). However, as the methylation reaction progresses, the number of R^* states increases (2.79), and after a sufficient amount of time, the sum of R_i^* restores to the prestimulus level. Similarly, when S_b increases, R_i transiently decreases at each i, but the sum of R states recovers the prestimulus level as the demethylation reaction proceeds. Thus, $\tilde{R}^*{}_{\mathrm{tot}} = \sum_i \tilde{R}_i^*$ is anticipated to adaptively respond to changes in ligand concentration (Fig. 2.12).

Let us study this adaptation of the receptor states. The total number of receptors that were methylated i times is

$$P_i = \tilde{R}_i + \tilde{R}_i^*. \tag{2.80}$$

Then, the total number of both R and R^* states is calculated by using P_i as follows:

$$\tilde{R}_{\mathrm{tot}} = \sum_{i=0}^{N_{\mathrm{met}}} \tilde{R}_i = \sum_{i=0}^{N_{\mathrm{met}}} P_i \frac{L_i}{1 + L_i}, \tag{2.81}$$

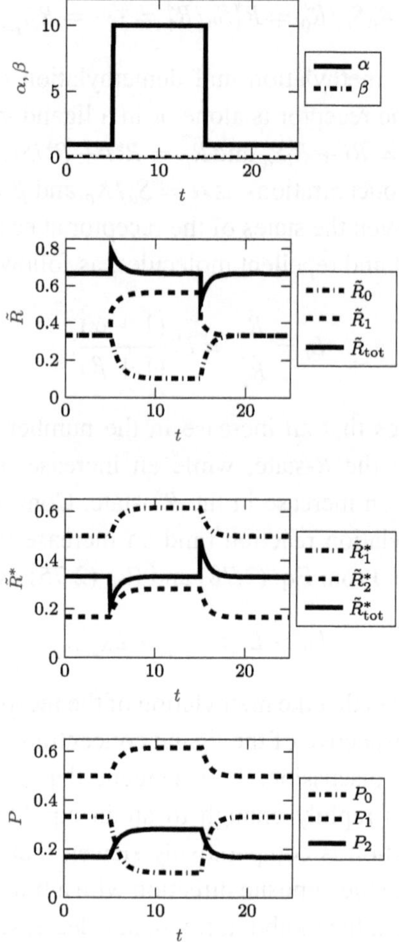

Figure 2.12 Asakura–Honda model. Response to the step input. $N_{\mathrm{met}} = 2$, $L_0 = 10^5$, $L_1 = 10^0$, $L_2 = 10^{-5}$, $k_i = 1$, $\gamma_i = 2$.

$$\tilde{R}^*_{\mathrm{tot}} = \sum_{i=0}^{N_{\mathrm{met}}} \tilde{R}^*_i = \sum_{i=0}^{N_{\mathrm{met}}} P_i \frac{1}{1 + L_i}. \tag{2.82}$$

Time evolution of P_i is written as

$$\frac{\mathrm{d}P_i}{\mathrm{d}t} = k_i \tilde{R}_{i-1} - k_{i+1} \tilde{R}_i + \gamma_{i+1} \tilde{R}^*_{i+1} - \gamma_i \tilde{R}^*_i, \tag{2.83}$$

where we assume $k_0 = \gamma_0 = k_{N_{\mathrm{met}}+1} = \gamma_{N_{\mathrm{met}}+1} = 0$. In the steady state,

$$\mu_i \equiv \frac{k_i}{\gamma_i} = \frac{\tilde{R}^*_i}{\tilde{R}_{i-1}} \tag{2.84}$$

holds. If $\mu_i = \mu$ (= constant) holds regardless of i, then

$$\mu = \tilde{R}_1^* / \tilde{R}_0 = \tilde{R}_2^* / \tilde{R}_1 = \cdots = \tilde{R}_{N_{\mathrm{met}}}^* / \tilde{R}_{N_{\mathrm{met}}-1}. \tag{2.85}$$

Here, if $L_0^0 \gg 1$ and $L_{N_{\mathrm{met}}}^0 \ll 1$, $\tilde{R}_0^*$ and $\tilde{R}_{N_{\mathrm{met}}}$ are negligible over a wide range of ligand concentrations so that

$$\mu \simeq \tilde{R}_{\mathrm{tot}}^* / \tilde{R}_{\mathrm{tot}}. \tag{2.86}$$

In summary, if $\mu_i = \mu$ (= constant) and $L_0^0 \gg 1$ and $L_{N_{\mathrm{met}}}^0 \ll 1$, the steady state is

$$\tilde{R}_{\mathrm{tot}}^* = \frac{\mu}{1+\mu} \tag{2.87}$$

and is independent of the ligand concentration, where we normalized the total amount of the receptor as $\sum_{i=0}^{N_{\mathrm{met}}} P_i = \{\tilde{R}_{\mathrm{tot}} + \tilde{R}_{\mathrm{tot}}^*\} = 1$.

Exercise 2.4.1 Derive the reaction rate equations for R_i, R_i^*, $R_i S_a$, and $R_i^* S_b$, respectively. Then confirm Eq. (2.83). Assuming that the binding and the dissociation between the ligand and the receptor is fast enough to reach equilibrium, using Eq. (2.76), Eq. (2.77), and Eq. (2.82), replace the right-hand side of Eq. (2.83) with a function of P_i to obtain a simultaneous differential equation for P_i.

Exercise 2.4.2 If $N_{\mathrm{met}} = 2$ and $k = (1+\alpha)/(1+\beta)$, then show that the steady-state solution is written as

$$\tilde{R}_{\mathrm{tot}}^* = \frac{1 + \mu_1 L_0^0 k + \mu_2 L_1^0 \mu_1 L_0^0 k^2}{1 + (1+\mu_1) L_0^0 k + (1+\mu_2) L_1^0 \mu_1 L_0^0 k^2 + L_2^0 \mu_2 L_1^0 \mu_1 L_0^0 k^3}. \tag{2.88}$$

If $L_0^0 k \gg 1$ and $k \ll 1$, then show that $\tilde{R}_{\mathrm{tot}}^* = \mu_1/(1 + \mu_1)$. Also, if $L_2^0 k \ll 1$ and $k \gg 1$, then show that $\tilde{R}_{\mathrm{tot}}^* = \mu_2/(1 + \mu_2)$ [14].

Exercise 2.4.3 Negative Feedback System [9] By comparing the results of the model

$$\frac{\mathrm{d}R}{\mathrm{d}t} = k_1 S(R_{\mathrm{T}} - R) - k_2 RI, \tag{2.89}$$

$$\frac{\mathrm{d}I}{\mathrm{d}t} = k_3 R \frac{I_{\mathrm{T}} - I}{K_3 + (I_{\mathrm{T}} - I)} - k_4 \frac{I}{K_4 + I} \tag{2.90}$$

and the Barkai–Leibler model (Eq. (2.96)), consider that the response to input S becomes adaptive. Also, confirm the dynamics in response to the step input by numerical simulation. Note that all the parameters are positive constants.

Column II: Barkai–Leibler Model [4]

Barkai and Leibler proposed a simple model for bacterial chemotaxis, which was derived by coarse-graining the actual reaction network involving receptor and signal proteins.

The receptor protein can form a complex with the kinase CheA. Upon binding an attractant molecule to the receptor's extracellular region, CheA phosphorylates CheY, leading to a reversal in the flagellar rotation direction. Additionally, the receptor can undergo reversible methylation, which enhances CheA's kinase activity. It is hypothesized that the presence of a ligand enhances receptor methylation. In the absence of extracellular ligand, the methylated receptor binds and phosphorylates CheB. CheR facilitates receptor methylation, while phosphorylated CheB promotes receptor demethylation. Here, we use R, S, and B to denote the concentrations of receptor, ligand, and CheB, respectively. Methylation is represented by the symbol $*$. The reaction scheme of this model is illustrated in Fig. 2.13.

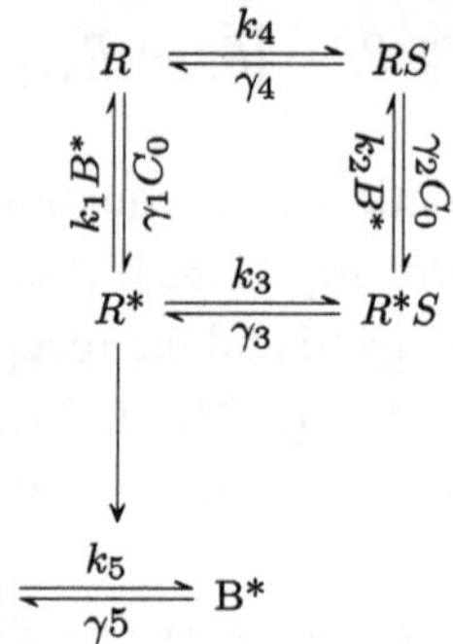

Figure 2.13 Reaction scheme of the Barkai–Leibler model.

Now, if the demethylation reaction of the receptor is assumed to be Michaelis–Menten type, and the methylation reaction of the receptor and the phosphorylation reaction of CheB are approximated as first-order reactions and the dephosphorylation reaction of CheB as zero-order, we obtain the following equations:

$$\frac{dR^*}{dt} = \gamma_1 C_0 - k_1 B^* \frac{R^*}{K_{M1} + R^*} - k_3 R^* S + \gamma_3 \tilde{R}^*, \tag{2.91}$$

$$\frac{d\tilde{R}^*}{dt} = \gamma_2 C_0 - k_2 B^* \frac{\tilde{R}^*}{K_{M2} + \tilde{R}^*} + k_3 R^* S - \gamma_3 \tilde{R}^*, \tag{2.92}$$

$$\frac{dR}{dt} = -\gamma_1 C_0 + k_1 B^* \frac{R^*}{K_{M1} + R^*} - k_4 RS + \gamma_4 \tilde{R}, \tag{2.93}$$

$$\frac{d\tilde{R}}{dt} = -\gamma_2 C_0 + k_2 B^* \frac{\tilde{R}^*}{K_{M2} + \tilde{R}^*} + k_4 RS - \gamma_4 \tilde{R}, \tag{2.94}$$

$$\frac{dB}{dt} = -k_5 R^* B + \gamma_5 B^*, \tag{2.95}$$

$$\frac{dB^*}{dt} = k_5 R^* B - \gamma_5 B^*. \tag{2.96}$$

Note that $R = [R], \tilde{R} = [RS], R^* = [R^*], \tilde{R}^* = [R^*S]$, C_0 is the concentration of CheR, and S is the concentration of the extra-cellular ligand.

Assuming a situation in which the receptor–ligand complex is more prone to methylation ($k_2 < k_1 = $ large) than the free receptor (as depicted in Fig. 2.14), an increase in ligand concentration results in a rapid decline in the number of methylated free receptors (R^*), leading to a reduction in tumbling frequency. In the case of slow CheB phosphorylation ($k_5 = $ small), the decrease in phosphorylated CheB (B^*) is followed by an increase in free methylated receptors (R^*), which eventually recovers to or near the original amount. In this negative feedback adaptation, the supply of free methylation receptors (R^*) in response to the complex of methylation receptor–ligands ($\tilde{R}^*$), which is amplified by the ligand increase, plays a significant role.

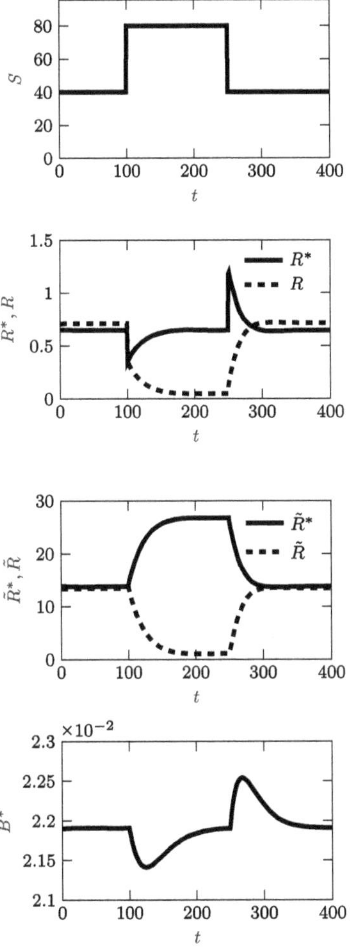

Figure 2.14 Barkai–Leibler model. The response to the step input. $k_1 = 200$, $k_5 = 0.05$, $\gamma_5 = 0.005$.

References

[1] G. Adair. The hemoglobin system. V. The relation of hemoglobin and bases. *Journal of Biological Chemistry*, **63**(2): 517–527, 1925.

[2] U. Alon *et al.* Response regulator output in bacterial chemotaxis. *The EMBO Journal*, **17**(15): 4238–4248, 1998.

[3] S. Asakura and H. Honda. Two-state model for bacterial chemoreceptor proteins: The role of multiple methylation. *Journal of Molecular Biology*, **176**(3): 349–367, 1984.

[4] N. Barkai and S. Leibler. Robustness in simple biochemical networks. *Nature*, **387**(6636): 913–917, 1997.

[5] H. Berg. Motile behavior of bacteria. *Physics Today*, **53**(1): 24–29, 2000.

[6] J.-P. Changeux and S. J. Edelstein. Allosteric mechanisms of signal transduction. *Science*, **308**(5727): 1424–1428, 2005.

[7] J. E. Ferrell. Tripping the switch fantastic: How a protein kinase cascade can convert graded inputs into switch-like outputs. *Trends in Biochemical Sciences*, **21**(12): 460–466, 1996.

[8] J. E. Ferrell and W. Xiong. Bistability in cell signaling: How to make continuous processes discontinuous, and reversible processes irreversible. *Chaos (Woodbury, NY)*, **11**(1): 227–236, 2001.

[9] J. E. Ferrell. Perfect and near-perfect adaptation in cell signaling. *Cell Systems*, **2**(2): 62–67, 2016.

[10] L. Goentoro *et al.* The incoherent feed-forward loop can provide fold-change detection in gene regulation. *Molecular Cell*, **36**(5): 894–899, 2009.

[11] A. Goldbeter and D. E. Koshland. An amplified sensitivity arising from covalent modification in biological systems. *Proceedings of the National Academy of Sciences of the United States of America*, **78**(11): 6840–6844, 1981.

[12] A. V. Hill. The possible effects of the aggregation of the molecules of hemoglobin on its dissociation curves. *The Journal of Physiology*, **40**: iv–vii, 1910.

[13] P. A. Iglesias and C. Shi. Comparison of adaptation motifs: Temporal, stochastic and spatial responses. *IET Systems Biology*, **8**(6): 268–281, 2014.

[14] M. Inoue. Dynamics of Adaptive Response in Biological Systems. PhD dissertation. The University of Tokyo, 2010.

[15] S. Ishihara *et al.* Cross talking of network motifs in gene regulation that generates temporal pulses and spatial stripes. *Genes to Cells*, **10**(11): 1025–1038, 2005.

[16] Y. Kamino and K. Kondo. Rescaling of spatio-temporal sensing in eukaryotic chemotaxis. *PLoS ONE*, **11**(10): e0164674, 2016.

[17] D. E. Koshland *et al.* Amplification and adaptation in regulatory and sensory systems. *Science*, **217**(4556): 220–225, 1982.

[18] D. E. Koshland *et al.* Comparison of experimental binding data and theoretical models in proteins containing subunits. *Biochemistry*, **5**(1): 365–385, 1966.

[19] A. Levchenko and P. A. Iglesias. Models of eukaryotic gradient sensing: Application to chemotaxis of amoebae and neutrophils. *Biophysical Journal*, **82**(1): 50–63, 2002.

[20] W. Ma *et al.* Defining network topologies that can achieve biochemical adaptation. *Cell*, **138**(4): 760–773, 2009.

[21] S. Mangan and U. Alon. Structure and function of the feed-forward loop network motif. *Proceedings of the National Academy of Sciences of the United States of America*, **100**(21): 11980–11985, 2003.

[22] J. Monod *et al.* On the nature of allosteric transitions: A plausible model. *Journal of Molecular Biology*, **12**: 88–118, 1965.

[23] Y. Shi. Effects of thermal fluctuation and the receptor–receptor interaction in bacterial chemotactic signaling and adaptation. *Physical Review E*, **64**(2): 021910, 2001.

[24] O. Shoval *et al.* Fold-change detection and scalar symmetry of sensory input fields. *Proceedings of the National Academy of Sciences*, **107**(36): 15995–16000, 2010.

[25] M. J. Tindall *et al.* Overview of mathematical approaches used to model bacterial chemotaxis I: The single cell. *Bulletin of Mathematical Biology*, **70**(6): 1525–1569, 2008.

[26] Y. Tu. Quantitative modeling of bacterial chemotaxis: Signal amplification and accurate adaptation. *Annual Review of Biophysics*, **42**(1): 337–359, 2013.

[27] J. J. Tyson *et al.* Sniffers, buzzers, toggles and blinkers: Dynamics of regulatory and signaling pathways in the cell. *Current Opinion in Cell Biology*, **15**(2): 221–231, 2003.

3

Oscillation and Excitability in Cellular Dynamics

3.1 Interplay between Positive and Negative Feedbacks

Large temporal fluctuations or oscillations in cellular states are widely observed in biological systems, for instance, in neural firing, circadian rhythms, and collective motion of amoebae. These phenomena arise from the interplay between positive and negative feedback mechanisms, as discussed in Chapter 2. In this section, we will focus on such dynamic changes in cellular states. To illustrate them, we will extend the model described in Section 2.4.2 by incorporating both positive and negative feedback. The model equation is provided here [8, 10]:

$$\frac{dR}{dt} = k_S S - \gamma_R R + k_{RR} \frac{R^m}{K^m + R^m} - k_{RI} I,$$

$$\frac{dI}{dt} = k_{IR} R - \gamma_I I. \tag{3.1}$$

In a similar manner to Section 2.4.2, we introduced an input, denoted as S, which changes in a stepwise manner, and waited for a sufficient time for the system to return to the original state (Fig. 3.1). When the input S is small, both the output R and the intermediate molecules I exhibit a sharp rise followed by oscillatory decay (Fig. 3.1, left; upper and middle panels). By considering the steady state of Eq. (3.1), it is apparent that the response does not make perfect adaptation. As the input S is further increased, oscillatory dynamics emerge (Fig. 3.1, right; upper and middle panels). We can examine this transition in the phase plane (Fig. 3.1, bottom row). Unlike the linear nullcline of variable I, the nullcline of variable R exhibits two turning points in the phase plane, representing a local maximum and a local minimum, due to the sigmoid function governing the time development of R. As the input S increases, the nullcline of R shifts upwards in the phase plane, and the fixed point goes beyond the local minimum. What is the relationship between this behavior and the onset of oscillatory dynamics? We will discuss this topic extensively in the following section.

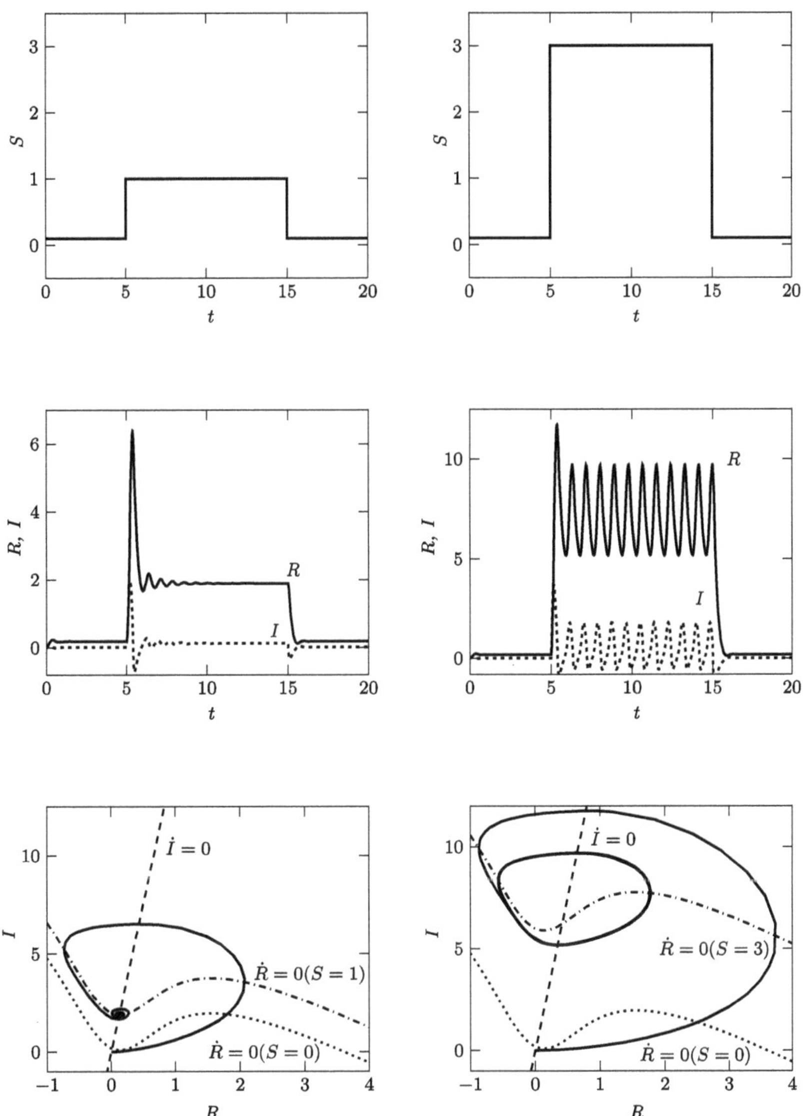

Figure 3.1 Oscillations emerge due to a combination of positive and negative feedback. (left) small input case ($S = 1$) and (right) large input case ($S = 3$).

3.2 Trajectory of Oscillatory Dynamics in Phase Plane

Oscillatory dynamics in biological systems involve diverse timescales, ranging from milliseconds in the neuron action potentials to seconds in heartbeats and hours in the circadian rhythm. While these dynamics may appear distinct at first glance, they share a common characteristic of being limit-cycles. As an illustration of a limit-cycle, we briefly review the Brusselator presented in Section 1.9. In this section, we will further explore its properties in detail.

For the Brusselator model:

$$\frac{dX}{dt} = 1 - (b+1)X + aX^2Y, \tag{3.2}$$

$$\frac{dY}{dt} = bX - aX^2Y, \tag{3.3}$$

the nullclines $dX/dt = f(X,Y) = 0$ and $dY/dt = g(X,Y) = 0$ are given as follows:

$$Y_f = \frac{(b+1)X - 1}{aX^2}, \tag{3.4}$$

$$Y_g = \frac{b}{aX} \quad \text{or} \tag{3.5}$$

$$X_g = 0. \tag{3.6}$$

There is a fixed point at $(X^*, Y^*) = (1, b/a)$, which corresponds to the intersection point of Y_f and Y_g. As discussed in Section 1.15, we can judge whether a limit-cycle attractor emerges by using the linear stability analysis. The Jacobi matrix at the fixed point $(X^*, Y^*) = (1, b/a)$ is

$$J = \begin{pmatrix} \frac{\delta f}{\delta X}\big|_{(X^*, Y^*)} & \frac{\delta f}{\delta Y}\big|_{(X^*, Y^*)} \\ \frac{\delta g}{\delta X}\big|_{(X^*, Y^*)} & \frac{\delta g}{\delta Y}\big|_{(X^*, Y^*)} \end{pmatrix} = \begin{pmatrix} b-1 & a \\ -b & -a \end{pmatrix} \tag{3.7}$$

Therefore, the eigenvalue of the Jacobi matrix can be calculated by

$$|J - \lambda I| = \begin{vmatrix} b-1-\lambda & a \\ -b & -a-\lambda \end{vmatrix} = 0, \tag{3.8}$$

where I is the identity matrix. Thus,

$$\lambda^2 - (b - 1 - a)\lambda + a = 0. \tag{3.9}$$

Let the solutions of the preceding equation be λ_1 and λ_2, which satisfies

$$\lambda_1 \lambda_2 = \begin{vmatrix} b-1 & a \\ -b & -a \end{vmatrix} = a > 0. \tag{3.10}$$

This means that the eigenvalues have the same sign. In other words, this fixed point cannot be a saddle point. In addition, the relationship

$$\lambda_1 + \lambda_2 = \mathrm{tr}\begin{pmatrix} b-1 & a \\ -b & -a \end{pmatrix} = b - 1 - a \tag{3.11}$$

indicates that the attractor is stable when $(b - 1 - a) < 0$ and is unstable when $(b - 1 - a) > 0$. Furthermore, if there is an oscillatory behavior, the eigenvalues have imaginary parts. From Eq. (3.9), this condition corresponds to

$$(b - 1 - a)^2 - 4a < 0. \tag{3.12}$$

Since $a > 0$, the relationship can be converted to

$$(b - 1 - a + 2\sqrt{a})(b - 1 - a - 2\sqrt{a}) < 0. \tag{3.13}$$

Then, by assuming the condition for unstable fixed point $(b - 1 - a) > 0$, we obtain $b - 1 - a - 2\sqrt{a} < 0$. This means that the oscillatory dynamics emerge when

$$0 < b - 1 - a < 2\sqrt{a}. \tag{3.14}$$

Exercise 3.1 Let us consider the van der Pol equation:

$$\frac{d^2 V}{dt^2} - \mu(1 - V^2)\frac{dV}{dt} + V = 0. \tag{3.15}$$

This equation describes an oscillator with a nonlinear damping term, which is studied as a model for oscillatory dynamics in electric circuits containing vacuum tubes. The constant parameter μ represents the strength of the damping term. By introducing the variable $m(t) = \dot{V}$, we can convert the preceding equation into a two-variable system involving $V(t)$ and $m(t)$. Then, by the method used earlier, we obtain the condition for oscillatory dynamics.

3.3 Conditions for Oscillation

The conditions for oscillatory dynamics discussed in the previous section can be generalized as follows. We consider a scenario where N variables $X_i(i = 1, 2, \ldots, N)$ satisfy the following system of differential equations:

$$\frac{dX_i}{dt} = f_i(\{X_j\}), \tag{3.16}$$

where the variables in the parentheses, $\{X_j\}$, represents the set of variables $(X_1, X_2, \ldots, X_N)$. Here, the stability of the fixed point X_j^*, satisfying $f_i(\{X_j\}) = 0$, is determined by the sign of eigenvalues of the Jacobi matrix

$$J = \begin{pmatrix} \frac{\delta f_1}{\delta X_1} & \cdots & \frac{\delta f_1}{\delta X_N} \\ \vdots & \ddots & \vdots \\ \frac{\delta f_N}{\delta X_1} & \cdots & \frac{\delta f_N}{\delta X_N} \end{pmatrix}_{\{X_j^*\}}. \tag{3.17}$$

In the case of $N = 2$, the eigen equation $|J - \lambda I| = 0$ is represented as

$$\lambda^2 - (\mathrm{tr}J)\lambda + |J| = 0. \tag{3.18}$$

Therefore, we can judge the stability by the signs of eigenvalues

$$\lambda_{\pm} = \frac{1}{2}\left\{\mathrm{tr}J \pm \sqrt{(\mathrm{tr}J)^2 - 4|J|}\right\}, \tag{3.19}$$

as already explained in Section 1.7.

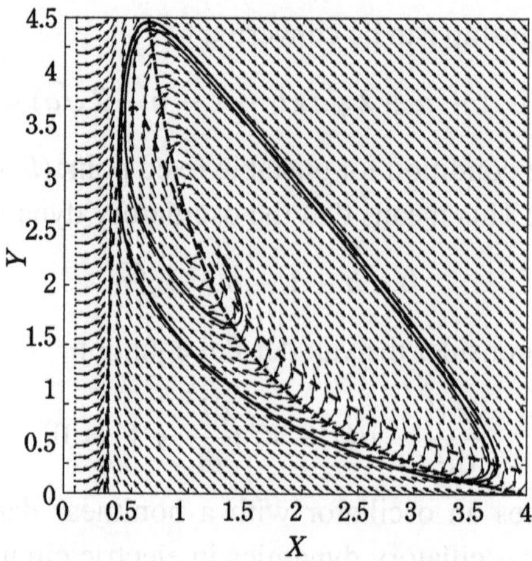

Figure 3.2 Nullcline (dashed line) and limit-cycle (solid line) of Brusselator.

Let us consider the stability conditions at the fixed point. When the content of the root in Eq. (3.19) is negative, we obtain

$$\mathrm{Re}\lambda_{\pm} = \frac{1}{2}\mathrm{tr}J. \tag{3.20}$$

Therefore, the fixed point is stable if $\mathrm{tr}J < 0$. When the content of the root is positive, one of the eigenvalues

$$\lambda_{-} = \frac{1}{2}\left\{ \mathrm{tr}J - \sqrt{(\mathrm{tr}J)^2 - 4|J|} \right\} \tag{3.21}$$

is negative if $\mathrm{tr}J < 0$. The other eigenvalue

$$\lambda_{+} = \frac{1}{2}\left\{ \mathrm{tr}J + \sqrt{(\mathrm{tr}J)^2 - 4|J|} \right\} \tag{3.22}$$

is negative if $\sqrt{(\mathrm{tr}J)^2 - 4|J|} < -\mathrm{tr}J$. Summing up, the stability conditions in the case of $N = 2$ are:

$$\mathrm{tr}J = J_{11} + J_{22} < 0, \tag{3.23}$$
$$|J| = J_{11}J_{22} - J_{12}J_{21} > 0. \tag{3.24}$$

In contrast, when the fixed point is unstable, and it is not a saddle point, the following conditions should be satisfied:

$$\mathrm{tr}J = J_{11} + J_{22} > 0, \tag{3.25}$$
$$|J| = J_{11}J_{22} - J_{12}J_{21} > 0. \tag{3.26}$$

In what follows, we discuss the dynamics with the preceding conditions by using the nullclines around the fixed point. Let we assume $dX_2/dX_1 > 0$ on the nullclines

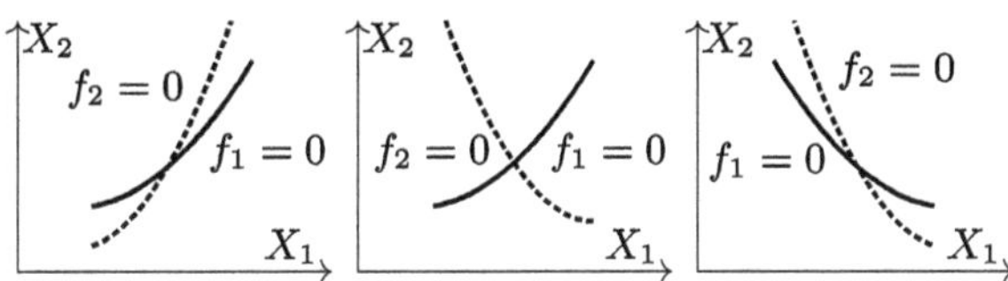

Figure 3.3 Intersection between nullclines.

$f_1 = 0$ and $f_2 = 0$, and the gradient of the nullcline $f_1 = 0$ is smaller than the nullcline $f_2 = 0$ (see the left panel in Fig. 3.3), that is,

$$\left.\frac{\mathrm{d}X_2}{\mathrm{d}X_1}\right|_{f_2=0} > \left.\frac{\mathrm{d}X_2}{\mathrm{d}X_1}\right|_{f_1=0} > 0. \tag{3.27}$$

From the implicit function theorem, the following relationships are derived:

$$\left.\frac{\mathrm{d}X_2}{\mathrm{d}X_1}\right|_{f_1=0} = -\frac{\frac{\mathrm{d}f_1}{\mathrm{d}X_1}}{\frac{\mathrm{d}f_1}{\mathrm{d}X_2}} = -\frac{J_{11}}{J_{12}}, \tag{3.28}$$

$$\left.\frac{\mathrm{d}X_2}{\mathrm{d}X_1}\right|_{f_2=0} = -\frac{\frac{\mathrm{d}f_2}{\mathrm{d}X_1}}{\frac{\mathrm{d}f_2}{\mathrm{d}X_2}} = -\frac{J_{21}}{J_{22}}. \tag{3.29}$$

Therefore, from the preceding relationships, we obtain

$$-\frac{J_{21}}{J_{22}} > -\frac{J_{11}}{J_{12}} > 0. \tag{3.30}$$

When J_{12} and J_{22} have the same sign, by multiplying $J_{12}J_{22}$ on both sides of Eq. (3.32), we obtain

$$-J_{21}J_{12} > -J_{11}J_{22} > 0, \tag{3.31}$$

where the condition $|J| > 0$ is satisfied, which means that the fixed point is not a saddle point. At the same time, the different signs between J_{11} and J_{12}, and between J_{21} and J_{22} are required, respectively. Furthermore, from the condition $\mathrm{tr}J > 0$ (Eq. (3.25)), one of J_{11} or J_{22} must be positive. That is, if J_{11} is positive, then J_{12} must be negative, while if J_{22} is positive, then J_{21} must be negative. If both J_{11} and J_{22} are positive, then J_{12} must also be positive because J_{12} and J_{22} have the same sign, and J_{11} and J_{12} cannot take different signs at the same time. To sum up, the combinations of signs satisfying the preceding conditions are:

$$J = \begin{pmatrix} + & - \\ + & - \end{pmatrix} \tag{3.32}$$

or

$$J = \begin{pmatrix} - & + \\ - & + \end{pmatrix}. \tag{3.33}$$

The preceding Jacobi matrices represent that one variable activates or inhibits the other. Dynamical systems satisfying these relationships are often called "activator-inhibitor" systems. The feedback system represented by Eq. (3.1) and the FitzHugh–Nagumo equations explained later are examples of the activator-inhibitor system.

Next, let us discuss the dynamics of the activator-inhibitor system using the trajectory on the phase space. Around the fixed point, the dynamics are approximated as follows:

$$f_1(X_1^* + dX_1, X_2^* + dX_2) \sim J_{11}dX_1 + J_{12}dX_2, \qquad (3.34)$$
$$f_2(X_1^* + dX_1, X_2^* + dX_2) \sim J_{21}dX_1 + J_{22}dX_2. \qquad (3.35)$$

Under the conditions $J_{11} > 0$ and $J_{12} < 0$, f_1 turns from negative to positive when the state crosses the nullcline $f_1 = 0$ by increasing X_1 (see the upper figure of Fig. 3.4(a)), while f_1 turns from positive to negative by increasing X_2. When $J_{21} > 0$ and $J_{22} < 0$, the sign of f_2 changes in the same way (the middle figure of Fig. 3.4(a)). These changes in the signs of f_1 and f_2 enable us to describe the dynamics around the fixed point as illustrated in the bottom figure of Fig. 3.4. In this configuration, the phase space is divided into four regions by the two nullclines. The signs of f_1 and f_2 vary across these regions, thereby determining the trajectory direction. As depicted, a circular trajectory is observed around the intersection of the nullclines, which corresponds to the fixed point. This is the case of the Jacobi matrix in Eq. (3.32).

In the case of the Jacobi matrix described in Eq. (3.33), the value of f_1 turns from positive to negative as the state crosses the nullcline by increasing X_1, while it turns from negative to positive with an increase in X_2 (Fig. 3.4(b)). Similarly, the sign of f_2 undergoes the same change. Consequently, a clockwise circular trajectory emerges around the fixed point. The preceding two cases correspond to the activator-inhibitor system.

Next, let us consider the case that $dX_2/dX_1 < 0$ on the nullclines $f_1 = 0$ and $f_2 = 0$ (see the right panel in Fig. 3.3). We also assume that the gradient of the nullcline $f_1 = 0$ is larger than the nullcline $f_2 = 0$, that is,

$$\left. \frac{dX_2}{dX_1} \right|_{f_2=0} < \left. \frac{dX_2}{dX_1} \right|_{f_1=0} < 0. \qquad (3.36)$$

Therefore, the following relationships are satisfied:

$$-\frac{J_{21}}{J_{22}} < -\frac{J_{11}}{J_{12}} < 0. \qquad (3.37)$$

To satisfy the preceding conditions and the condition that the fixed point is not a saddle point simultaneously, J_{12} and J_{22} must have different signs. At the same

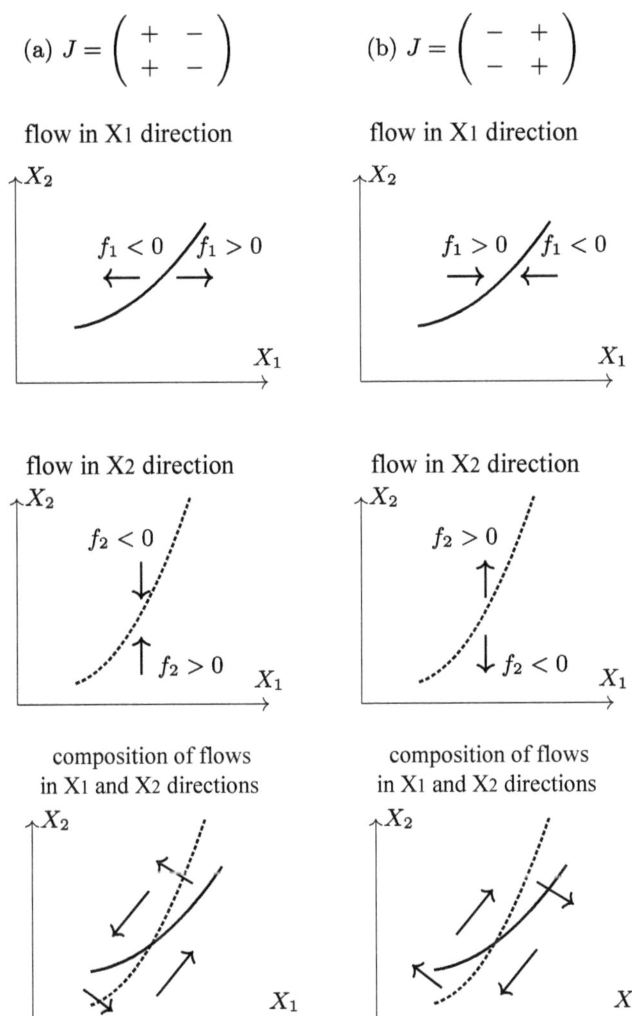

Figure 3.4 The relationship between the sign of the Jacobi matrix and the orientation of the orbit around a fixed point in an activator-inhibitor system. Intersection of nullclines in an in the upper-right direction causes both variables to oscillate in-phase.

time, from Eq. (3.37), it is required that J_{11} and J_{12} have the same sign, and J_{21} and J_{22} also have the same sign. Then, the combinations of signs satisfying the preceding conditions are:

$$J = \begin{pmatrix} + & + \\ - & - \end{pmatrix} \tag{3.38}$$

or

$$J = \begin{pmatrix} - & - \\ + & + \end{pmatrix} \tag{3.39}$$

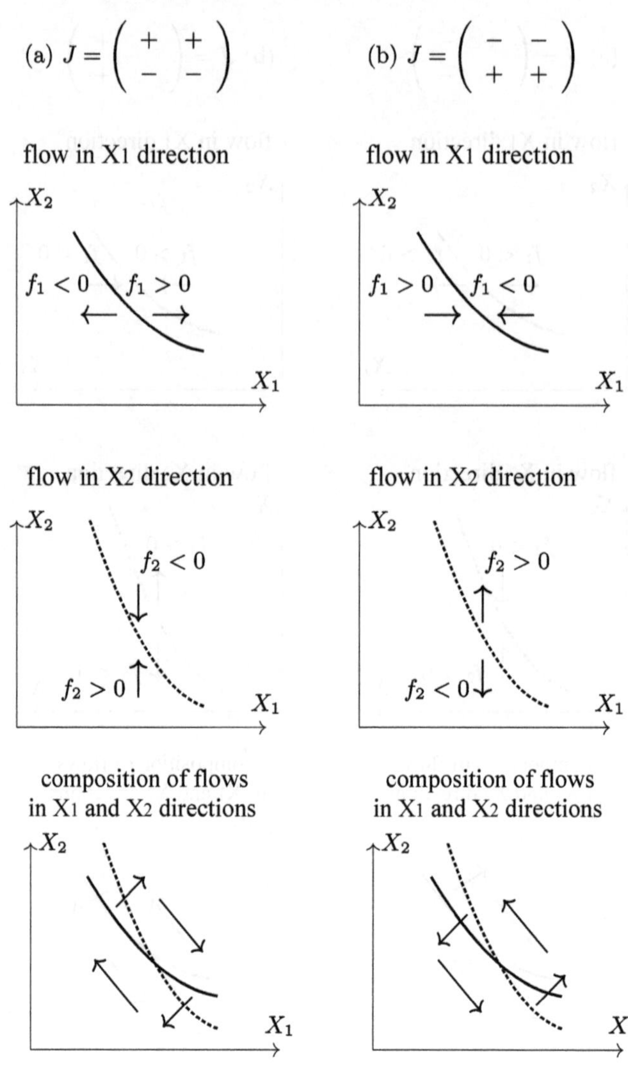

Figure 3.5 The relationship between the sign of the Jacobi matrix and the orientation of the trajectories around a fixed point in a substrate-depletion system. When the nullcline intersection occurs in a downward-right direction, both variables oscillate in anti-phase.

In the aforementioned scenario, when one variable is increasing, the other is decreasing. This characterizes a situation where a factor is generated by the substrate and consumes the substrate leading to an inhibitory effect. Such a system is referred to as a "substrate-depletion" system (Fig. 3.5). The Brusselator gives an example, in which inhibition is triggered by substrate depletion. In this system, the emergence of a decreasing trend at the intersection of nullclines, as depicted in Fig. 3.5, leads to the development of a limit-cycle around that intersection (see Fig. 3.2).

Exercise 3.2   Oscillation in Gene Expression Dynamics (Goodwin model [2, 3])
Let us examine the dynamics of a system consisting of mRNA, protein, and substrate, denoted as X, Y, and Z respectively. In this system, mRNA X encodes protein Y, and the protein produces substrate Z. Concurrently, the substrate suppresses mRNA transcription. The dynamics of this system can be described by the following set of differential equations:

$$\frac{\mathrm{d}X}{\mathrm{d}t} = \frac{V}{K^n + Z^n} - aX, \tag{3.40}$$

$$\frac{\mathrm{d}Y}{\mathrm{d}t} = bX - cY, \tag{3.41}$$

$$\frac{\mathrm{d}Z}{\mathrm{d}t} = dY - eZ. \tag{3.42}$$

Here, a, b, c, d, e are positive constants. Analyze how the system dynamics change depending on the Hill coefficient n and verify that oscillatory dynamics tend to emerge when n is large.

Exercise 3.3   A Self-Regulatory System with Two Variables
Depict the nullclines of the following self-regulatory system:

$$\frac{\mathrm{d}y}{\mathrm{d}t} = \frac{\alpha_1 y^m}{K_1^m + y^m} + \frac{\alpha_3}{K_3^n + z^n} - y, \tag{3.43}$$

$$\frac{\mathrm{d}z}{\mathrm{d}t} = \frac{\alpha_2 y^m}{K_2^m + y^m} - z, \tag{3.44}$$

and analyze how the emergence of limit-cycle depends on the Hill coefficients m and n.

3.4   Hodgkin–Huxley Equations

In neurons, dynamic changes in the membrane potential, known as nerve impulses or action potentials, propagate through the axon with almost no attenuation. This alteration in membrane potential gives a basis for the dynamics of brain activity. Hodgkin, Huxley, Katz, and their colleagues performed experimental analysis of the relationships between voltage and current across the membrane using axons from crabs and cuttlefish. The development of electrophysiological techniques, such as the voltage clamp method, has facilitated a comprehensive and quantitative description of experimental data, leading to the establishment of the Hodgkin–Huxley equations [4]. The equations predicted the behavior of unidentified ion channels and provided insights into excitability and oscillation in biological systems. Consequently, this modeling approach marked a significant milestone in the fields of biophysics, theoretical biology, and electrophysiology.

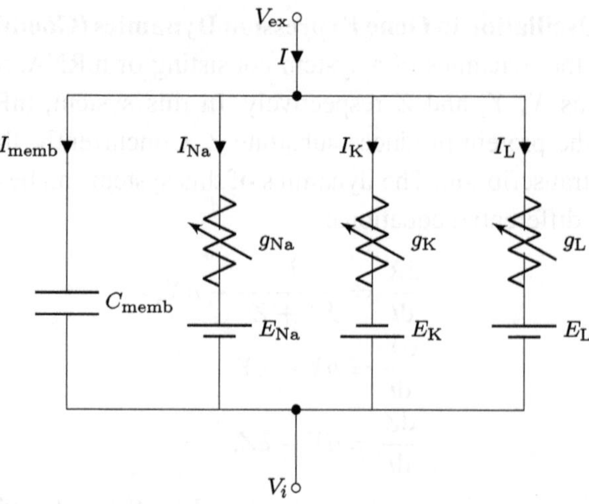

Figure 3.6 Electric circuit equivalent to a system consisting of a cellular membrane and ion channels.

The cell membrane primarily consists of a phospholipid bilayer with embedded membrane proteins. Due to the hydrophobic nature of the phospholipid membrane, only a small number of water molecules can traverse it. Likewise, ions have limited ability to pass through, rendering the membrane a dielectric. On the membrane, there are proteins called ion channels that facilitate the passive movement of ions such as sodium, potassium, and calcium. In this context, we treat the membrane as a capacitor and the ion channels as conductors with variable resistances. We analyze the relationship between the temporal change in membrane potential, denoted as V per unit area, and the current by modeling a small portion of the membrane as an equivalent circuit in which these elements are connected in parallel (see Fig. 3.6). Specifically, we consider the potassium channel and sodium channel, with electrical conductivities of g_{K} and g_{Na} respectively, while the electrical capacitance of the membrane is represented as C_{memb}. Additionally, E represents the potential generated by the ion concentration difference between the intracellular and extracellular environments, g_{L} corresponds to the leak current, and I denotes the externally applied current in experimental settings.

Let us consider the case that the current I is applied from outside the cell. The currents flowing through each pathway, namely membrane, Na, K, and L, are represented as I_{memb}, I_{Na}, I_{K}, and I_{L} respectively. According to Kirchhoff's laws, we obtain

$$I - I_{\mathrm{memb}} - I_{\mathrm{Na}} - I_{\mathrm{K}} - I_{\mathrm{L}} = 0, \qquad (3.45)$$

where the current is positive in the direction of flow into the contact (the black filled circle in Fig. 3.6).

Here, we assume that the external environment of the cell is sufficiently large and grounded, so that we can set $V_{ex} = 0$. For convenience, we denote the electrical potential inside the cell near the membrane as V_i, which we can simply refer to as V. Considering the pathway through the cell membrane as a capacitor with a constant electrical capacitance C_{memb}, we can derive the following equation:

$$I_{memb} = C_{memb}\frac{dV}{dt}. \tag{3.46}$$

On the other hand, the pathways through $i = $ Na, K, and L are considered to be conductors, and we can apply Ohm's law to derive the following equation:

$$I_i = g_i(V - E_i), \tag{3.47}$$

where E_i represents the equilibrium potential obtained by the Nernst equation, which provides the electromotive force from inside to outside the cell. Therefore, from Eqs. (3.45) to (3.47), we obtain

$$C_{memb}\frac{dV}{dt} = -g_{Na}(V - E_{Na}) - g_K(V - E_K) - g_L(V - E_L) + I. \tag{3.48}$$

Now we need a relationship between g_{Na}, g_K, and g_L, which determine the characteristics of the electrical circuit across the cell membrane. Let us consider a hypothetical ion channel consisting of two types of "gates." One is the activation gate, which opens with an increase in membrane potential, and the other is the inhibitory gate, which closes as the membrane potential increases. We assume that the hypothetical ion channel has multiple activation and inhibitory gates, whose numbers are p_i and q_i, respectively. The activation of these gates, represented by $m(V)$ and $h(V)$ respectively, changes independently. Assuming that the membrane potential dependency of electrical conductivity is determined by the activation of these gates, we can establish the following relationship:

$$g_i(V) = G_i m_i^{p_i}(V) h_i^{q_i}(V) \tag{3.49}$$

for $i = $ Na, K, and L. Here, m_i and h_i are functions of the domain $[0,1]$, where the values 0 and 1 correspond to the close (inactivate) and the open (activate) gate, respectively. The G_i is the maximum conductivity, which represents the number of channels. Note that g_L is so small that the leak current is almost negligible, and the voltage behavior of the neuron is mainly described by the two synaptic currents of sodium and potassium ions.

For the gating variables m_i and h_i of the pathways $i = \{$Na, K$\}$, let us denote the transition rates from closing state to opening state per unit time using $\alpha_i^{(m)}(V), \alpha_i^{(h)}(V)$, respectively. In the same way, the transition rates from opening state to closing state are given by $\beta_i^{(m)}(V), \beta_i^{(h)}(V)$, respectively. By using these transition rates, the time evolution of gate activity is represented as

$$\frac{dm_i}{dt} = \alpha_i^{(m)}(1 - m_i) - \beta_i^{(m)}m_i, \tag{3.50}$$

$$\frac{dh_i}{dt} = \alpha_i^{(h)}(1 - h_i) - \beta_i^{(h)}h_i. \tag{3.51}$$

By assuming fixed V and then α and β can be considered as constants, the solutions of the preceding equations are

$$m_i(t) = m_{i,\infty} + (m_i(0) - m_{i,\infty})e^{-t/\tau_i^{(m)}}, \tag{3.52}$$

$$h_i(t) = h_{i,\infty} + (h_i(0) - h_{i,\infty})e^{-t/\tau_i^{(h)}}. \tag{3.53}$$

Here, $\tau_i^{(m)} = 1/(\alpha_i^{(m)} + \beta_i^{(m)})$ and $\tau_i^{(h)} = 1/(\alpha_i^{(h)} + \beta_i^{(h)})$ are the time constants, and m_∞, h_∞ represent asymptotic values after sufficient time, that is,

$$m_{i,\infty} = \frac{\alpha_i^{(m)}}{\alpha_i^{(m)} + \beta_i^{(m)}}, \tag{3.54}$$

$$h_{i,\infty} = \frac{\alpha_i^{(h)}}{\alpha_i^{(h)} + \beta_i^{(h)}}. \tag{3.55}$$

Hodgkin, Huxley, and their colleagues analyzed the dynamics of electrical conductivity under various membrane potentials using the voltage clamp method, which enables the maintenance of a constant membrane potential. In the case of the potassium channel, they postulated the presence of four activation gates ($p_K = 4$) and no inhibitory gate ($q_K = 0$). For the sodium channel, they assumed $p_{Na} = 3$ and $q_{Na} = 1$. Under these assumptions, the conductivities can be expressed as follows:

$$g_K(V) = G_K m_K^4, \tag{3.56}$$

$$g_{Na}(V) = G_{Na} m_{Na}^3 h_{Na}. \tag{3.57}$$

Hereafter, h_K is not necessary because there is no inhibitory gate in the potassium channel ($q_K = 0$).

By fitting the parameters to the conductivity data of relaxation processes to the steady state, $m_{i,\infty}, h_{i,\infty}, \tau_i^{(m)}$, and $\tau_i^{(h)}$ can be obtained. Then, from Eqs. (3.54) and (3.55), the following relationships are satisfied:

$$\alpha_i^{(m)} = \frac{m_{i,\infty}}{\tau_i^{(m)}}, \tag{3.58}$$

$$\beta_i^{(m)} = \frac{(1 - m_{i,\infty})}{\tau_i^{(m)}}, \tag{3.59}$$

$$\alpha_i^{(h)} = \frac{h_{i,\infty}}{\tau_i^{(h)}}, \tag{3.60}$$

$$\beta_i^{(h)} = \frac{(1 - h_{i,\infty})}{\tau_i^{(h)}}. \tag{3.61}$$

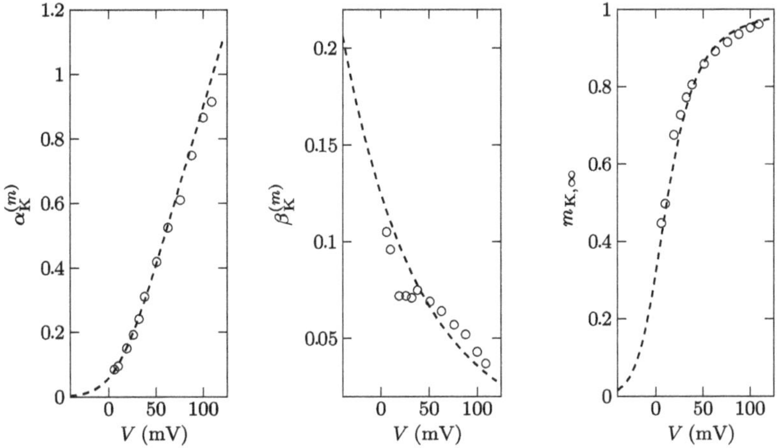

Figure 3.7 Membrane potential dependence of potassium channel gating variables. Plotted based on [4].

By getting these parameters of experimental data obtained under various membrane potentials, they determined the dependence of these parameters on the potential V. In the case of the potassium channel, they obtained the following fitting functions (Fig. 3.7):

$$\alpha_K^{(m)} = \frac{0.01(10 - V)}{e^{(10-V)/10} - 1}, \tag{3.62}$$

$$\beta_K^{(m)} = 0.125e^{-V/80}, \tag{3.63}$$

while for the sodium channel (Fig. 3.8) the fitted functions are as follows:

$$\alpha_{Na}^{(m)} = \frac{0.1(25 - V)}{e^{(25-V)/10} - 1}, \tag{3.64}$$

$$\beta_{Na}^{(m)} = 4e^{-V/18}, \tag{3.65}$$

$$\alpha_{Na}^{(h)} = 0.07e^{-V/20}, \tag{3.66}$$

$$\beta_{Na}^{(h)} = \frac{1}{e^{(30-V)/10} + 1}. \tag{3.67}$$

These functions were built phenomenologically, which cannot be obtained directly from physical and chemical processes in neurons.

Based on the preceding discussions, the differential equations that describe four variables, V, m_{Na}, h_{Na}, and m_K, are derived. These equations are commonly referred to as the Hodgkin–Huxley equations. Since these four variables interact with each other in a complex manner, it is difficult to understand their dynamics immediately from the equations. To facilitate our understanding, we first investigate the relationship between membrane potential and electrical conductivity through numerical

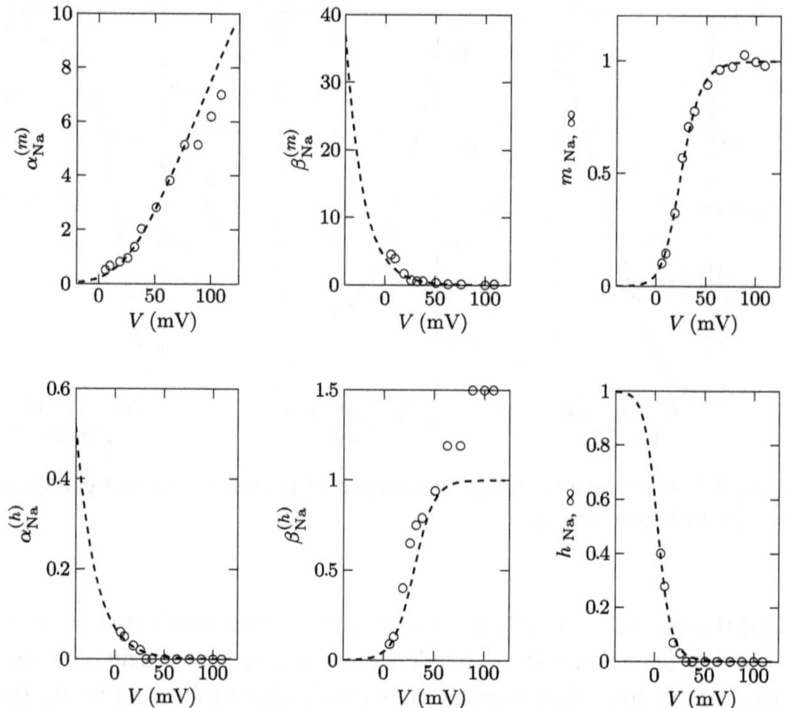

Figure 3.8 Membrane potential dependence of sodium channel gating variables. Plotted based on [4].

simulations. Figure 3.9 presents the numerical results obtained from calculating the dynamics of membrane potential in response to transient external current pulses.

Following the increment of the external current, the membrane potential V rises, resulting in the opening of the m_{Na} gate and allowing the influx of sodium ions. (In the resting state, due to the function of the ion pumps in the membrane, the cell possesses a lower concentration of sodium ions and a higher concentration of potassium ions compared to the extracellular environment.) This further elevates the membrane potential and triggers positive feedback, causing an increase in the influx of sodium ions. Subsequently, the m_{K} gate opens, allowing potassium ions to start flowing out. However, the influx of sodium ions remains dominant, sustaining the upward trend of the membrane potential. When the membrane potential reaches approximately 30 mV, the inhibitory gate eventually closes (resulting in a decrease of h), leading to the cessation of sodium ion influx (accompanied by a decrease in m_{Na}). Consequently, the increase in membrane potential halts, and the membrane potential begins to decline due to the efflux of potassium ions. Eventually, the membrane potential returns to its resting state, with a slight deviation toward hyperpolarization. Here, the variables V and m_{Na} change faster, while the dynamics of h_{Na} and m_{K} have slower time-scale.

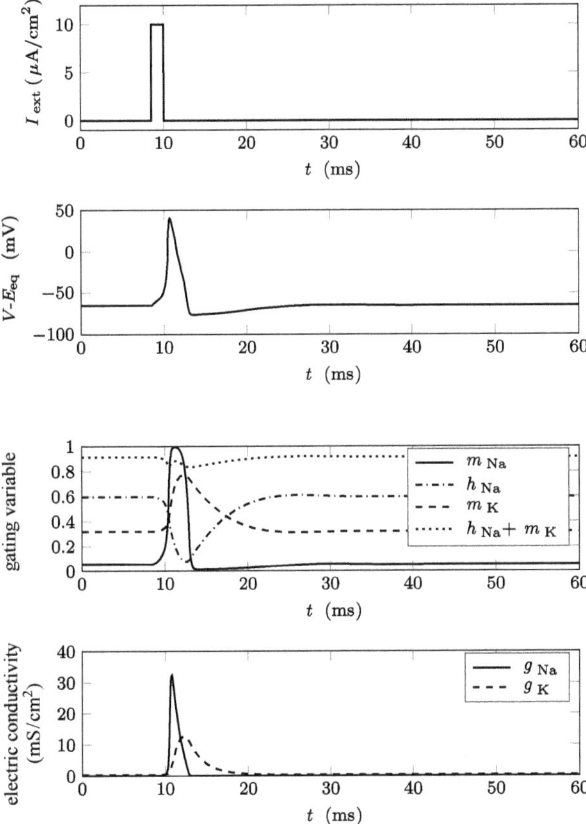

Figure 3.9 Temporal change in membrane potential in response to pulsed external current. In the second row of the figure, the vertical axis shows the values subtracted by Nernst potential. $E_{Na} = 115$ (mV), $E_K = 12$ (mV), $E_L = 10.6$ (mV).

The behavior of these variables can be understood by examining the membrane potential dependence of the gating processes. We illustrate the relationship between V and the gating variables, as well as the time constants, in Fig. 3.10. After a sufficient amount of time has passed, the membrane potential dependence of the sodium and potassium channels exhibits a similar trend (shown in the left and middle panels of Fig. 3.10). However, it is important to note that the value of $\tau_{Na}^{(m)}$ is less than 1 millisecond, significantly smaller than $\tau_{Na}^{(h)}$ and $\tau_{K}^{(m)}$. Consequently, when the membrane potential changes, sodium's m immediately follows the steady-state electrical conductivity. When the influx of sodium ions elevates the membrane potential for several milliseconds, the current carried by potassium ions becomes inhibitory. This pattern of rapid positive action followed by relatively slower inhibitory action is commonly observed in excitatory systems.

Next, let us examine how the behavior of the membrane potential changes in response to the external current I_{ext} (Fig. 3.11). First, when I_{ext} with a small

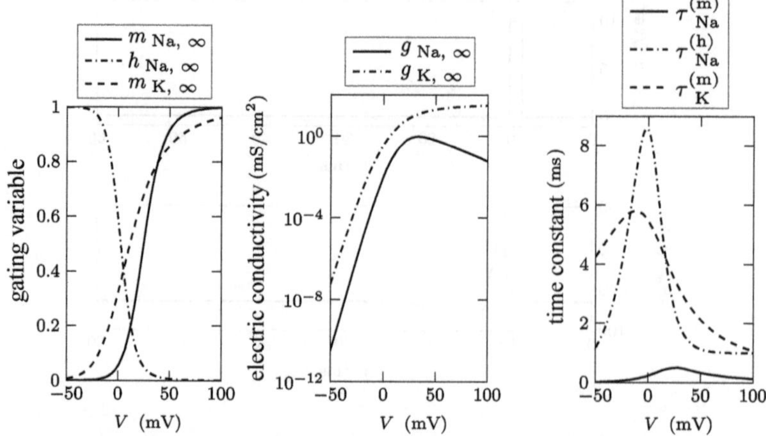

Figure 3.10 Membrane potential dependence of gating variables in steady state.

magnitude is applied stepwise for a specific duration, the membrane potential exhibits very small deviation from the resting potential and rapidly returns to its original state. Even upon the cessation of the stepwise input, the membrane potential gradually returns to the resting potential, after briefly exhibiting a slight negative value. Second, the stepwise application of moderate external currents transiently depolarizes the membrane potential, followed by a return to the resting potential after hyperpolarization. In this case, the response exhibits a large amplitude, a steep rise, and relatively slow relaxation. Such a waveform is commonly referred to as a pulse. It will be explained later that this response occurs only when the external current surpasses a certain threshold. Hence, the response follows the all-or-none rule that occurs when the input exceeds a threshold. Furthermore, continuous application of external input yields transient changes in the membrane potential. This type of response is known as an adaptive response. The presence of a threshold and an adaptive response characterizes excitability, and the system displaying such behavior is referred to as an excitable system. Last, a larger external current induces alternating depolarization and hyperpolarization of the membrane potential. This response exhibits oscillatory behavior even when the external current remains constant, which is termed spontaneous oscillation. In this case, the waveform resembles a sawtooth pattern, in contrast to the symmetrical waveform of a sinusoidal function.

How can we mathematically comprehend such excitability and oscillation? Due to the intricate interactions among the four variables in the Hodgkin–Huxley equations, conducting nullcline analysis, as discussed in previous sections, is not easy. Therefore, let us explore the possibility of reducing the number of variables without sacrificing the essence of the model. Firstly, considering the short relaxation time of m_{Na}, we approximate m_{Na} as $m_{Na,\infty}$ (similar to the adiabatic elimination in Section 1.12). As observed in Fig. 3.9, during the generation of action potential

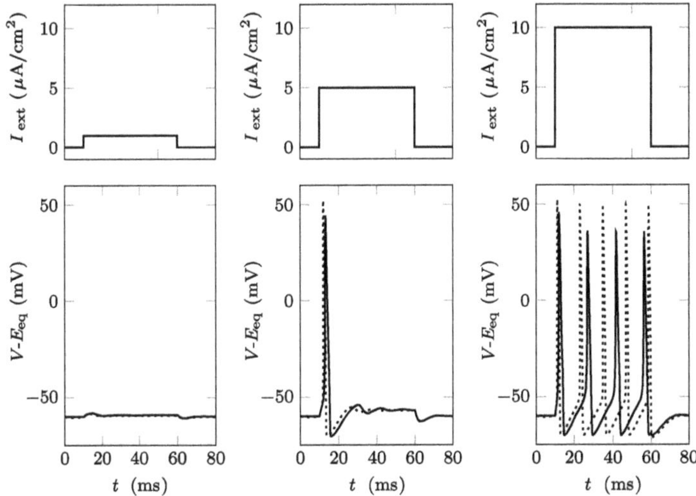

Figure 3.11 Excitation and oscillation in Hodgkin–Huxley equations. Examples of numerical calculations in the case of external current added in $10 < t < 60$: (left) $I_{\text{ext}} = 1.0$, (center) $I_{\text{ext}} = 5.0$, (right) $I_{\text{ext}} = 10.0$. Transient excitation and oscillation emerge depending on the strength of the external current. Dotted lines show the behaviors of the equation reduced to two dimensions.

impulses, the variation of $h_{\text{Na}} + m_{\text{K}}$ remains relatively small. Because m_{Na} is fast, it follows the change in I_{ex} and opens immediately, causing V to rise. The equilibrium potential of sodium ions is high, and V continues to rise up to that point. Then h_{Na} gradually takes effect and the sodium ion's contribution decreases. The equilibrium potential of potassium ions is low, and m_{K} causes V to gradually decrease, resulting in the formation of the neural spike. In this excitatory phenomenon, the movement of sodium and potassium ions in and out plays a major role, so that the state is practically close to $h_{\text{Na}} + m_{\text{K}} = 0$. Hence, we make the approximation that $h_{\text{Na}} + m_{\text{K}}$ is constant. This approximation allows us to reduce the original four-variable dynamical system to a system with only two variables, namely, V and m_{K}, recalling that h_{K} is always zero:

$$C_{\text{memb}}\frac{\mathrm{d}V}{\mathrm{d}t} = -\, G_{\text{Na}}m_{\text{Na},\infty}^{3}(0.8 - m_{\text{K}})(V - E_{\text{Na}})$$

$$-\, G_{\text{K}}m_{\text{K}}^{4}(V - E_{\text{K}}) - g_{\text{L}}(V - E_{\text{L}}) + I, \tag{3.68}$$

$$\frac{\mathrm{d}m_{\text{K}}}{\mathrm{d}t} = \alpha_{\text{K}}^{(m)}(1 - m_{\text{K}}) - \beta_{\text{K}}^{(m)}m_{\text{K}}, \tag{3.69}$$

where we assume $h_{\text{Na}} + m_{\text{K}} = 0.8$. In the reduced model with two variables, excitability and oscillatory behavior can also be observed, depending on the input current (dotted lines in Fig. 3.11). By plotting the nullclines of this reduced system in the $m_{\text{K}} - V$ plane, we can observe that the dynamics are characterized by the

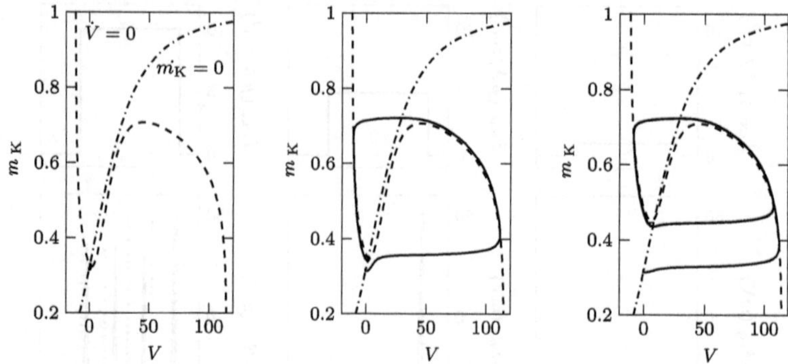

Figure 3.12 Behavior on the phase plane in the two-variable reduced system of Hodgkin–Huxley equations. (left) $I_{\text{ext}} = 1.0$, (center) $I_{\text{ext}} = 3.0$, (right) $I_{\text{ext}} = 20.0$. The solid line shows the trajectory of numerical simulation on the phase plane, and the dashed and single dotted lines show nullclines, respectively.

intersection of nullclines (Fig. 3.12). When I_{ext} is small, the nullcline $dm_{\text{K}}/dt = 0$ (which is increasing over V) intersects with the nullcline $dV/dt = 0$ (which is decreasing over V). Although the fixed point at this intersection is linearly stable, a slight increase in the external input I_{ext} by the application of a small perturbation leads to a large increase in V. Consequently, the state returns to the fixed point through a trajectory close to the $dV/dt = 0$ nullcline (middle panel of Fig. 3.12). With a further increase in I_{ext}, however, the intersection of nullclines shifts, causing the nullcline dV/dt to increase over V. In this case, the fixed point becomes unstable, and a limit-cycle emerges (right panel of Fig. 3.12).

Exercise 3.4   Morris–Lecar Equations

The firing behavior of neurons can also be described by a simpler model than the Hodgkin–Huxley equation. Morris and Lecar, among others, proposed the following model to explain the excitability of barnacle muscle fibers [7]. In this model, the membrane potential is determined by the influx of calcium ions through a voltage-dependent calcium channel and the efflux of potassium ions through a potassium channel. The dynamics of the membrane potential and the gating variables of the channels are governed by the following set of differential equations:

$$C_{\text{memb}} \frac{dV}{dt} = -g_{\text{Ca}} m_K (V - E_{\text{Ca}}) - g_K m_{\text{Ca}} (V - E_K)$$

$$+ g_{\text{L}}(V - E_{\text{L}}) + I_{\text{ext}}, \tag{3.70}$$

$$\tau_K \frac{dm_K}{dt} = m_{K,\infty}(V) - m_K, \tag{3.71}$$

$$\tau_{\text{Ca}} \frac{dm_{\text{Ca}}}{dt} = m_{\text{Ca},\infty}(V) - m_{\text{Ca}}. \tag{3.72}$$

Based on experimental observations, in the steady state, the gating function is known to be threshold functions. Here, they are assumed to be the following relationships:

$$m_{K,\infty} = \frac{1 + \tanh((V - V_K^*)/\theta_K)}{2}, \tag{3.73}$$

$$m_{Ca,\infty} = \frac{1 + \tanh((V - V_{Ca}^*)/\theta_{Ca})}{2}, \tag{3.74}$$

where $V_K^*, V_{Ca}^*, \theta_K$, and θ_{Ca} are fitting parameters. By assuming that the gating dynamics of the sodium channel are sufficiently faster than those of the potassium channel, depict the nullclines of the dynamical system of $V(t)$ and $m_{Ca}(t)$, and discuss its dynamic behavior by comparing the Hodgkin–Huxley equations.[1]

3.5 FitzHugh–Nagumo Equations

To understand the behavior of the reduced Hodgkin–Huxley equations (Eqs. (3.68) and (3.69)) on the phase space in more detail, let us consider the following dynamical system, which is an extension of the van der Pol equation (Eq. (3.15)):

$$\epsilon \frac{dv}{dt} = v(a - v)(v - 1) - m + I_{ext}, \tag{3.75}$$

$$\frac{dm}{dt} = bv - cm, \tag{3.76}$$

where a, b, c, and ϵ are positive parameters. They are called FitzHugh–Nagumo equations, and they enable us to extract the essence of the Hodgkin–Huxley model. The nullclines of this model are

$$m|_{dv/dt=0} = v(a - v)(v - 1) + I_{ext}, \tag{3.77}$$

$$m|_{dm/dt=0} = \frac{b}{c}v. \tag{3.78}$$

As depicted in Figs. 3.12 and 3.13, the shape of the nullclines and the manner in which they intersect are similar to those observed in the Hodgkin–Huxley model.

The behavior of the nullclines on the phase plane effectively captures the characteristics of excitable systems. The fixed point corresponds to the resting state. When a perturbation of sufficient magnitude is applied to the resting state, it elicits a strong response and transitions into an excited state characterized by transient dynamics. During the transient dynamics, the variable v rapidly returns to its

---

[1] In neurons, there are two classes of cells (class I and class II) with different firing properties, which correspond to different bifurcation dynamics. Class I cells exhibit saddle-node bifurcation and class II cells exhibit Hopf bifurcation. The Hodgkin–Huxley model corresponds to class II, while the leaky integrate-and-fire model (LIF model) corresponds to class I, and Morris–Lecar model corresponds to class I or class II depending on its parameters. Models corresponding to class I can show more flexible frequency response [11].

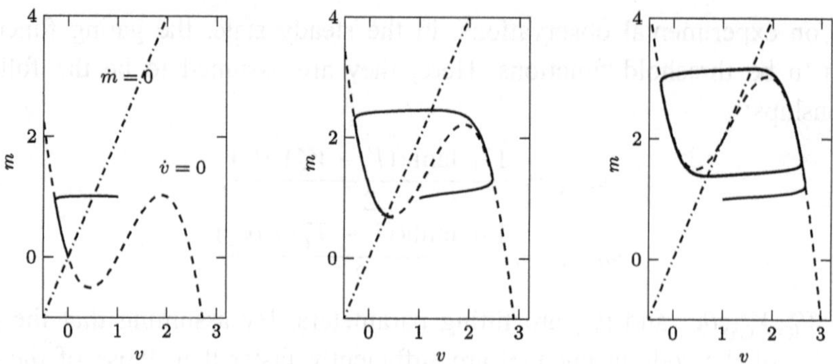

Figure 3.13 Behaviors of the FitzHugh–Nagumo equation on the phase plane. $a = 2.5$, $b = 2.0$, $c = 1.0$, $\epsilon = 0.05$. (left) $I_{\text{ext}} = 0.0$, (center) $I_{\text{ext}} = 1.2$, (right) $I_{\text{ext}} = 2.0$.

original level, followed by a slower return of m to its original level, eventually returning to the resting state. This period of slow state change is known as the refractory period, as the state exhibits no response to further stimulation. Excitability is observed not only in the action potentials of nerve cells but also in various phenomena, such as the beating of cardiac muscle cells, chemotactic motion of amoeba, differentiation of bacteria (e.g., *Bacillus subtilis*) [9], and pigment expression in hair follicle cells. The time constant associated with these excitability phenomena varies across a wide range, from milliseconds to minutes.

Exercise 3.5   Excitability in Regulation of Gene Expression [9].
The competence of *Bacillus subtilis* only arises in a small fraction of cells, a phenomenon regulated by the expression of the *comK* gene. The expression of the *comK* gene is controlled through a positive feedback mechanism, while its expression indirectly represses the expression of the *comS* gene. Additionally, the ComS protein inhibits the degradation of the ComK protein by the MecA protein. By using the following variables, $y_1 = [\text{ComK}]$, $y_2 = [\text{ComS}]$, $m_1 = [\text{MecA}]$, $m_2 = [\text{MecA-ComK}]$, $m_3 = [\text{MecA-ComS}]$, and assuming $m_1 + m_2 + m_3 = \text{constant}$, the following model is introduced:

$$\frac{dy_1}{dt} = a_k + b_k \frac{y_1^n}{y_1^n + K_1^n} - \gamma_a m_1 y_1 + \gamma_{-a} m_2, \tag{3.79}$$

$$\frac{dy_2}{dt} = b_s \frac{K_2^p}{y_1^p + K_2^p} - \gamma_b m_1 y_2 + \gamma_{-b} m_3, \tag{3.80}$$

$$\frac{dm_2}{dt} = -(\gamma_{-a} + \gamma_1) m_2 + \gamma_a m_1 y_1, \tag{3.81}$$

$$\frac{dm_3}{dt} = -(\gamma_{-b} + \gamma_1) m_3 + \gamma_b m_1 y_2. \tag{3.82}$$

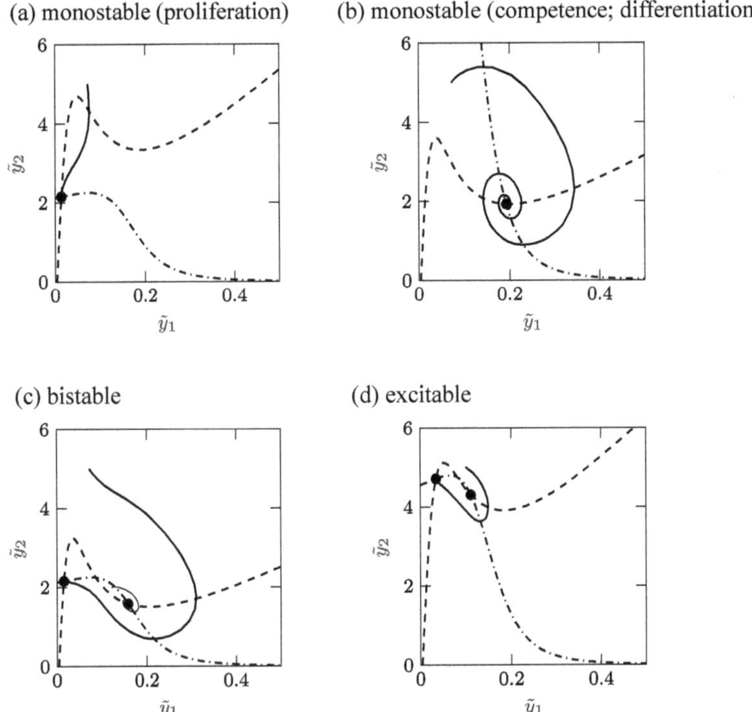

Figure 3.14 Behavior of *Bacillus subtilis* differentiation model. Plotted based on [9]. $a_k = 0.004$, $K_0 = 0.2$, $K_1 = 0.222$, $n = 2$, $p = 5$. (a) $b_k = 0.08$, $b_s = 0.68$, (b) $b_k = 0.12$, $b_s = 0.92$, (c) $b_k = 0.14$, $b_s = 0.68$, (d) $b_k = 0.07$, $b_s = 0.82$. The solid line shows the trajectory on the phase plane, whereas the dashed and single dotted lines show nullclines corresponding to $d\tilde{y}_1/dt = 0$ and $d\tilde{y}_2/dt = 0$, respectively.

Here assume that m_2 and m_3 reach steady levels because the degradations of complexes are fast. Then, show that the model can be reduced to the following:

$$\frac{d\tilde{y}_1}{dt} = \tilde{a}_k + \tilde{b}_k \frac{\tilde{y}_1^n}{\tilde{y}_1^n + \tilde{K}_1^n} - \frac{\tilde{y}_1}{1 + \tilde{y}_1 + \tilde{y}_2}, \tag{3.83}$$

$$\frac{d\tilde{y}_2}{dt} = \tilde{b}_s \frac{\tilde{y}_1^p}{\tilde{y}_2^p + \tilde{K}_2^p} - \frac{\tilde{y}_2}{1 + \tilde{y}_1 + \tilde{y}_2}. \tag{3.84}$$

Furthermore, depict changes in the nullclines by changing b_s, and analyze the characteristics of the fixed point (see Fig. 3.14).

Example 3.1 Martiel–Goldbeter Equations

When cellular slime mold undergoes starvation, cyclic AMP (cAMP) is produced and subsequently released, leading to an increase in extracellular cAMP concentration. The binding of extracellular cAMP to membrane receptors stimulates further cAMP production, which is then secreted into the extracellular space. Concurrently, there exists a mechanism where continuous cAMP input suppresses intracellular cAMP production. During this phase, extracellular cAMP is degraded by an extracellular enzyme. As the extracellular cAMP concentration decreases, cellular responsiveness is restored, resulting in a subsequent rise in cAMP concentration. This cyclic process leads to oscillations in both intracellular and extracellular cAMP levels, propagating as waves that determine the direction of cell migration (refer to Section 8.2). The Martiel–Goldbeter equations [1, 6] provide a well-known model for explaining cAMP oscillations in a suspension of slime mold cells:

$$\frac{d\rho}{dt} = -\rho\frac{k_1 + k_2\gamma}{1 + \gamma} + (1 - \rho)\frac{k_{-1} + k_{-2}c\gamma}{1 + c\gamma}, \tag{3.85}$$

$$\frac{d\beta}{dt} = k_\beta\phi(\rho, \gamma, \alpha) - k_d\beta, \tag{3.86}$$

$$\frac{d\gamma}{dt} = k_\gamma\beta - k_e\gamma. \tag{3.87}$$

Here, β and γ are the dimensionless intracellular and extracellular concentrations of cAMP, and ρ represents the ratio of activated receptors. The function ϕ is

$$\phi = \frac{\alpha(\lambda\theta + \epsilon Y^2)}{1 + \alpha\theta + \epsilon Y^2(1 + \alpha)}, \tag{3.88}$$

$$Y = \rho\frac{\gamma}{1 + \gamma}. \tag{3.89}$$

The cAMP receptor exists in two states: activated (R) and inactivated (D). The right-hand side of Eq. (3.85) represents the transition between these two states. The activated and inactivated receptors have affinities for cAMP represented by K_R and K_D, respectively, with their ratio denoted as $c = K_R/K_D$. Intracellular cAMP is produced from ATP when adenylate cyclase is activated by the complex of the activated receptor and cAMP. Adenylate cyclase also exists in two states. The parameter θ in the first term of the right-hand side of Eq. (3.86) represents the ratio of the Michaelis–Menten constants between these two states. The parameter γ denotes the ratio of production rate constants, and α represents the normalized concentration of ATP. A portion of intracellular cAMP is released into the extracellular environment, while the remainder is degraded within the cell. The parameter k_d in the second term of the right-hand side of Eq. (3.86) signifies the rate of intracellular degradation. The first and second terms of the right-hand side of Eq. (3.87) represent the increase in extracellular cAMP concentration through secretion from the cell and degradation of cAMP in the extracellular environment.

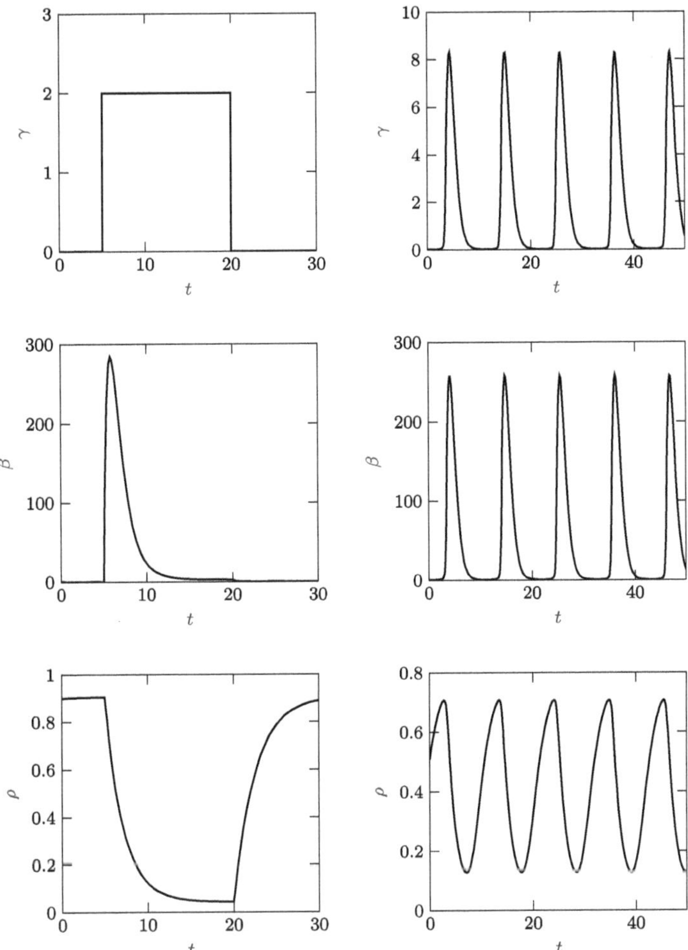

Figure 3.15 Martiel–Goldbeter equation. cAMP relay response (left) and oscillation (right) in cellular slime mold. Intracellular cAMP exhibits a sharp elevation (middle left) in response to a step input of extracellular cAMP (upper left), followed by a gradual decrease in the percentage of active receptors (lower left), and a subsequent reduction in intracellular cAMP levels. When the feedback mechanism is sufficiently strong, spontaneous oscillation occurs (right). See [1, 6] for details.

In a clamp experiment where extracellular cAMP concentration can be controlled through perfusion, an adaptive response to a step input is observed. In the aforementioned model, a transient increase in intracellular cAMP also occurs when γ is fixed, and the extracellular cAMP concentration is changed stepwise (left panel in Fig. 3.15). As illustrated, following a transient elevation of intracellular cAMP concentration, it begins to decrease. This decline is attributed to the suppression of cAMP production, induced by receptor inactivation under continuous stimulation. Under the same parameter values, when the time evolution of extracellular cAMP is allowed autonomously, the extracellular cAMP concentration decreases due to extracellular degradation. In this scenario, oscillatory dynamics emerge by reactivating the receptors (right panel in Fig. 3.15).

Exercise 3.6   Show that the following activator-inhibitor system [5] exhibits excitability and oscillatory dynamics depending on the input S, by describing the nullclines on the phase space:

$$\frac{\mathrm{d}u}{\mathrm{d}t} = S + a_u \frac{u^2}{K_u + (u^2 + v)} - \gamma_u u,$$

$$\frac{\mathrm{d}v}{\mathrm{d}t} = a_v \frac{u^2}{K_v + u^2} - \gamma_v v.$$

References

[1] A. Goldbeter. *Biochemical Oscillations and Cellular Rhythms: The Molecular Bases of Periodic and Chaotic Behaviour*. Cambridge, Cambridge University Press, 1996.

[2] B. C. Goodwin. Oscillatory behavior in enzymatic control processes. *Advances in Enzyme Regulation*, **3**: 425–438, 1965.

[3] J. S. Griffith. Mathematics of cellular control processes. I. Negative feedback to one gene. *Journal of Theoretical Biology*, **20**(2): 202–208, 1968.

[4] A. L. Hodgkin and A. F. Huxley. A quantitative description of membrane current and its application to conduction and excitation in nerve. *The Journal of Physiology*, **117**(4): 500–544, 1952.

[5] K. Kamino, K. Fujimoto and S. Sawai. Collective oscillations in developing cells: Insights from simple systems. *Development Growth & Differentiation*, **53**(4): 503–517, 2011.

[6] J. L. Martiel and A. Goldbeter. A model based on receptor desensitization for cyclic AMP signaling in Dictyosterium cells. *Biophysical Journal*, **52**(5): 807–828, 1987.

[7] C. Morris and H. Lecar. Voltage oscillations in the barnacle giant muscle fiber. *Biophysical Journal*, **35**(1): 193–213, 1981.

[8] C. Shi *et al.* Interaction of motility, directional sensing, and polarity modules recreates the behaviors of chemotaxing cells. *PLoS Computational Biology*, **9**(7): e1003122, 2013.

[9] G. M. Süel *et al.* An excitable gene regulatory circuit induces transient cellular differentiation. *Nature*, **440**(7083): 545–550, 2006.

[10] M. Tang *et al.* Evolutionarily conserved coupling of adaptive and excitable networks mediates eukaryotic chemotaxis. *Nature Communications*, **5**: 5175, 2014.

[11] I. Tsuda *et al.* Chaotic itinerancy as a mechanism of irregular changes between synchronization and desynchronization in a neural network. *Journal of Integrative Neuroscience*, **3**(2): 159–182, 2004.

4

Spatiotemporal Pattern Formation

4.1 Diffusion Equation

In cells and tissues, substances undergo continuous spatial reorganization. The displacement of molecules within cells and tissues can be attributed to various factors, including pressure-driven flows like protoplasmic flow and blood flow, as well as diffusion, wherein molecules move from regions of high concentration to regions of low concentration. Additionally, transport systems employing motor proteins contribute to molecular relocation. In the present chapter, we shall direct our attention toward the process of diffusion.

Let us consider the change in the amount of a substance in a one-dimensional (1D) space. If the concentration of a molecule at a specific position x and time t is denoted as $c(x, t)$, then the alteration in the amount of substance over a time interval Δt within the range from x to $x + \Delta x$ can be expressed using the following equation:

$$(c(x, t + \Delta t) - c(x, t))\Delta x. \tag{4.1}$$

Alternatively, if we define the number of molecules traversing a unit cross-sectional area per unit time as the flux density $J_x(x)$, the change in the quantity of substance over a time interval Δt within a small region from x to $x + \Delta x$ is given by

$$(J_x(x) - J_x(x + \Delta x))\Delta t. \tag{4.2}$$

Here, it is important to note that the second term carries a negative sign due to the definition of flux being positive in the x direction. As the equilibrium between inflow and outflow must be maintained alongside the temporal concentration changes, the following equations are derived to account for this balance:

$$\frac{c(x, t + \Delta t) - c(x, t)}{\Delta t} = \frac{J_x(x) - J_x(x + \Delta x)}{\Delta x}. \tag{4.3}$$

This can be written in differential form as

$$\frac{\partial c(x,t)}{\partial t} = -\frac{\partial J_x(x)}{\partial x}. \tag{4.4}$$

Expanding the balance to three-dimensional space (x, y, z), we obtain

$$\frac{\partial c(x,y,z,t)}{\partial t} = -\frac{\partial J_x(x,y,z)}{\partial x} - \frac{\partial J_y(x,y,z)}{\partial y} - \frac{\partial J_z(x,y,z)}{\partial z}. \tag{4.5}$$

By assuming that the magnitude of the flux is proportional to the concentration gradient (in accordance with Fick's first law) and denoting the proportionality coefficient as D, we can derive the following equations:

$$J_x = -D\frac{\partial c(x,y,z,t)}{\partial x}, J_y = -D\frac{\partial c(x,y,z,t)}{\partial y}, J_z = -D\frac{\partial c(x,y,z,t)}{\partial z}. \tag{4.6}$$

Therefore, from Eq. (4.5) and Eq. (4.6), we obtain:

$$\frac{\partial c(x,y,z,t)}{\partial t} = \frac{\partial}{\partial x}\left(D\frac{\partial c(x,y,z,t)}{\partial x}\right) + \frac{\partial}{\partial y}\left(D\frac{\partial c(x,y,z,t)}{\partial y}\right)$$
$$+ \frac{\partial}{\partial z}\left(D\frac{\partial c(x,y,z,t)}{\partial z}\right). \tag{4.7}$$

Under the assumption of no anisotropy or heterogeneity in the coefficient D, where D is a constant, we can establish the following partial differential equation that relates the temporal and spatial variations of the concentration c:

$$\frac{\partial c(x,y,z,t)}{\partial t} = D\left(\frac{\partial^2}{\partial x^2} + \frac{\partial^2}{\partial y^2} + \frac{\partial^2}{\partial z^2}\right)c(x,y,z,t) \tag{4.8}$$
$$= D\nabla^2 c(x,y,z,t). \tag{4.9}$$

Here, ∇ is a vector of differential operation, and in three dimensions, it is the following expression:

$$\nabla = \begin{pmatrix} \dfrac{\partial}{\partial x} \\[2mm] \dfrac{\partial}{\partial y} \\[2mm] \dfrac{\partial}{\partial z} \end{pmatrix}.$$

Eq. (4.9) corresponds to the well-known diffusion equation (refer to Section 5.8), and the constant D is commonly referred to as the diffusion constant.

4.1.1 Characteristics of the Diffusion Equation

The form of the diffusion equation is similar to that of the linear wave equation. In one dimension, the linear wave equation can be expressed as follows, where W represents the amplitude:

$$\frac{\partial^2 W(x,y,z,t)}{\partial t^2} = v^2 \frac{\partial^2 W(x,y,z,t)}{\partial x^2}. \tag{4.10}$$

Here, the coefficient v is a parameter that represents the phase velocity of the wave. In comparison, the one-dimensional diffusion equation is

$$\frac{\partial c(x,y,z,t)}{\partial t} = D \frac{\partial^2 c(x,y,z,t)}{\partial x^2}. \tag{4.11}$$

While the wave equation involves a second-order derivative with respect to time t, the diffusion equation, involves a first-order derivative.

Now, if we apply the time-reversing variable transformation $\tau = -t$ to the equation, we obtain

$$\frac{\partial}{\partial t} = \frac{\partial \tau}{\partial t}\frac{\partial}{\partial \tau} = -\frac{\partial}{\partial \tau}. \tag{4.12}$$

Indeed, when we apply the time-reversing variable transformation, the left-hand side of the wave equation maintains its original form, while the left-hand side of the diffusion equation undergoes a sign reversal. Consequently, the wave equation exhibits symmetry with respect to time reversal, whereas the diffusion equation breaks the time reversal symmetry. This implies that the physical process represented by the solution of the wave equation is reversible (i.e., the motion can be observed as if the video were played backwards), whereas the process described by the diffusion equation is irreversible.

4.1.2 The Solution of the Diffusion Equation

Let us consider a scenario where a droplet of milk is carefully poured onto the surface of a coffee-filled mug, causing the milk to gradually spread out. For purposes of simplification, we can envision an idealized one-dimensional space. In this scenario, at the initial time $t=0$, all the milk molecules, denoted as c_0, are concentrated at position $x=0$; then we obtain

$$\frac{\partial c(x,t)}{\partial t} = D \frac{\partial^2 c(x,t)}{\partial x^2} + c_0 \delta(x,t). \tag{4.13}$$

The second term on the right-hand side, $\delta(x,t)$, represents the delta function, which signifies the application of c_0 milk molecules specifically at $x=0$ and $t=0$. This term is nonzero only at $x=0$ and $t=0$. Consequently, Eq. (4.13) simplifies to Eq. (4.11) for all instances except for the moment when the milk is initially poured.

In order to determine the solution to this equation, we can take the Fourier transform of the entire Eq. (4.13) using the following formula:

$$\tilde{c}(k, t) = F[c(x, t)] = \int_{-\infty}^{\infty} e^{-ikx} c(x, t) dx. \tag{4.14}$$

Assuming that the order of differentiation and integration is interchangeable, we obtain

$$(\text{LHS}) = \frac{\partial}{\partial t} \tilde{c}(k, t), \tag{4.15}$$

$$(\text{RHS}) = -Dk^2 \tilde{c}(k, t) + c_0 \delta(t) e^{ikt}. \tag{4.16}$$

Therefore,

$$\tilde{c}(k, t) = e^{-Dk^2 t} c_0 H(t), \tag{4.17}$$

where H is the Heaviside step function ($H(t) = 0$ for $t < 0$, $H(t) = 1$ for $t \geq 0$). Then, applying the inverse Fourier transform,

$$c(x, t) = \frac{H(t)}{2\pi} \int_{-\infty}^{\infty} e^{-Dk^2 t + ikx} dk \tag{4.18}$$

$$= \frac{c_0 H(t)}{2\pi} \int_{-\infty}^{\infty} e^{-Dt\left(k - \frac{ix}{2Dt}\right)^2 - \frac{x^2}{4Dt}} dk \tag{4.19}$$

$$= \frac{c_0 H(t)}{\sqrt{4\pi Dt}} e^{-\frac{x^2}{4Dt}}. \tag{4.20}$$

Therefore, if the time domain is restricted to $t > 0$, the concentration $c(x, t)$ satisfies the following equation:

$$\frac{c(x, t)}{c_0} = \frac{1}{\sqrt{4\pi Dt}} e^{-\frac{x^2}{4Dt}}. \tag{4.21}$$

It is worth noting that this equation shares the same form as the normal distribution, also known as the Gaussian distribution, which is well-known in probability and statistics (see also Section 5.8).

If the random variable X has a mean value of 0 and a variance of σ^2, and follows a normal distribution, the probability density function $n(X)$ can be expressed as follows:

$$n(X) = \frac{1}{\sqrt{2\pi\sigma^2}} e^{-\frac{X^2}{2\sigma^2}}. \tag{4.22}$$

Now, let's consider the realization of the random variable X as the position variable x. By comparing Eq. (4.21) and Eq. (4.22), we observe that they coincide when the variance $\sigma^2 = 2Dt$. Consequently, the ratio $c(x)/c_0$ in Eq. (4.21) becomes normalized, exhibiting a symmetrical bell shape centered at $x = 0$. As t approaches zero, it approaches a delta function, and for sufficiently large distances ($x \rightarrow \pm\infty$),

$c(x, t) \rightarrow 0$. As time progresses, the distribution of particles is broader around $x = 0$ with a variance of $2Dt$, eventually reaching a spatially uniform state. This intuitive nature of the solution aligns with the principles of the second law of thermodynamics. The microscopic details of the diffusion phenomenon described in the previous subsection will be further discussed in Section 5.8.

4.1.3 Concentration Gradients under Fixed Boundary Conditions and the Tricolor Flag Problem: Exploring Positional Information

Diffusion processes play a crucial role in the dispersion of ions, gases, and proteins, including chemokines (signaling factors involved in chemotaxis) and morphogens (factors influencing morphogenesis) within cells and tissues. For example, morphogen concentration determines cell differentiation according to where cells are within an organism or tissue. This concept, known as positional information, is a fundamental principle for the spatial organization of multicellular systems. The precise localization of cell differentiation, guided by diffusible factors, has a significant impact on pattern formation and its robustness. It is important to note that the concentration distribution of molecules, as governed by the diffusion equation, is influenced by specific boundary conditions.

Let us consider an idealized scenario of a uniform one-dimensional space, where molecules (represented by concentration c) diffuse through the extracellular spaces between cells in a tissue. At the boundaries of this tissue, specifically at the left end $(x = 0)$ and the right end $(x = L)$, special cells are present, and homeostatic mechanisms operate to maintain a constant concentration of molecules around these cells:

$$\frac{\partial c(x, t)}{\partial t} = D \frac{\partial^2 c(x, t)}{\partial x^2}, \tag{4.23}$$

$$c(0, t) = c_{\max}, \tag{4.24}$$

$$c(L, t) = c_{\min}. \tag{4.25}$$

If the system is in a steady state, indicated by $\partial c / \partial t = 0$, we can obtain the equation

$$c(x) = ax + b \tag{4.26}$$

from Eq. (4.23). Here, a and b are constants determined by the boundary conditions. By applying the boundary conditions, we then obtain

$$c(x) = (c_{\min} - c_{\max})\frac{x}{L} + c_{\max}. \tag{4.27}$$

This equation represents a linear gradient, appearing as a straight line connecting $c(0) = c_{\max}$ and $c(L) = c_{\min}$ at both ends. While this equation may seem simple, it conveys that the concentration is exclusively determined by the position normalized by the system size L.

To illustrate this point, let us consider a scenario where cells differentiate into different states in the respective regions of molecular concentration $c_{max} > c(x) > c_1$, $c_1 > c(x) > c_2$, and $c_2 > c(x) > c_{min}$. This situation is often referred to as the French flag problem (refs. [5, 29]). Notably, it becomes evident that the differentiation pattern scales in proportion to the spatial size, maintaining a conserved ratio. During development, as cells proliferate and individuals grow in size over time, inevitable variations in size emerge due to both environmental and endogenous fluctuations. To enable the scaling of the morphogen gradient, a mechanism that incorporates a dependence on x/L would be required.

4.1.4   Chemical Gradient with Degradation Reaction

Indeed, the distribution of diffusible factors in tissues is influenced by various complex factors. One significant factor is degradation. When the diffusion process involves degradation through first-order reactions, the time evolution equation can be expressed as:

$$\frac{\partial c(x, t)}{\partial t} = -kc(x, t) + D\frac{\partial^2 c(x, t)}{\partial x^2}, \tag{4.28}$$

where k represents the rate constant of degradation. This formulation, where the reaction term is incorporated into the diffusion equation, is generally referred to as the reaction-diffusion equation.

Similar to the previous section, when the system reaches a steady state with $\frac{\partial c}{\partial t} = 0$, the right-hand side of Eq. (4.28) becomes zero, resulting in the equation

$$\frac{\partial^2 c(x)}{\partial x^2} = \frac{k}{D}c(x). \tag{4.29}$$

Setting the boundary conditions as $c(x = 0) = c_{max}$ and $c(x = +\infty) \to 0$, we find that

$$c(x) = c_{max} \exp\left(-\frac{x}{\lambda}\right) \tag{4.30}$$

holds, where $\lambda = \sqrt{\frac{D}{k}}$. Here, it is important to note that the concentration gradient, when degradation is involved, takes the form of an exponential decrease. While the previous section discussed a linear gradient normalized by x/L, in this case, it is normalized by x/λ. The parameter λ represents the characteristic length scale of the system.

By considering the rate of concentration change between positions x and $x + \Delta x$, we can observe that

$$\frac{\Delta c}{c} = \frac{c(x) - c(x + \Delta x)}{c(x)} \tag{4.31}$$

$$\begin{aligned} &= 1 - \frac{c_{\max} \exp(-(x + \Delta x)/\lambda)}{c_{\max} \exp(-x/\lambda)} \\ &= 1 - \exp(-\Delta x/\lambda) \\ &\approx 1 - (1 - \Delta x/\lambda) = \frac{\Delta x}{\lambda}. \end{aligned} \qquad (4.32)$$

This indicates that $\frac{1}{\lambda}$ expresses the slope of the concentration gradient per unit length. Exponentially distributed morphogens, characterized by this exponential gradient, are well-known in Bicoid [10] and Dpp [17] in *Drosophila* (fruit fly) development. The characteristic length, λ, exhibits robustness against perturbations, which is believed to be a crucial property for the reproducibility of development. Exploring the mechanisms underlying such robustness is an intriguing area of study.

Exercise 4.1.1 Concentration Gradient with Association, Dissociation, and Degradation In cases where both the binding of the ligand to the receptor and the reverse dissociation reaction are not negligible (with a sufficiently large dissociation rate constant k_{off}) [20], the following equations describe the system dynamics:

$$\frac{\partial c}{\partial t} = -k_{\mathrm{on}} M c (1 - b) + k_{\mathrm{off}} b + D \frac{\partial^2 c}{\partial x^2} \qquad (4.33)$$

$$\frac{\partial b}{\partial t} = k_{\mathrm{on}} M c (1 - b) - k_{\mathrm{off}} b - k_{\mathrm{deg}} b. \qquad (4.34)$$

Here, c represents the concentration of the free ligand, b represents the concentration of the ligand-receptor complex, k_{deg} is the rate constant for degradation, and both c and b are normalized by the total number of receptors M. Consider what kind of behavior appears in the preceding equations.

4.2 Transformation of Pattern: Feed-Forward Loop

4.2.1 Pattern Formation by Feed-Forward Loop: Transforming a Gradient to Another Gradient

In Section 2.4.1, we observed that adaptive responses to step inputs arise through feed-forward response loops. In chemotaxis of cellular slime molds and neutrophils, feed-forward response pathways are believed to detect temporal and spatial changes in the concentration of extracellular attractants, which control the formation of the leading edge of cell elongation.

Consider a 1D space where the input $S(x)$ is stationary at each position x on the cell membrane. The following equations describe the dynamics of the system:

$$\frac{\partial A(x,t)}{\partial t} = k_A S(x) - \gamma_A A, \tag{4.35}$$

$$\frac{\partial I(x,t)}{\partial t} = k_I S(x) - \gamma_I I + D\frac{\partial^2 I}{\partial x^2}, \tag{4.36}$$

$$\frac{\partial R(x,t)}{\partial t} = k_R A(R_T - R) - \gamma_R IR. \tag{4.37}$$

The diffusion term is included only in the time evolution equation of I; assuming that the activator A acts locally in space, while the inhibitor acts globally. For this reason, the model is called LEGI (Local Excitation Global Inhibition). Consider first the steady state, without the diffusion term in the evolution equation. In this case, the output R is written, by using $Q(x) = A(x)/I(x)$ and $\tilde{\gamma} = \gamma_R/k_R$, as

$$R(x) = R_T \frac{Q(x)}{\tilde{\gamma} + Q(x)}. \tag{4.38}$$

This equation has the same form as the spatially independent Eq. (2.73) discussed in Section 2.4.1. If the diffusion of I is sufficiently fast, $I(x)$ is represented by the intracellular mean $\overline{I(x)}$, so that from Eq. (4.36), $\overline{I(x)} \propto \overline{S(x)}$, with the spatial mean of $S(x)$. Noting from Eq. (4.35), $A(x) \propto S(x)$, we get

$$Q(x) \propto \frac{S(x)}{\overline{S(x)}}. \tag{4.39}$$

Hence $Q(x)$ represents the deviation of the input S from its mean $\overline{S(x)}$; in other words, the concentration gradient. When $Q(x)$ is sufficiently small and $R(x) \propto Q(x)$, we obtain

$$\frac{1}{R_T}\frac{\partial R}{\partial x} \propto \frac{1}{\overline{S}}\frac{\partial S}{\partial x}. \tag{4.40}$$

This shows that the gradient of the input S is converted into the gradient of the cell's output R. The simulation shown in Fig. 4.1 illustrates the output R when the input S transitions from a spatially uniform state to a linear gradient. It is evident that the gradient of the input S is converted into the gradient of the output R once a steady state is reached over a sufficient period of time.

In cellular slime molds, quantitative measurements have revealed a notable correlation between variations in the relative concentration of extracellular cAMP, a chemotaxis inducer, and the relative intensity of intracellular signals associated with leading edge formation [14].

Exercise 4.2.1 Even when the spatial distribution of the attractant exhibits a gradual concentration gradient, chemotactic cells may exhibit distinct localization of actin and upstream small G-proteins which can be considered as a measure of R at the front and back regions. As we have seen in the previous section, there can be various nonlinearities in the mechanisms underlying such

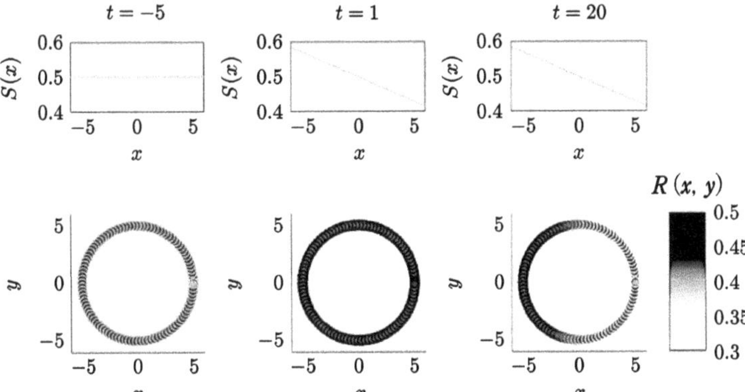

Figure 4.1 Gradient detection using the LEGI model. When switching from $S(x) = 0.5$ to a linear gradient of $S(x) = -kx + 0.5$ at $t = 0$, an abrupt change is introduced. Initially ($t = 1$), the output R rises uniformly due to the elevation of A, which occurs before spatial variations in I emerge. However, as time progresses, the output R gradually forms a gradient from the high side to the low side of the $S(x)$ values ($t = 20$) as the system reaches a steady state. The figure by Daisuke Imoto (University of Tokyo) is partially modified.

all-or-none changes. Let us consider the simplest extension by rewriting Eq. (4.36) into a Michaelis–Menten-type Eq. [23, 28]:

$$\frac{\partial R(x, t)}{\partial t} = k_R A \frac{(R_T - R)}{K_{\mathrm{MA}} + (R_T - R)} - \gamma_R I \frac{R}{K_{\mathrm{MR}} + R}. \tag{4.41}$$

In the steady state, R is obtained as a solution to a quadratic equation. Show that as the Michaelis–Menten coefficients K_{MR} and K_{MA} decrease, the solution exhibits pronounced nonlinearity with respect to A/I. Furthermore, describe how the transient response to temporal changes in $S(x)$ differs depending on the position of the fixed point on the input-output curve.

4.2.2 Pattern Formation by Feed-Forward Loops: Transformation of Chemical Gradient to Stripe Pattern

When analyzing gene regulatory networks across different organisms, it has been widely observed that the number of feed-forward loops present in these networks is significantly higher compared to randomly generated networks (Fig. 4.2). Building upon this established observation, here we aim to investigate the distinctive characteristics associated with the combination of multiple feed-forward loops within these networks.

During *Drosophila* embryogenesis, the transcriptional regulation of genes encoding transcription factors, including Bicoid, Hunchback, and Krüppel, is

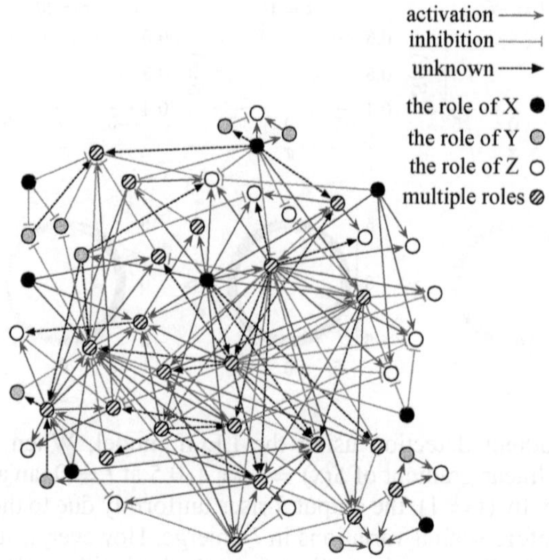

Figure 4.2 *Drosophila* gene regulatory network. Fifty-four genes and 167 controls are illustrated (cited from [11]). The nature of interactions within the feed-forward loop can be discerned by the representation of the lines: black lines indicate activation, gray lines indicate inhibition, and dotted lines indicate unknown interactions. In the circuit, the black, white, and gray nodes correspond to the roles of X, Y, and Z, respectively. Furthermore, shaded nodes represent genes that participate in multiple roles within the circuit.

thought to be operated in a feed-forward loop. Notably, Bicoid proteins establish a concentration gradient along the anterior–posterior axis of the early embryo. This gradient is formed through the diffusion and degradation of Bicoid proteins, which are transcribed and translated from mRNAs localized in the anterior region (Eq. (4.30)). Hunchback proteins, in turn, exhibit anterior expression patterns in response to the concentration gradient of Bicoid proteins. The resulting concentration gradient of Hunchback proteins induces the expression of gap genes, such as Krüppel. These gap genes, in turn, further induce the expression of pair-rule genes and segment polarity genes, leading to their expression as spatially organized stripes. The number of stripes is determined by the presence of specific genes in the control region. As development progresses, the number of these stripes increases, ultimately determining the formation of larval body segments.

In the following analysis, we will explore the developmental dynamics responsible for generating a stripe pattern based on an initial concentration gradient of a morphogen. Specifically, we assume that a gradient of Z_0 is established along the x axis, which is then subjected to a conversion process by a series of N feed-forward networks consisting of factors Y_i and Z_i with $i = 1, 2, \ldots, N$. The model equations are as follows:

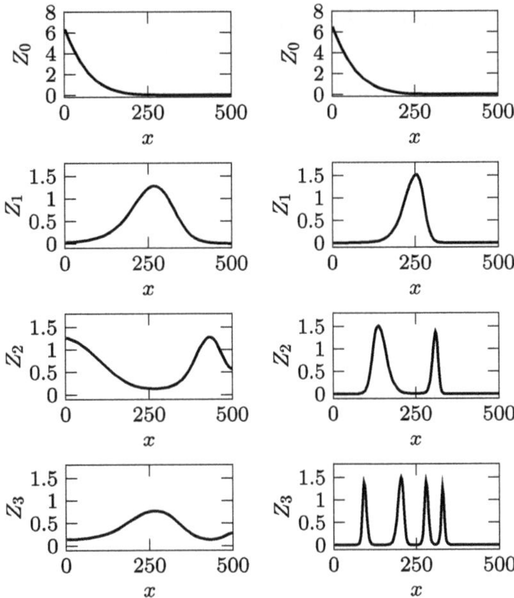

Figure 4.3 Stripe formation with a three-layer feed-forward loop. Hill coefficients correspond to 1 for each layer on the left and to 2 or 3 for each layer on the right. See paper for other details [11].

$$\frac{\partial Z_0}{\partial t} = -\gamma_{Z_0} Z_0 + D_{Z_0} \frac{\partial^2 Z_0}{\partial x^2},$$

$$\frac{\partial Y_i}{\partial t} = -\gamma_{Y_i}(Y_i - f_{A_{Y_i}}(Z_{i-1})) + D_{Y_i} \frac{\partial^2 Y_i}{\partial x^2},$$

$$\frac{\partial Z_i}{\partial t} = -\gamma_{Z_i}(Y_i - g_{Y_i}(Z_{i-1}, Y_i)) + D_{Z_i} \frac{\partial^2 Z_i}{\partial x^2},$$

where the index i denotes the circuit number. As described in Section 2.3.4, a feed-forward loop is governed by a combination of activation and inhibition reactions (e.g., Eqs. (2.69) and (2.70)). In the present model, f_A and f_I represent the activation and inhibition sigmoid functions. The specific form of g depends on the actual reaction mode; here we adopt $g = f_A \cdot f_I$ as shown in Section 2.3.4. Figure 4.3 shows a numerical example for $N = 3$.

Based on the gradient of Z_0, it is evident that a peak of Z_1 is formed. The activation of gene expression in the mid-point of the gradient is attributed to the feed-forward loop combination, which realizes an input–output relationship that responds specifically at the intermediate concentration range (Fig. 2.8). To generate multiple stripes, specific network architectures with many feed-forward loops and parameter settings are required [11]. In addition, the creation of multiple peaks relies on appropriate settings of the K_d (dissociation constant) values for the activation and inhibition sigmoid curves, and also the Hill coefficients. Failure to achieve suitable K_d values for each curve would impede the formation of multiple distinct peaks. By comparing Fig. 4.3 (left column) and Fig. 4.3 (right column), it is evident that the formation of the third peak is hindered when the Hill coefficient is small. While it is possible to create a localized peak on a portion of the slope with a large Hill coefficient, excessively large values result in overly sharp peaks, reducing the available area for peak formation in the downstream layer. This limitation is undesirable when aiming to generate multiple repeating peaks.

These observations indicate that increasing the number of stripes is not a straightforward task and is contingent on parameter settings. Notably, as the number of desired stripes increases, the number of feed-forward loops would be required to increase. This observation aligns with the findings that genetic defects in certain regulatory elements can lead to the disappearance of specific peaks in fly embryos.

Conversely, the formation of a large number of repeated segments, such as in arthropods like centipedes, poses a challenge when solely relying on feed-forward loops. Such intricate patterns are considered difficult to explain solely through the mechanisms of feed-forward loops [8]. In the subsequent section, we will explore an alternative method capable of generating stripe patterns.

4.3   Self-Organizing Pattern Formation

4.3.1   *Linear Stability Analysis of Turing Pattern*

In the preceding sections, we have examined the formation of gradients resulting from the outflow or absorption of specific molecules under certain local boundary or initial conditions. By interpreting these gradients, we have observed the emergence of repetitive stripe patterns along the embryonic axis. Another mechanism for the pattern formation is one that does not require initial gradient. In such a scenario, reactions occur in any location within a spatial domain, and the uniform solution of chemical concentrations can become unstable regardless of the specific boundary conditions. Such instability is referred to as Turing instability, and it can give rise to the formation of periodic spatial patterns. Turing instability offers valuable insights into the mechanisms that underlie the emergence of diverse patterns observed in living organisms.

Let c_i be the concentration of $i = 1, \ldots, N$ kinds of chemicals, the reaction-diffusion equation can generally be written for each ith component as follows:

$$\frac{\partial c_i(x, t)}{\partial t} = f_i\left(\{c\}\right) + D_i \frac{\partial^2 c_i(x, t)}{\partial x^2}, \tag{4.42}$$

where $\{c\}$ means the set $\{c_1, c_2, \ldots, c_N\}$. Considering $c_i(x, t) = c_i^*$ satisfying a uniform solution $\partial c_i/\partial x = 0$ as the steady state $\partial c_i/\partial t = 0$, the following equation is satisfied:

$$f_i\left(\{c^*\}\right) = 0. \tag{4.43}$$

Let us now investigate the conditions that lead to the destabilization of the uniform solution using linear stability analysis. By considering small perturbations on c_i^*, denoted as $c_i(x, t) = c_i^* + \delta c_i(x, t)$, we can derive the following equation:

$$\frac{\partial(\delta c_i)}{\partial t} = \sum_{j=1}^{N} J_{ij}\delta c_j + D\frac{\partial(\delta c_i)}{\partial x^2}. \tag{4.44}$$

Here, J_{ij} is the Jacobi matrix defined as follows:

$$J_{ij} = \frac{\partial f_i}{\partial c_j}\bigg|_{c^*}. \tag{4.45}$$

Recalling the linear stability analysis for systems without spatial degrees of freedom, we observe that the diffusion term is not present. However, when spatial degrees of freedom are considered, the diffusion term is included in the analysis.

By applying the Fourier transform

$$\delta\hat{c}_i(k, t) = \int_{-\infty}^{\infty} e^{-ikx}\delta c_i(x, t)\mathrm{d}x, \tag{4.46}$$

the preceding linearized equation becomes the following ordinary differential equation:

$$\frac{\partial(\delta\hat{c}_i(k, t))}{\partial t} = \sum_{j=1}^{N}(J_{ij} - D_i k^2 \delta_{ij})\delta\hat{c}_j(k, t) \tag{4.47}$$

$$= \sum_{j=1}^{N} M_{ij}\delta\hat{c}_j(k, t), \tag{4.48}$$

where δ_{ij} is Kronecker's delta function and

$$M_{ij} = \begin{pmatrix} J_{11} - D_1 k^2 & J_{12} & J_{13} & \cdots & J_{1N} \\ J_{21} & J_{22} - D_2 k^2 & J_{23} & \cdots & J_{2N} \\ J_{31} & J_{32} & J_{33} - D_3 k^2 & \cdots & J_{3N} \\ \vdots & \vdots & \vdots & \ddots & \vdots \\ J_{N1} & J_{N2} & J_{N3} & \cdots & J_{NN} - D_N k^2 \end{pmatrix}. \tag{4.49}$$

Similar to the characteristic equation for the Jacobi matrix J_{ij} in the linear stability analysis without spatial degrees of freedom, we can perform a similar analysis for the matrix M_{ij}. Let $\lambda_i(k)$ denote the ith eigenvalues of the matrix M_{ij} for any i and k. If the real part $\text{Re}(\lambda_i(k)) < 0$ for all i and k, then the fixed point c^* is linearly stable.

Conversely, if $\text{Re}(\lambda_i(k)) > 0$ for some k, it indicates that the corresponding wavelength undergoes exponential growth. Additionally, when the imaginary part $\text{Im}(\lambda_i(k)) = \omega_i$ takes a nonzero value, it implies the presence of time-varying waves or oscillations in the concentration dynamics, which will be discussed in more detail later. Turing instability is defined as the case where $\text{Im}(\lambda_i(k)) = 0$, indicating the emergence of a stationary (time-invariant) spatially periodic pattern.

4.3.2 Condition of Turing Instability

Let us consider the matrix M_{ij} in the previous section for $N = 2$ components, that is,

$$M_{ij} = \begin{pmatrix} J_{11} - D_1 k^2 & J_{12} \\ J_{21} & J_{22} - D_2 k^2 \end{pmatrix}. \tag{4.50}$$

To simplify the notation, let's denote $a = J_{11}$, $b = J_{12}$, $c = J_{21}$, and $d = J_{22}$, respectively. Furthermore, we introduce $\tilde{a} = a - D_1 k^2$ and $\tilde{d} = d - D_2 k^2$. Then, the eigenvalues λ should satisfy the following quadratic equation:

$$\lambda^2 - (\tilde{a} + \tilde{d})\lambda + \tilde{a}\tilde{d} - bc = 0. \tag{4.51}$$

For the space-uniform solution to be linearly stable, the real part of λ for both roots should be negative. This condition can be expressed as follows:

$$(\tilde{a} + \tilde{d}) < 0, \tag{4.52}$$

$$(\tilde{a}\tilde{d} - bc) > 0. \tag{4.53}$$

Therefore, the inequalities

$$(a + d) < (D_1 + D_2)k^2, \tag{4.54}$$

$$bc < (a - D_1 k^2)(d - D_2 k^2) \tag{4.55}$$

are satisfied. Note that the condition for linear stability is given as a function of wavelength k.

Suppose that a perturbation of the chemical concentration is added around the spatially uniform solution. Then, consider the case where the spatially uniform state ($k = 0$) is stable and some $k \neq 0$ is destabilized and grows. The stability criteria for a spatially uniform state can be derived by substituting $k = 0$ into the inequalities (4.54) and (4.55) as

$$(a + d) < 0, \tag{4.56}$$

$$bc < ad. \tag{4.57}$$

Now Eq. (4.56) excludes the case that both a and d are positive. Next, to have an instability at some k, Eqs. (4.54) or (4.55) has to be violated at some $k > 0$. This excludes the case that both a and d are negative. Hence the sign of a and d should be the opposite. Without losing generality, we can choose $a > 0 > d$. Then, the right-hand side of Eq. (4.57) is negative, so that bc should be negative. Accordingly, we can consider the following two cases (i) $a, b > 0$ and $c, d < 0$ or (ii) $a, c > 0$ and $b, d < 0$. The first case is given by the following Jacobi matrix:

$$J_{ij} = \begin{pmatrix} + & + \\ - & - \end{pmatrix}. \tag{4.58}$$

The second case is given by the following Jacobi matrix:

$$J_{ij} = \begin{pmatrix} + & - \\ + & - \end{pmatrix}. \tag{4.59}$$

The former is a so-called "substrate-depletion" system, as seen in the Brusselator, and the latter is an "activator-inhibitor" system such as the FitzHugh–Nagumo equation.

The first stability condition in inequality (4.54) is always satisfied for any k when the inequality (4.56) holds. Hence, should instability arise, the second stability condition given by Eq. (4.55) would be violated for $k \neq 0$. This condition arises when $\tilde{a}d - bc = 0$, corresponding to a transition of an eigenvalue from a negative to a positive value, and $\lambda(k^2) = 0$.

Let us rewrite (4.55) as

$$b \times (-c) > (a - D_1 k^2)((-d) + D_2 k^2). \tag{4.60}$$

In the cases of Eqs. (4.58) and (4.59), the left-hand side of this inequality is positive. Furthermore, the content of the first bracket on the right-hand side decreases from the value $a(> 0)$ as k increases, and the content of the second bracket increases from the value $(-d)(> 0)$ as k increases. Conversely, we can see that this inequality is readily violated depending on k by decreasing D_1 and increasing D_2 so that the content of the first bracket does not decrease much, but the content of the second bracket increases.

In light of these observations, it can be inferred that Turing instability is more likely to manifest itself when the diffusion rate of the inhibitor c_2 (or substrate) is significantly greater than that of the activator c_1 (or product), that is,

$$D_2 \gg D_1. \tag{4.61}$$

This relationship can be expressed in simple terms as follows: activators act locally and both activators and inhibitors are increased simultaneously. The larger D_2 means the inhibitor diffuses more rapidly, preventing the activator from rising in the vicinity. A single concentration peak is formed by local activation and global inhibition. Similarly, in the case of the substrate-depletion system, the substrate is locally depleted, but the substrate itself is transmitted ahead of the others, resulting in a situation where there is more substrate in the surrounding area. In the activator-inhibitor system, the peaks of the activator and the peaks of the inhibitor coincide in space, whereas in the substrate-depletion system, the peaks and valleys of the product and the substrate coincide.

4.3.3 Characteristic Length of Turing Pattern

When local activation and global inhibition occur simultaneously, the peaks of concentration become aligned at certain intervals. Let us discuss the characteristic length of the pattern based on the Turing instability of the two-component system. When the inequality (4.55) is violated, a perturbation of wavelength k satisfying the following equation grows:

$$k^2 = \frac{D_2 a + D_1 d \pm \sqrt{(D_2 a + D_1 d)^2 - 4D_1 D_2(ad - bc)}}{2D_1 D_2}. \tag{4.62}$$

Let us assume that when the spatially uniform system becomes unstable, the inequality fails at a unique value of k $(= k_c)$. This scenario corresponds to the case where Eq. (4.62) possesses a double root. Thus, we obtain

$$k_c = \left(\frac{D_2 a + D_1 d}{2D_1 D_2}\right)^{\frac{1}{2}} \tag{4.63}$$

$$= \left(\frac{ad - bc}{D_1 D_2}\right)^{\frac{1}{4}}. \tag{4.64}$$

In analyzing the Turing instability, one must note that the characteristic lengths of emergent patterns remain unaffected by the boundary conditions. The characteristic length in question is directly proportional to the ratio of the effective means of both the diffusion coefficient and the reaction rate coefficient. Furthermore, the onset of spatial symmetry breaking arises from the amplification of initial perturbations, and it is not contingent upon any inherent asymmetry within the system, a phenomenon often termed as "spontaneous symmetry breaking." This behavior stands in stark contrast to the examples of the concentration gradients induced by diffusion, as discussed in Section 4.1.3, and other self-organizing patterns arising from mechanisms such as convection. However, in models integrating a global variable dependent on the system size, a consistent ratio between the system size and stripe spacing is evident, as elucidated in reference [12].

As discussed previously, the conditions for Turing instability are associated with the reaction mechanisms of activator-inhibitor and substrate-depletion systems. Notably, while systems such as the FitzHugh–Nagumo equation and the BZ reaction meet these conditions, it remains a considerable challenge to experimentally induce a significant difference in the diffusion coefficients of chemicals. This is particularly the case in reaction systems characterized by relatively dilute solutions. The experimental validation of the Turing pattern was conclusively demonstrated approximately four decades after Turing's death, using the CIMA (Chlorite-Iodite-Malonic Acid) reaction [1, 24]. A key to this success was the strategic reduction of the diffusion coefficient for certain reactants by introducing starch, which serves as a color indicator for iodine.

The presence of Turing patterns within genuine biological systems remains a matter of ongoing investigation. For instance, it has been observed that as the marine angelfish *Pomacanthus* grows, the interstripe intervals on their epidermis accommodate additional stripes to compensate for increased spacing [19]. Notably, the constancy of the stripe intervals, irrespective of the overall size of the fish surface, aligns well with the characteristics postulated by Turing instability. Other manifestations reminiscent of Turing patterns have been delineated in populations of slime molds exhibiting limited mobility, in the murine palate [6], and in tetrapods. Challenges in interpreting these phenomena arise from the intricacies of the underlying reaction dynamics and the consideration that cellular motility, as well as the inherent morphology and dimensions of the cells, play pivotal roles. A more comprehensive view will be presented in Chapter 8.

Example 4.1 Characteristic Length of Brusselator in One-Dimensional Space Let us analyze the Brusselator model equations

$$\frac{\partial X}{\partial t} = 1 - (B+1)X + AX^2Y + D_1\frac{\partial^2 X}{\partial x^2}$$

$$\frac{\partial Y}{\partial t} = BX - AX^2Y + D_2\frac{\partial^2 Y}{\partial x^2}.$$

First, let us confirm Turing instability of this system. Around the fixed point,

$$M_{ij} = \begin{pmatrix} B - 1 - D_1 k^2 & A \\ -B & -A - D_2 k^2 \end{pmatrix}, \tag{4.65}$$

the characteristic equation is written as

$$(B - 1 - D_1 k^2 - \lambda)(-A - D_2 k^2 - \lambda) + AB = 0. \tag{4.66}$$

It is rearranged into

$$\lambda^2 + \left\{ A - B + 1 + (D_1 + D_2)k^2 \right\}\lambda$$
$$+ \left\{ A + (1 - B)D_2 k^2 + AD_1 k^2 + D_1 D_2 k^4 \right\} = 0. \tag{4.67}$$

From the condition of linear stability (4.55),

$$k^4 + \left(\frac{A}{D_2} + \frac{1-B}{D_1} \right)k^2 + \frac{A}{D_1 D_2} > 0. \tag{4.68}$$

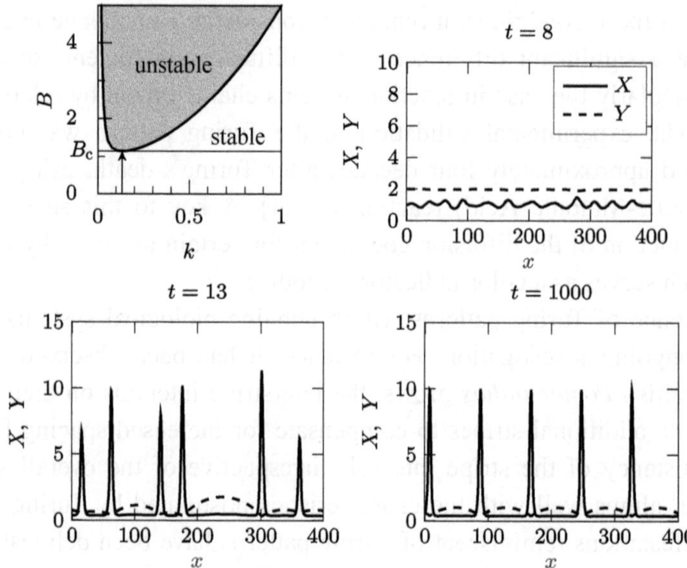

Figure 4.4 Phase diagram of a Brusselator (1D space) (upper left) and the formation process of the Turing pattern. At first, peaks of roughly spaced characteristic lengths appear ($t = 8, t = 13$), and as time proceeds, they are arranged at equal intervals ($t = 1000$). $A = 1, B = 2, D_X = 4, D_Y = 800$.

Then, let us look at the situation where this inequality breaks when B is varied as a bifurcation parameter. Writing the destabilization condition as an inequality for B, we obtain

$$B \geq 1 + \frac{D_1}{D_2}A + \frac{A}{D_2 k^2} + D_1 k^2. \tag{4.69}$$

Suppose we increase the value of B from zero, as shown in Fig. 4.4. When the bend point of the boundary indicated by this inequality is reached ($B = B_c$), the wavelength of $k = k_c \left(= \left(\frac{A}{D_1 D_2} \right)^{1/4} \right)$ is destabilized.

Exercise 4.3.1 The Schnackenberg equation is an example of a substrate-depletion system:

$$\frac{\partial u}{\partial t} = \gamma(a - u + u^2 v) + \nabla^2 u,$$
$$\frac{\partial v}{\partial t} = \gamma(b - u^2 v) + D\nabla^2 v. \tag{4.70}$$

Obtain the instability condition and the characteristic length of the system.

Exercise 4.3.2 The Gierer–Meinhardt equation [9] is the representative example of an activator-inhibitor system:

$$\frac{\partial u}{\partial t} = \left(-au + \frac{u^2}{v}\right) + \nabla^2 u,$$

$$\frac{\partial v}{\partial t} = \left(-cv + u^2\right) + D\nabla^2 v. \tag{4.71}$$

Obtain the instability condition and the characteristic length of the system.

4.3.4 Turing Pattern in a Two-Dimensional Plane

The animal skin patterns, characterized by a diverse array of stripes and spots, present intriguing questions regarding the emergence of patterns with such configurations [21, 27]. As a case in point, we examine the Turing pattern maintained by the Schnackenberg equation (Eq. (4.70)) on a two-dimensional plane. The computational outcomes of this system are depicted in Fig. 4.5. Rough peaks and valleys emerge from the process of flattening random initial conditions. Over time, these patterns become regular and evenly spaced. In this example, a spot pattern with u characterized as a valley is displayed; however, contingent on parameter selection, configurations with u as a peak or exhibiting a striped morphology can also be achievable. Selection between stripes and spots is determined by the order of the nonlinear term. Specifically, in the absence of a second-order nonlinear term, stripes emerge as the predominant pattern [7].

In morphogenesis, repetitive patterns such as spots and stripes are often seen on body segments, limb bones, palate, and body surface. As discussed in the previous section, one method of forming repetitive structures is the feed-forward input-output conversion, as in the *Drosophila* embryo. In this scenario, each successive stripe necessitates the integration of a subsequent feed-forward loop downstream. Conversely, the reiterated structure observed in mouse embryonic limb buds, which intrinsically exhibits five digits, can expand to 14 digits due to Hox mutations [26]. The assumption of pre-existing supplementary feed-forward loops seems implausible, leading to the hypothesis of a Turing pattern–like mechanism in operation. Interestingly, a combination of feed-forward and feedback has been found in patterns created by evolving response networks on a computer. Notably, evolutionary simulation demonstrated that a combination of feed-forward and feedback loops are selected to maintain periodic patterns. It has been suggested that these patterns are related to the striped patterns seen in short-germ beetles [8].

The following simple activator-inhibitor may well explain the relationship between the formation of the vertebrate limb skeletal pattern and the regulation by Hox genes. Let us discuss the following simple model that can generate a two-dimensional pattern [26]:

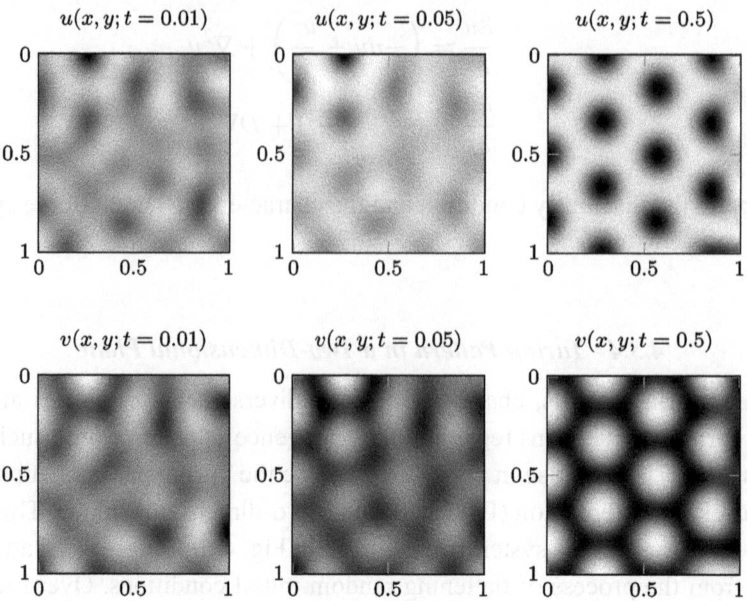

Figure 4.5 Two-dimensional pattern formation of the Schnackenberg model (Eq. (4.70)). At first, roughly spaced peaks of characteristic lengths appear, and as time proceeds, they appear to be evenly spaced.

$$\frac{\partial u}{\partial t} = au - bv - u^3 + D_u \nabla^2 u,$$
$$\frac{\partial v}{\partial t} = cu - dv + D_v \nabla^2 v,$$

$$(4.72)$$

where u and v are the concentrations of activators and inhibitors, a, b, c, d are positive constants, and D_u, D_v are the diffusion coefficients of u, v respectively.

Typically, the stripe patterns maintained by the Turing instability appear in orientations dictated by stochastic initial conditions. Nonetheless, the expression of the Sox9 gene, pivotal for digit specification in limb bud morphogenesis, invariably emerges as stripes aligned with the axis of growth elongation. Can Turing dynamics account for the orientation specificity of these stripes? Indeed, if initial conditions were ordered periodically rather than random, they could provide a consistent direction. However, this merely postpones the core issue, as the occurrence of such ordered initial conditions remains a mystery. Morphogenetic pattern formation is postulated to be orchestrated by the germinal axis, which is defined by the morphogen gradient. In the case of limb formation, regulatory factors such as FGF and BMP are distributed with varying concentrations in the tissue, potentially conveying positional and directional cues to cells through gradient establishment. Fig. 4.6 (bottom) shows an example of a pattern that appears when c is regarded as such an input that forms a linear gradient. The higher the value of c is, the

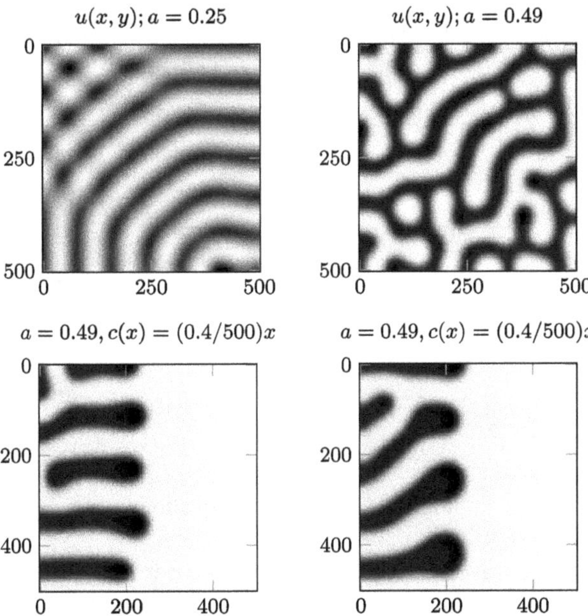

Figure 4.6 Stripe and labyrinth patterns that appear in the activator-inhibitor system (Eq. (4.72)). The left and right figures below show results from different initial conditions with the same parameters. The following parameters were used: $D_u = 17.5, D_v = 218.75, b = 0.5, c = 0.5, d = 0.5$.

earlier destabilization proceeds, thus the patterns tend to line up along the gradient of c. Consequently, while the morphogen gradient may not delineate positional information directly, it could influence pattern orientation indirectly.

4.4 Traveling Wave Pattern

4.4.1 Propagation of Membrane Potential: Cable Equation

The propagation of changes in the membrane potential can be described mathematically by a model analogous to diffusion. To address such systems, let us consider intracellular and extracellular compartments separated by the cell membrane, as depicted in Fig. 4.7. Here, let $V_{in}(x)$ denote the potential inside the membrane, that is, the cytoplasmic side, and $V_{ex}(x)$ denote the potential outside the cell at a certain position x. As described in the Hodgkin–Huxley equation in Section 3.4, the membrane microinterval $[x, x + dx]$ is conceptualized as an equivalent circuit consisting of a capacitor (with capacitance C_{memb}) and conductances representing ion channels. In Fig. 4.7, let $I_{in}(x)$ and $I_{ex}(x)$ represent the currents propagating along the x axis within and outside the cell, respectively. Additionally, currents traverse the membrane both intracellularly and extracellularly. The current is characterized

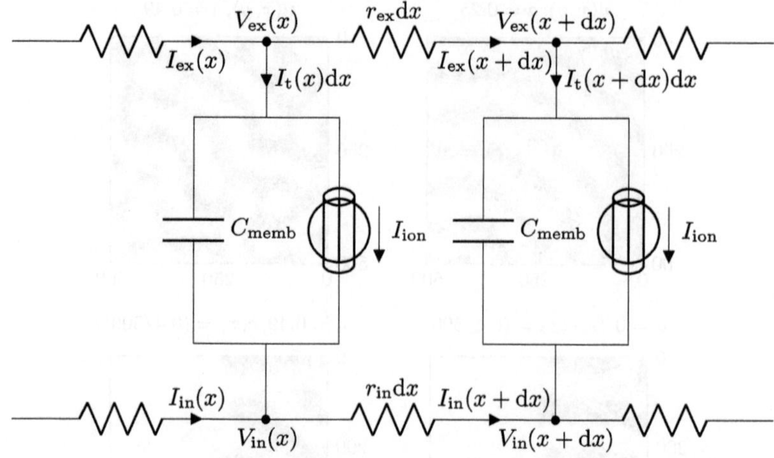

Figure 4.7 The cable equation.

as the electric charge traversing the cell membrane per unit area and per unit time. For the purpose of this analysis, consider a strip of area with unit depth orthogonal to the plane of the depiction and denote the current as $I_t(x)\mathrm{d}x$.

The change in charge per unit time in the interval $[x, x + \mathrm{d}x]$ is described as follows:

$$-I_{\mathrm{ex}}(x + \mathrm{d}x) + I_{\mathrm{ex}}(x) - I_t(x)\mathrm{d}x = 0,$$
$$-I_{\mathrm{in}}(x + \mathrm{d}x) + I_{\mathrm{in}}(x) + I_t(x)\mathrm{d}x = 0. \tag{4.73}$$

Here, the current assumes a positive value when flowing into the contacts. Consequently, as $\mathrm{d}x$ approaches zero, we can express this in differential form as follows:

$$\frac{\partial I_{\mathrm{ex}}}{\partial x} = -I_t,$$
$$\frac{\partial I_{\mathrm{in}}}{\partial x} = I_t. \tag{4.74}$$

On the other hand, when the potentials V_{in} and V_{ex} vary in the x direction, the current flows in response to the potential gradient as governed by Ohm's law. Let r_{ex} and r_{in} denote the resistances per unit length in the x direction externally and internally to the cell, respectively. Then, we derive:

$$V_{\mathrm{ex}}(x + \mathrm{d}x) - V_{\mathrm{ex}}(x) = -I_{\mathrm{ex}}r_{\mathrm{ex}}\mathrm{d}x, \tag{4.75}$$
$$V_{\mathrm{in}}(x + \mathrm{d}x) - V_{\mathrm{in}}(x) = -I_{\mathrm{in}}r_{\mathrm{in}}\mathrm{d}x. \tag{4.76}$$

Thus, we obtain the following equations in the differential form:

$$\frac{\partial V_{\mathrm{ex}}}{\partial x} = -I_{\mathrm{ex}}r_{\mathrm{ex}},$$
$$\frac{\partial V_{\mathrm{in}}}{\partial x} = -I_{\mathrm{in}}r_{\mathrm{in}}. \tag{4.77}$$

Let $V = V_{ex} - V_{in}$ be the potential difference between the inside and outside of the cell. If the sum of the currents flowing in the x-direction inside and outside the cell is denoted as $I_{total} = I_{ex} + I_{in}$, then we obtain the following:

$$
\begin{aligned}
I_{total} &= -\frac{1}{r_{ex}}\frac{\partial V_{ex}}{\partial x} - \frac{1}{r_{in}}\frac{\partial V_{in}}{\partial x} \\
&= -\frac{1}{r_{ex}}\frac{\partial (V + V_{in})}{\partial x} - \frac{1}{r_{in}}\frac{\partial V_{in}}{\partial x} \\
&= -\frac{1}{r_{ex}}\frac{\partial V}{\partial x} + \left(\frac{r_{in} + r_{ex}}{r_{ex}}\right) I_{in},
\end{aligned}
\tag{4.78}
$$

where Eq. (4.77) is used. Since I_t on the outside of the cell and I_t on the inside of the cell must be equal, and from Eq. (4.74)

$$
\frac{\partial I_{total}}{\partial x} = 0
\tag{4.79}
$$

must be satisfied, then by substituting Eq. (4.78) for I_{in} in Eq. (4.74), we obtain

$$
I_t(x) = \frac{\partial I_{in}}{\partial x} = \frac{\partial}{\partial x}\left(\frac{1}{r_{in} + r_{ex}}\frac{\partial V}{\partial x}\right).
\tag{4.80}
$$

Letting I_{ion} denote the current through the ion channel, and substituting Eq. (4.80) for the equivalent circuit for a unit-length membrane element

$$
C_{memb}\frac{\partial V}{\partial t} = -I_{ion} + I_t(x),
\tag{4.81}
$$

then we obtain

$$
C_{memb}\frac{\partial V}{\partial t} = -I_{ion} + \frac{\partial}{\partial x}\left(\frac{1}{r_{in} + r_{ex}}\frac{\partial V}{\partial x}\right).
\tag{4.82}
$$

This equation has the same form as the reaction-diffusion equation where $1/(C_{memb}(r_{in} + r_{ex}))$ corresponds to the diffusion coefficient.

The previous framework does not consider the morphology of the membrane in the depth direction. However, in a real biological context, the resistance depends on the morphology of the membrane. For example, consider a nerve axon modeled as a cylinder of radius a extending along the x axis. The surface of this cylinder is composed of membrane strands, while the interior and exterior of the cylinder represent the intracellular and extracellular regions, respectively. Here, let $R_{ex} = \pi a^2 r_{ex}$ and $R_{in} = \pi a^2 r_{in}$ denote the resistivities outside and inside the cylinder, respectively. Moreover, let $C_{memb'} = C_{memb}/2\pi a$ represent the capacitance both per unit length and per unit circumference in the axial direction and $I_{ion'} = I_{ion}/2\pi a$ denotes the current through the ion channel on the cylindrical membrane surface, respectively. Then, from Eq. (4.82), we obtain

$$
C_{memb'}\frac{\partial V}{\partial t} = -I_{ion'} + \frac{\partial}{\partial x}\left(\frac{a}{2(R_{in} + R_{ex})}\frac{\partial V}{\partial x}\right),
\tag{4.83}
$$

which is called the cable equation in electrophysiology. By applying the aforementioned relation, we derive the spatially dependent form of the Hodgkin–Huxley equation, which is presented as follows:

$$
C_{\text{memb}} \cdot \frac{\partial V}{\partial t} = -g_{\text{Na}}(V - E_{\text{Na}}) - g_{\text{K}}(V - E_{\text{K}}) - g_{\text{L}}(V - E_{\text{L}}) + I_{\text{ext}}
$$

$$
+ \frac{a}{2(R_{\text{in}} + R_{\text{ex}})} \frac{\partial^2 V}{\partial x^2}, \tag{4.84}
$$

where the space inside and outside of the cell is assumed to be uniform, and $R_{\text{in}}, R_{\text{ex}}$ are set to be constant.

4.4.2 *One-Dimensional Traveling Wave: Fisher Equation*

In the previous section, we saw the similarities between the Hodgkin–Huxley saw the cable equation for membrane potential, and a standard reaction-diffusion system. In the following, we will explore the traveling wave solutions emerging from these equations. Let us consider the following equation:

$$
\frac{\partial u}{\partial t} = kf(u) + \frac{\partial}{\partial x}\left(D\frac{\partial u}{\partial x}\right), \tag{4.85}
$$

where we assume the wave form $u(x, t)$ propagates at a constant speed c without changing its shape in time. This is called a traveling wave. Since the function u depends only on the position on the coordinate $z = x - ct$ moving with velocity c,

$$
u(x, t) = u(z), \tag{4.86}
$$

that is, the function u must be written as a function of a uniquely determined value z. Rewriting Eq. (4.85) with the variable z, the left-hand side becomes

$$
\frac{\partial u}{\partial t} = \frac{\partial z}{\partial t}\frac{\partial u}{\partial z} = -c\frac{du}{dz}. \tag{4.87}
$$

Also, the derivative for x is

$$
\frac{\partial^2 u}{\partial x^2} = \frac{d^2 u}{dz^2}, \tag{4.88}
$$

and when D is constant, Eq. (4.85) can be rewritten as an ordinary differential equation as follows:

$$
D\frac{d^2 u}{dz^2} + c\frac{du}{dz} + kf(z) = 0. \tag{4.89}
$$

Let us now consider $f = ku(1 - u)$. In this case, Eq. (4.85) corresponds to the Fisher equation that describes the spatial propagation of population fluctuations. The Fisher equation is a model that describes the spatial spread of a single invasive species through diffusion and multiplication. It is called a logistic equation because f represents the increase in population ku with an increase rate k and the fact that the rate of increase is suppressed by $1 - u$ as the density increases. As can be seen from the fact that $u > 0$ increases exponentially for small values of u, the pace of increase slows down as the solution approaches $u = 1$.

$$D\frac{d^2u}{dz^2} + c\frac{du}{dz} + ku(1 - u) = 0 \tag{4.90}$$

Here, we rewrite the preceding equation into the following first-order simultaneous differential equations:

$$\begin{aligned} \frac{du}{dz} &= v, \\ \frac{dv}{dz} &= -\frac{k}{D}u(1 - u) - \frac{c}{D}v. \end{aligned} \tag{4.91}$$

Subsequently, let us conduct a linear stability analysis around the fixed points $(u, v) = (0, 0)$ and $(u, v) = (1, 0)$. The Jacobi matrix is

$$J_{ij} = \begin{pmatrix} 0 & 1 \\ -\frac{k}{D}(1 - 2u) & -\frac{c}{D} \end{pmatrix} \tag{4.92}$$

and in the cases of $(u, v) = (0, 0)$, $(u, v) = (1, 0)$, the eigenvalues λ_0, λ_1 are

$$\begin{aligned} \lambda_0 &= \frac{-c/D \pm \sqrt{(c/D)^2 - 4k/D}}{2}, \\ \lambda_1 &= \frac{-c/D \pm \sqrt{(c/D)^2 + 4k/D}}{2}. \end{aligned} \tag{4.93}$$

Thus, in the (u, v) plane, $(0, 0)$ is a stable fixed point and $(1, 0)$ is a saddle point. If $(c/D)^2 < 4k/D$, then u will be negative because of damped oscillations around $u = 0$ due to the negative value in the root. If u is the population density, u is required to be positive, so $(c/D)^2 > 4k/D$. In other words, the propagation speed of the traveling wave is expected to be $c > 2\sqrt{kD}$.

Recalling that the trajectory on the (u, v) plane gives the behavior of the system in the coordinate system moving at the propagation speed c of the traveling wave, it is evident that this trajectory represents the characteristics of the wavefront. Consider a scenario where the population density experiences a slight increase from $u = 0$. At the wavefront, which exhibits exponential growth, u is spatially nearly homogeneous, making the reaction term $f(u)$ predominant. Therefore, u increases, and the increased population of individuals is transmitted to the surroundings by diffusion, causing the increase of u in such region. Throughout this phase, by definition, v remains positive. As u nears the value of $u = 1$, its growth rate decelerates, converging toward $v = 0$. Given that $f(u) = 0$ when $u = 1$, the diffusion component

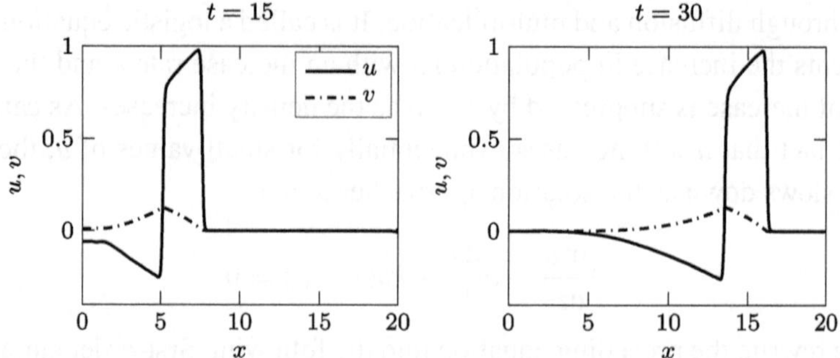

Figure 4.8 Traveling waves in the FitzHugh–Nagumo equation. The propagation of the excited wave is in the positive direction of the x-axis. Initial values are $u(0) = 0$, $v(0) = 0.04$, and $u(0) = 0.9(x < 1)$ is given as a perturbation. $\epsilon = 0.03$, $a = 0.1, b = 0.1, \gamma = 0.04, D = 1 \times 10^{-3}$.

assumes a principal role in governing the system's dynamics. As a result, the state at $u = 1$ cannot be maintained, asymptotically moving toward $u = 0$. In this phase, v assumes negative values. This provides an intuitive rationale for $u = 1$ manifesting as a saddle point in a moving coordinate system. It elucidates the intrinsic tug-of-war between reaction and diffusion in the wave propagation of a reaction-diffusion system.

4.4.3 One-Dimensional Traveling Wave: FitzHugh–Nagumo Equation

Let us now turn our attention to the traveling wave solution of the FitzHugh–Nagumo equation. Analogous to the cable equation, the membrane potential is posited to propagate in line with the one-dimensional diffusion equation. Conversely, given that the channel is embedded within the membrane, its diffusion is negligible compared to the membrane potential. In Section 3.5, we discussed the FitzHugh–Nagumo equation using the same variables with the Hodgkin–Huxley equation. In this chapter, for analytical clarity, we rewrite the independent variables of the FitzHugh–Nagumo equation, u and v, as follows:

$$\epsilon \frac{\partial u}{\partial t} = -u(u - a)(u - 1) - v + I + D \frac{\partial^2 u}{\partial x^2}, \tag{4.94}$$

$$\frac{\partial v}{\partial t} = bu - \gamma v, \tag{4.95}$$

where $0 < a < 1$. Now consider the case $I = 0$ with no external current. The stable fixed point is $(u, v) = (0, 0)$. Then, consider the initial condition with a large perturbation at the left boundary of x. We subsequently analyze its temporal evolution, as illustrated in Fig. 4.8. As seen in Section 3.5, when the value of u is sufficiently far away from the stable fixed point, u and v takes a large trajectory before returning to the fixed point. With the inclusion of a diffusion term, the initial perturbation to u is transmitted to the periphery. As a result, a traveling wave is formed. In an excitable system, this transient response can be transmitted without spatial decay.

In Section 3.5, we examined the qualitative behavior of the FitzHugh–Nagumo equation as a function of the external current I and other parameters. To elucidate the differences in behavior between the cases without spatial degrees of freedom and those incorporating diffusion, we depict the dynamics on the phase plane in Fig. 4.9. As illustrated in Fig. 4.9 (a, left), when there is excitability at $I = 0$, $(u, v) = (0, 0)$ is a stable fixed point. When a large perturbation is applied from this point, it propagates spatially, as seen above (Fig. 4.8). When the v nullcline is tilted as shown in Fig. 4.9 (a, middle), another stable fixed point u^+ appears and the system becomes bistable. Even when every spatial region is initially at a stable fixed point of $(0, 0)$, if some of the region transitions to u^+, the travelling wave front will propagate from there by diffusion, leading the entire region to eventually transition to u^+ (Fig. 4.9(b)). In another case, when I increases as shown in Fig. 4.9 (a, right), the nullcline of u shifts upward, which makes the fixed point unstable. Regardless of the initial conditions being set randomly, limit-cycle oscillations synchronize their phases in proximate spatial regions, and eventually periodic traveling waves emerge (Fig. 4.9(c)).

4.4.4 One-Dimensional Traveling Wave: Nagumo Equation

Let us consider the characteristics of travelling waves in more detail. In the numerical example earlier, we saw a wave (called a pulse wave) that appears isolated in a certain spatial domain. In other words, u and v are at a stable fixed point regardless of whether it is positive or negative at a sufficiently distant location. At $x \to \pm\infty$, the following equations are obtained: $u \to 0, v \to 0, \frac{du}{dt} \to 0$. If we focus only on the edge of the pulse wave, we can approximate $v = 0$ at the beginning of the excitation, as we also assumed the case without spatial variation. Therefore, Eq. (4.94) can be written as follows:

$$\epsilon\frac{\partial u}{\partial t} = -u(u - a)(u - 1) + D\frac{\partial^2 u}{\partial x^2}. \tag{4.96}$$

When diffusion is not considered, $u = 0, 1$ is a stable fixed point and $u = a$ is an unstable fixed point. If we focus on the back of the wave where u drops sharply, the change in v is slow, so that v may be approximated as $v = $ (constant). Then, let us consider a more generalized form that is valid for both the front and back of the wave as follows:

$$\frac{\partial u}{\partial t} = -k(u - u_1)(u - u_2)(u - u_3) + D\frac{\partial^2 u}{\partial x^2}, \tag{4.97}$$

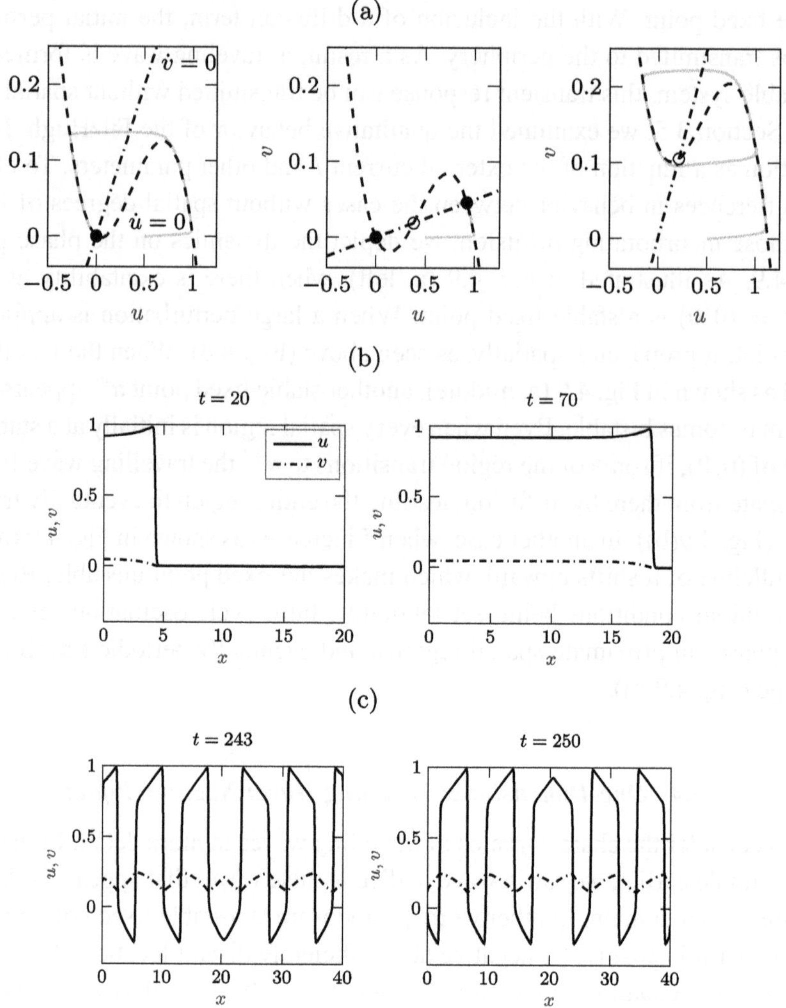

Figure 4.9 Behavior of the FitzHugh–Nagumo equation. (a) The phase plane in the case of no spatial dependence. The black and white circles represent stable and unstable fixed points, respectively, and the dashed line represents the null-cline. The gray line shows the trajectory of time evolution from the initial value of $(0.5, 0)$. (b) Front wave. Front wave propagating in the positive direction of the x-axis. $\epsilon = 0.03$, $a = 0.3$, $b = 0.2$, $\gamma = 0.01$, $I = 0$, $D = 1 \times 10^{-3}$. (c) Coupled wave. Periodically propagating waves in the positive and negative directions of the x-axis collide and disappear at the center. $\epsilon = 0.03$, $a = 0.1$, $b = 0.1$, $\gamma = 0.04$, $I = 0.12$, $D = 1 \times 10^{-3}$.

which is called the Nagumo equation. The excitability of a nullcline on a phase plane is characterized by the trajectory around the nullcline, which consists of valleys and peaks. Let k be a positive real number and $0 < u_1 < u_2 < u_3$. When diffusion is not considered, $u = u_1, u_3$ are stable fixed points and $u = u_2$ is an unstable fixed point. As in the discussion in Section 4.4.2, for a coordinate moving with a traveling wave and propagation velocity c, we can rewrite the equation as follows:

$$D\frac{d^2u}{dz^2} + c\frac{du}{dz} - k(u - u_1)(u - u_2)(u - u_3) = 0. \tag{4.98}$$

Exercise 4.4.1 Suppose that the boundary conditions in Eq. (4.98) are $u(-\infty) = u_3, u(+\infty) = u_1$ and the slope of the wave is zero, that is, $du/dz|_{x\to\pm\infty} = 0$, at sufficiently distant locations. Based on these conditions, Consider the following waveform:

$$\frac{du}{dz} = a(u - u_1)(u - u_3). \tag{4.99}$$

show that the propagation velocity is given by

$$c = \sqrt{\frac{kD}{2}}(u_1 - 2u_2 + u_3). \tag{4.100}$$

Furthermore, let A be an appropriate constant and show that the analytical solution is

$$u(z) = \frac{u_1 - Au_3 \exp\{a(u_1 - u_3)z\}}{1 - A\exp\{a(u_1 - u_3)z\}}. \tag{4.101}$$

4.4.5 Two-Dimensional Pattern Formation of Excitable Systems

Excited elements transmit stimuli to their surroundings, and when a chain of firings occurs, this chain propagates as an "excitatory wave." Excitatory waves often propagate in two- or three-dimensional space, as in the case of calcium waves in fertilized eggs and cAMP waves in cellular slime molds. Concentric waves (target waves) are formed by repeated transmission of excitation from a single location. It is easy to understand that a concentric pattern is formed when the center of the wave is excitable or oscillatory for some reason. Fig. 4.10 (left) shows the wave that appears when such an initiation center is placed at a single location in the two-dimensional plane in the FitzHugh–Nagumo equation. On the other hand, when there are multiple random firing centers, spiral waves tend to appear, as shown in Fig. 4.10 (right). What constitutes this spiral structure, and through which mechanism does it emerge?

Recall that an excitable system can be understood as transitions between three basic states, namely, the resting state, the excited state, and the refractory period, as discussed in Section 3.5. Given this understanding, one anticipates that the characteristics of the spatial patterns observed in the FitzHugh–Nagumo equation would

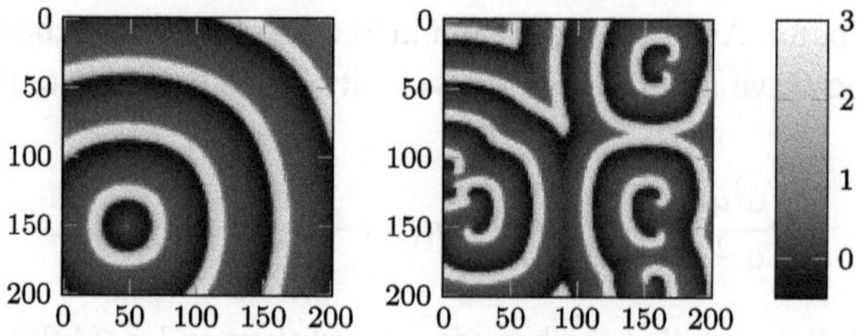

Figure 4.10 Traveling waves on a two-dimensional plane for the FitzHugh–Nagumo equation. $a = 2.5$, $b = 2.0$, $\epsilon = 0.05$, $\gamma = 1.0$, $I_{ext} = 1.2$, $D = 0.5$. (left) Concentric wave propagating from an ignition source near $(x, y) = (50, 150)$, (right) spiral wave formed from multiple randomly placed ignition sources. The grayscale represents the value of u.

also apply to state transition in discrete systems. In the following, let us express it in terms of a cellular automaton.

A cellular automaton is a model in which time, space, and the state of the system are all discrete, each cell (an element in a lattice) in space has a state, and each cell updates its state according to rules that incorporate the states of neighboring cells, thus representing the temporal evolution of the system. Suppose that excitable elements are arranged in a lattice in a two-dimensional plane. Each lattice element has a state variable that represents the "state of the element" and can be 0, 1, or 2. State 0 corresponds to a stable fixed point, which remains in this state unless an external stimulus enters. When an external stimulus enters (starting from this point; $t = 0$), the state transitions from 0 to 1 and remains in this state until a certain time ($t = \tau$). Then it moves to state 2 in the "refractory period," and at ($t = \tau_R$), it returns to the initial resting state 0. When three or more elements around the element in state 0 are in the active state, a stimulus of sufficient intensity is applied, and excitation (transition from state 0 to 1) is triggered. For example, if all initial conditions are set to zero and the center point is set to 1, a cyclic traveling wave is formed (Fig. 4.11(a)). Conversely, upon initially assigning states to the lattice at random, or starting with a configuration where 0, 1 and 2 are next to each other, and subsequently applying the aforementioned rules over time, we observe a recurring transition from state 0 to 2, resulting in the emergence of a spiral pattern (Fig. 4.11(b)).[1]

---

[1] Some may feel strange about the sudden change of the state taking only 0 or 1. In contrast, a class of dynamical systems in which time and space are discrete while the state remains a continuous variable is known as a coupled map lattice. For example, a state $x_n(i,j)$ (n is a discrete time step) is placed at each lattice point (i,j), and $x'(i,j) = f(x_n(i,j)) = sx_n(i,j) + c(\mathrm{mod}\ 1)$ ($0 < s \leq 1, 0 < c$ is constant). After performing the transformation, we use the discretized Laplacian Δ for this $x'(i,j)$ to represent the diffusion, and then calculate the values for the next time step as $x_{n+1}(i,j) = (1 - D)x'(i,j) + (D/N_{neighbor}) \sum_{i',j'} x'(i',j')$ ($\sum$ is the sum over i,j neighborhoods and $N_{neighbor}$ is the number of such neighbors). This model can also easily reproduce spiral waves in an excited system (e.g., Fig. 4.33 in [15]).

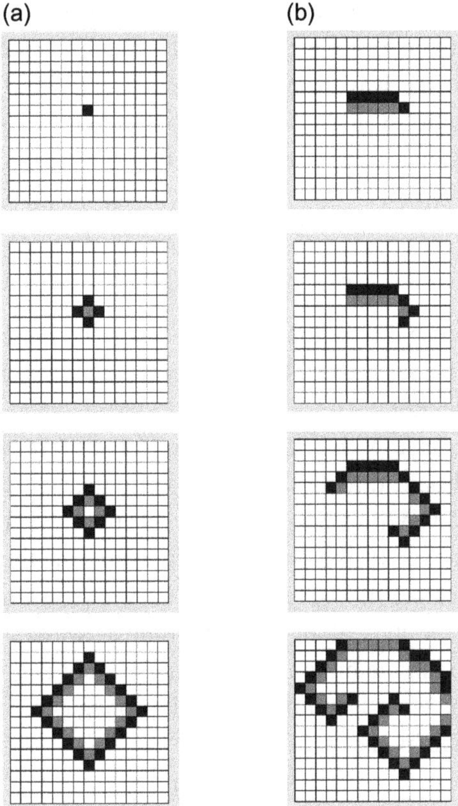

Figure 4.11 Behavior of an excitable automaton. States 0 (white), 1 (black), and 2 (gray) are shown. Panels show time evolution from top to bottom. (a) State 1 spreads outward. (b) Development of helical waves.

Considering this back to a continuous system, integrating the periodically varying phase ϕ of the variable at each point, then, it can be seen that only the spiral's center takes a nonzero value,

$$\oint \phi \cdot \mathrm{d}r = \pm 2\pi, \tag{4.102}$$

that is, called spatial phase singularity. In this model, such an arrangement is achieved by initially random assignment of states. Depending on the model, this can be noise that is always introduced as well as initial conditions, or it can be heterogeneity in the refractory period or sensitivity to stimuli.

4.4.6 Plane Wave, and Velocity of Wave with Curvature

At the spiral's center, the wave propagation is slower due to reduced neighbors for excitation transmission compared to plane waves. Conversely, toward the periphery, the curvature becomes more relaxed, and the wave propagation becomes faster. The degree of this property determines the extent to which the spiral wraps around. To consider the shape of the wavefront, we consider Eq. (4.85). While it originally addresses the propagation velocity of a traveling wave in one-dimensional space, this equation can be extended to encompass a two-dimensional plane as follows:

$$\frac{\partial u}{\partial t} = kf(u) + \frac{\partial}{\partial x}\left(D\frac{\partial u}{\partial x}\right) + \frac{\partial}{\partial y}\left(D\frac{\partial u}{\partial y}\right)$$
$$= kf(u) + \nabla \cdot (D\nabla u). \tag{4.103}$$

Here, the symbol ∇ is the differential operator described in Section 4.1, and in two-dimensional Cartesian coordinates. Note that writing $(\nabla\cdot)$ means the inner product, since it is a vector operator.

If there is a plane wave with a flat wavefront, the argument for waves in one-dimensional space should directly apply. To introduce a coordinate system that moves with a plane wave with velocity c as in 1D space, we need to define a vector of length 1 perpendicular to the wavefront to represent the direction of motion (called the normal vector) $\boldsymbol{n}$. Using the normal vector $\boldsymbol{n}$, the components of the position vector $\boldsymbol{x}$ of the axis along the direction of wave propagation are $\boldsymbol{n} \cdot \boldsymbol{x}$. Therefore, the position in the moving coordinate system is given by $z = \boldsymbol{n} \cdot \boldsymbol{x} - ct$. The shape of the plane wave is maintained as it propagates, that is, u is assumed to be uniquely determined by the variable z, then

$$\frac{\partial}{\partial t} = -c\frac{\partial}{\partial z} \tag{4.104}$$

$$\nabla = \boldsymbol{n}\frac{\partial}{\partial z}. \tag{4.105}$$

By using them to transform Eq. (4.103), we obtain

$$(\boldsymbol{n} \cdot D\boldsymbol{n})\frac{\mathrm{d}^2 u}{\mathrm{d}z^2} + c\frac{\mathrm{d}u}{\mathrm{d}z} + kf(u) = 0. \tag{4.106}$$

Therefore, comparing with the 1D expression (4.89), we can see that D is replaced by $(\boldsymbol{n} \cdot D\boldsymbol{n})$. Note that $(\boldsymbol{n} \cdot D\boldsymbol{n}) = D$ if there is no anisotropy in the diffusion coefficient (when D has the same value in the x-axis and y-axis directions). In general, the propagation speed of a traveling wave depends on the curvature of the wave front [16].

4.4.7 Normal Form and Traveling Wave of Limit-Cycle

Similar behavior to the coupled waves (Fig. 4.9(c)) of the oscillatory system is also seen in the excitable system when the system is sufficiently perturbed. Consequently, the behaviors of oscillatory and excitable systems cannot be distinguished based on the emergent spatiotemporal patterns. The emergence of a wave means that the oscillation phases at various points are organized in a spatiotemporal manner. We will examine this in the context of the standard type of oscillations near the Hopf bifurcation point.

Consider a reaction-diffusion system of two variables u, v in one-dimensional space x:

$$\frac{\partial u}{\partial t} = f(u, v) + D_u \frac{\partial^2 u}{\partial x^2},$$

$$\frac{\partial v}{\partial t} = g(u, v) + D_v \frac{\partial^2 v}{\partial x^2}.$$

Let us assume that in the absence of a diffusion term, the system has a fixed point at $(u, v) = (0, 0)$. The system is assumed to be destabilized when the parameter γ exceeds a critical value γ_c, and an oscillatory solution appears. By using a vector $U = (u, v)$ and a Jacobi matrix J, the linearized expression expanded about the fixed point can be represented as follows:

$$\frac{\partial U}{\partial t} = JU + \hat{D}\nabla^2 U. \tag{4.107}$$

Here, $\hat{D}$ is the following matrix:

$$\hat{D} = \begin{pmatrix} D_u & 0 \\ 0 & D_v \end{pmatrix}. \tag{4.108}$$

Here, we use the Fourier transform to analyze the Turing instability in the linearized equations with the diffusion term. Since we assume that the system is unstable when there is no diffusion term, we simply diagonalize the Jacobi matrix. We then perform a variable transformation $U = NW$ with a constant matrix N and assume that the resulting Jacobi matrix of equation W is diagonalized, that is,

$$\frac{\partial W}{\partial t} = N^{-1}JNW + N^{-1}\hat{D}N\nabla^2 W$$

$$= \tilde{J}W + \tilde{D}\nabla^2 W.$$

The transformed diffusion matrix $\tilde{D}$ is not a diagonal matrix. Here we consider the case $D_u = D_v \equiv D'$. In this case, the matrix $\tilde{D} = \hat{D} = D'I$, with I being a unit matrix. Below, we will analyze the system with this assumption. As seen in Sections 1.15 and 3.3, the condition for Hopf bifurcation is that the eigenvalue σ of Jacobi matrix satisfies $\text{Re}(\sigma) < 0$ in the case of $\gamma < \gamma_c$, $\text{Re}(\sigma) > 0$ in the case of $\gamma > \gamma_c$, and $\text{Re}(\sigma) = 0$ and $\text{Im}(\sigma) \neq 0$ at the bifurcation point of $\gamma = \gamma_c$. Therefore, from $\text{tr}\tilde{J} = 0$, $|\tilde{J}| > 0$, we obtain

$$\sigma(\gamma_c) = \pm i\sqrt{|\tilde{J}|} \tag{4.109}$$

as a function of the branching parameter γ. Here, with the eigenvalue $\sigma(\gamma_c) = \pm i\sqrt{|\tilde{J}|} = \pm i\omega_c$, we obtain

$$\tilde{J} = N^{-1}JN$$
$$= \begin{pmatrix} 0 & -\omega_c \\ \omega_c & 0 \end{pmatrix}.$$

This is the situation where the limit-cycle appears exactly at $\gamma = \gamma_c$. Then, the following equation is obtained as the equation for a slight deviation from the bifurcation point:

$$\frac{\partial W}{\partial t} = \begin{pmatrix} \lambda(\gamma) & -\omega(\gamma) \\ \omega(\gamma) & \lambda(\gamma) \end{pmatrix} W + \hat{D}\nabla^2 W. \tag{4.110}$$

By employing a complex number $W = u + iv$ and rewrite the preceding equation into

$$\frac{\partial W}{\partial t} = (\lambda(\gamma) + i\omega(\gamma))W + \hat{D}\nabla^2 W, \tag{4.111}$$

and by displaying polar coordinates, we obtain

$$\frac{dr}{dt} = \lambda(\gamma)r + D'\left[\frac{\partial^2 r}{\partial x^2} - r\left(\frac{\partial\theta}{\partial x}\right)^2\right]$$
$$\frac{d\theta}{dt} = \omega(\gamma) + D'\left[\frac{1}{r^2}\frac{\partial}{\partial x}\left(r^2\frac{\partial\theta}{\partial x}\right)\right].$$

It can be clearly seen that $D' = 0, \lambda = 0$ represents an orbit that rotates with an angular velocity ω_c while maintaining a radius r.

Assuming that u, v are 2π-periodic functions of phase θ, consider a traveling wave solution $W = r_1 \exp(i(\sigma t - \kappa x))$ with wavenumber κ. By substituting $\theta = \sigma t - \kappa x, r = r_1 = constant$, we obtain

$$\kappa^2 = \lambda(\gamma)/D',$$
$$\sigma = \omega(\gamma),$$

where $\gamma = \gamma(r_1)$, that is,

$$u(x, t) = r_1 \cos\left[-\sqrt{\frac{\lambda(\gamma)}{D'}}\{x - \frac{\omega(\gamma)}{\sqrt{\lambda(\gamma)/D'}}t\}\right],$$
$$v(x, t) = r_1 \sin\left[-\sqrt{\frac{\lambda(\gamma)}{D'}}\{x - \frac{\omega(\gamma)}{\sqrt{\lambda(\gamma)/D'}}t\}\right].$$

Thus, we obtain the propagation velocity c as

$$c = \omega(\gamma)\sqrt{\frac{D'}{\lambda(\gamma)}}. \tag{4.112}$$

Example 4.2 Stuart–Landau Equation

Let us consider the case of $\lambda(\gamma) = \gamma - r^2, \omega(\gamma) = \omega_0$ as a specific form of λ, which satisfies the condition (4.109) of the bifurcation parameter γ near the Hopf bifurcation point. By substituting them into Eq. (4.112), we obtain

$$\frac{\partial W}{\partial t} = (i\omega_0 + \gamma - |W|^2)W + \hat{D}\frac{\partial^2 W}{\partial x^2}. \tag{4.113}$$

This is called the Stuart–Landau equation. In polar coordinates,

$$\frac{dr}{dt} = (\gamma - r^2)r + D'\left[\frac{\partial^2 r}{\partial x^2} - r\left(\frac{\partial\theta}{\partial x}\right)^2\right]$$

$$\frac{d\theta}{dt} = \omega_0 + D'\left[\frac{1}{r^2}\frac{\partial}{\partial x}\left(r^2\frac{\partial\theta}{\partial x}\right)\right].$$

In the case of $D' = 0$, both solutions of $r = 0, r = \sqrt{\gamma}$ are linearly stable. When a diffusion term is added, r shifts from $r = \sqrt{\gamma}$, resulting in the emergence of a finite wavelength. In this case, assuming $r = r_1 = $ constant, we obtain

$$\kappa^2 = (\gamma - r_1{}^2)/D',$$

$$\sigma = \omega_0.$$

Here, we find that there is the relation $r_1 < \sqrt{\gamma}$ between the wave length κ and the bifurcation parameter γ. In addition, the shape of the propagating wave is represented as follows:

$$u(x,t) = r_1 \cos\left[-\sqrt{\frac{(\gamma - r_1^2)}{D'}}\left\{x - \frac{\omega_0}{\sqrt{(\gamma - r_1^2)/D'}}t\right\}\right],$$

$$v(x,t) = r_1 \sin\left[-\sqrt{\frac{(\gamma - r_1^2)}{D'}}\left\{x - \frac{\omega_0}{\sqrt{(\gamma - r_1^2)/D'}}t\right\}\right],$$

where the wave length is $\kappa = 2\pi\sqrt{\frac{D'}{\gamma - r_1^2}}$ and the propagating velocity is $c = \omega_0\sqrt{\frac{D'}{\gamma - r_1^2}}$, respectively.

4.4.8 Transformation of Temporal Oscillation into Spatial Pattern

As an alternative possibility for stripe pattern formation, let us revisit the reaction-diffusion system with limit-cycle attractors discussed in the preceding section. We posit that this system does not satisfy the Turing instability criterion, ensuring the stability of the spatially uniform oscillatory state. Focusing on a one-dimensional

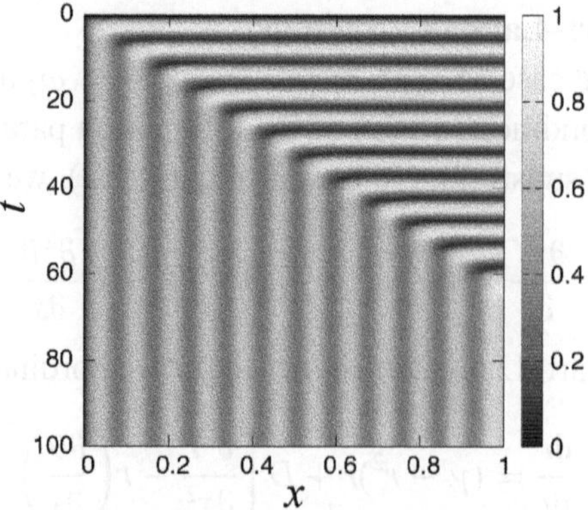

Figure 4.12 Transformation of spatially uniform oscillation into periodic spatial pattern. The shading represents the values of the variables in the bivariate reaction-diffusion system [18]. The peaks are successively fixed as time t advances from the left side of the spatial coordinate x (courtesy of Dr. Kohsokabe (BDR, Riken)).

space, we adopt a fixed boundary condition, wherein the component concentrations are held constant at the boundaries. This eliminates oscillations there. When the diffusion coefficients of the activator and inhibitor are almost the same, moving away from the edge, the intermediate region converges toward a uniform oscillatory state with entrainment in temporal oscillations. Conversely, if the diffusion coefficient of the inhibitor is significantly larger, temporal oscillations are suppressed, giving rise to a temporally stationary, spatially periodic pattern. When the oscillation transitions to a stable fixed point, it acts as a fixed boundary, and the next fixed pattern is formed successively in the direction away from the boundary [18] (Fig. 4.12). In other words, temporal oscillations are converted into spatially fixed waves to form patterns. The wavelength of the pattern produced in this process is approximately proportional to the period of the oscillation. This stripe formation, in some sense, is simpler, because to have oscillation, inhibitor does not require to inhibit itself (see Figs. 1.7 and 7.7). Hence it can be easily achieved by gene regulation networks [8, 18].

Experiments have shown that the expression of transcription factors that determine body segments oscillates in time in the tail bud during vertebrate spine development, and that this oscillation is translated into a repetitive pattern of somite (block of mesodermal cells) [25]. For example, the gene c-hairy1 is expressed and oscillates during chick embryo development and is the key to the formation of segmental repeat structures. As a model to explain this behavior, the "clock-and-wavefront" mechanism [2] has been proposed. In this mechanism, there is an oscillation of the gene expression of interest, or "segmental clock," and a separate global parameter which moves backward with the extension of the tail bud and forms a "wavefront." The phase of the clock as the wavefront passes through is fixed at that position, and a spatial periodic pattern is generated. For example, a model in which an external chemical component with a concentration gradient is input to an intracellular reaction and stops oscillating as the tissue grows has been proposed [3, 22]. On the other hand, the possibility that time oscillations are converted into spatial patterns by diffusion interactions, as mentioned above, has also been discussed [4].

In both the reaction-diffusion system and the clock-wavefront model in this section, neither can explain scaling [2, 13], in which the ratio of body-node spacing changes accordingly when tissue size changes. A typical example of scaling is when a fixed boundary is coupled to a simple diffusion, as seen in the tricolor flag problem in Section 4.1.3. Size scaling may be explained by combining the gradient that scales with the size of such a system with the clock and wavefront mechanism. On the other hand, in the case of the diffusion interaction described in this section, scaling can be explained if the wavelength changes in proportion to the size due to the existence of conserved quantities (e.g., [12]).

References

[1] V. Castets *et al.* Experimental evidence of a sustained standing Turing-type nonequilibrium chemical pattern. *Physical Review Letters*, **64**(24): 2953–2956, 1990.

[2] J. Cooke. Control of somite number during morphogenesis of a vertebrate, Xenopus laevis. *Nature*, **254**(5497): 196–199, 1975.

[3] J. Cooke and E. C. Zeeman. A clock and wavefront model for control of the number of repeated structures during animal morphogenesis. *Journal of Theoretical Biology*, **58**(2): 455–476, 1976.

[4] J. Cotterell, A. Robert-Moreno and J. Sharpe. A local, self-organizing reaction-diffusion model can explain somite patterning in embryos. *Cell Systems*, **1**(4): 257–269, 2015.

[5] F. Crick. Diffusion in embryogenesis. *Nature*, **225**: 420–422, 1970.

[6] A. D. Economou *et al.* Periodic stripe formation by a Turing mechanism operating at growth zones in the mammalian palate. *Nature Genetics*, **44**(3): 348–351, 2012.

[7] B. Ermentrout. Stripes or spots? Nonlinear effects in bifurcation of reaction–diffusion equations on the square. *Proceedings of the Royal Society A*, **434**(1891): 413–417, 1991.

[8] K. Fujimoto, S. Ishihara and K. Kaneko. Network evolution of body plans. *PLoS ONE*, **3**(7): e2772, 2008.

[9] A. Gierer and H. Meinhardt. A theory of biological pattern formation. *Kybernetik*, **12**(1): 30–39, 1972.

[10] T. Gregor *et al.* Diffusion and scaling during early embryonic pattern formation. *Proceedings of the National Academy of Sciences*, **102**(51): 18403–18407, 2005.

[11] S. Ishihara, K. Fujimoto and T. Shibata. Cross talking of network motifs in gene regulation that generates temporal pulses and spatial stripes. *Genes to Cells*, **10**(11): 1025–1038, 2005.

[12] S. Ishihara and K. Kaneko. Turing pattern with proportion preservation. *Journal of Theoretical Biology*, **238**(3): 683–693, 2006.

[13] K. Ishimatsu *et al*. Size-reduced embryos reveal a gradient scaling-based mechanism for zebrafish somite formation. *Development*, **145**(11): dev161257, 2018.

[14] C. Janetopoulos *et al*. Chemoattractant-induced phosphatidylinositol 3,4,5-trisphosphate accumulation is spatially amplified and adapts, independent of the actin cytoskeleton. *Proceedings of the National Academy of Sciences of the United States of America*, **101**(24): 8951–8956, 2004.

[15] K. Kaneko and I. Tsuda. *Complex Systems: Chaos and Beyond: A Constructive Approach with Applications in Life Sciences*. Berlin, Springer, 2000.

[16] J. P. Keener. A geometrical theory for spiral waves in excitable media. *SIAM Journal on Applied Mathematics*, **46**(6): 1039–1056, 1986.

[17] A. Kicheva *et al*. Kinetics of morphogen gradient formation. *Science*, **315**(5811): 521–525, 2007.

[18] T. Kohsokabe and K. Kaneko. Boundary-induced pattern formation from temporal oscillation: Spatial map analysis. *EPL (Europhysics Letters)*, **116**(4): 48005, 2016.

[19] S. Kondo and R. Asai. A reaction–diffusion wave on the skin of the marine angelfish Pomacanthus. *Nature*, **376**: 765–768, 1995.

[20] A. D. Lander, Q. Nie and F. Y. M. Wan. Do morphogen gradients arise by diffusion? *Developmental Cell*, **2**(6): 785–796, 2002.

[21] M. J. Lyons and L. G. Harrison. Stripe selection: An intrinsic property of some pattern-forming models with nonlinear dynamics. *Developmental Dynamics*, **195**(3): 201–215, 1992.

[22] H. Meinhardt. *Models of Biological Pattern Formation*. London, Academic Press, 1982.

[23] A. Nakajima *et al*. Rectified directional sensing in long-range cell migration. *Nature Communications*, **5**: 5367, 2014.

[24] Q. Ouyang and H. L. Swinney. Transition from a uniform state to hexagonal and striped Turing patterns. *Nature*, **352**(6336): 610–612, 1991.

[25] I. Palmeirim *et al*. Avian hairy gene expression identifies a molecular clock linked to vertebrate segmentation and somitogenesis. *Cell*, **91**: 639–648, 1997.

[26] R. Sheth *et al*. Hox genes regulate digit patterning by controlling the wavelength of a Turing-type mechanism. *Science*, **338**(6113): 1476–1480, 2012.

[27] H. Shoji, Y. Iwasa and S. Kondo. Stripes, spots, or reversed spots in two-dimensional Turing systems. *Journal of Theoretical Biology*, **224**(3): 339–350, 2003.

[28] M. Skoge *et al*. Cellular memory in eukaryotic chemotaxis. *Proceedings of the National Academy of Sciences*, **111**(40): 14448–14453, 2014.

[29] L. Wolpert. Positional information and the spatial pattern of cellular differentiation. *Journal of Theoretical Biology*, **25**(1): 1–47, 1969.

5

"Fluctuation" of Intracellular Dynamics

5.1 Conditions for the Existence of Cells

A cell lives in a world smaller than the spatiotemporal scale in which we humans operate. To understand the behavior of cells, it is necessary to deal with spatiotemporal scales that are more microscopic than thermodynamic systems, but more macroscopic than those of quantum mechanical systems. Such spatiotemporal scales are called "mesoscopic," which are between the macroscopic and microscopic scales. In thermodynamics, there are universal quantitative relationships among state variables such as temperature T, pressure P, volume V, and entropy S for equilibrium states. It usually involves a macroscopic system consisting of a homogeneous mixture and a very large number of particles (see Avogadro's number 6.02×10^{23}), and its formulation does not depend on the details of the system or on the type of material. On the other hand, a cell is in a nonequilibrium state as a whole, and consists of a huge variety of molecules, each of which is not as numerous. A cell is maintained in a spatiotemporally varying physicochemical environment, in which the fluctuations of the elements around the mean is not negligible. It should be noted, however, that even if the cell as a whole is in a nonequilibrium state, individual reactions in the cell are often close to equilibrium and have a local spatiotemporal scale that can be described thermodynamically. This is based on the fact that a process occurring at a relatively fast time scale can be regarded as an equilibrium state in terms of a process occurring at a slower time scale (see Section 1.12). In the following chapter, we will study numbers and reaction rates of molecules in a cell quantitatively, together with the thermal fluctuations, Brownian motion, in the mesoscopic perspective.

5.2 Thermal Fluctuations of Molecules

Let us consider thermal fluctuations around thermal equilibrium. If the system is in equilibrium with an environment of constant temperature T [K] (heat bath), the energy E [J] of the system will inevitably fluctuate (canonical distribution in statistical physics) [42, 21] with a probability according to the Boltzmann distribution. The Boltzmann distribution is a probability distribution proportional to $e^{-\frac{E}{k_{\mathrm{B}}T}}$ with the Boltzmann constant k_{B}, and if we take its expected value, it is of the order of $k_{\mathrm{B}}T$ [J]. This is a typical energy scale for thermal fluctuations, and the higher the temperature of the system, the larger the fluctuations. For example, at 25°C, the energy of $1\,k_{\mathrm{B}}T$ [J] is

$$4.1 \times 10^{-21}\,[\mathrm{J}] = 4.1\,[\mathrm{pN} \cdot \mathrm{nm}] = 2.5\,[\mathrm{kJ/mol}] = 0.60\,[\mathrm{kcal/mol}] = 24\,[\mathrm{meV}]$$

This is an order of magnitude naturally expressed as the product of *nm*, the typical size of a biomolecule, and *pN*, the typical magnitude of the force applied to a biomolecule. This reflects the fact that thermal fluctuations are not negligible at the scale of biomolecules (in fact, the driving force of a molecular motor is on the order of $1 \sim 10$ [pN], which was confirmed in a single-molecule measurement experiment using optical tweezers [46]).

The *nm* is the scale at which the energy of various noncovalent interactions governing living organisms, such as electrostatic interactions, hydrogen bonds, hydrophobic interactions, and van der Waals forces, is as large as the energy of thermal fluctuations, where biomolecules function [26, 31]. For example, the Bjerrum distance (or Onsager distance) l_{B} is the distance at which the electrostatic interaction and thermal fluctuations are energetically equal,

$$\frac{1}{4\pi\epsilon}\frac{e^2}{l_{\mathrm{B}}} \approx k_{\mathrm{B}}T, \tag{5.1}$$

where ϵ is the permittivity of the medium. This l_{B} is known to be about 1 [nm] in water. The van der Waals forces and hydrophobic interactions in water also have an interaction distance of approximately 1 [nm]. The bond strength (order of energy) of these noncovalent interactions is about $1 \sim 2\,k_{\mathrm{B}}T$ for van der Waals forces, about $4\,k_{\mathrm{B}}T$ for hydrophobic interactions (for methyl groups), about $4 \sim 12\,k_{\mathrm{B}}T$ for hydrogen bonds, about $2 \sim 20\,k_{\mathrm{B}}T$ for electrostatic interactions, whereas it is larger than $100\,k_{\mathrm{B}}T$ for covalent bonds. Therefore, on the scale of biomolecules such as proteins, multiple noncovalent interactions have energy scales that are not much different from the thermal fluctuations of the $k_{\mathrm{B}}T$ order, and can be affected by thermal fluctuations. In proteins, the peptide bonds in the main chain are covalent and stable, but the interactions between the side chains are noncovalent. Therefore, thermal fluctuations cause the structure to fluctuate, which in turn gives a "softness" to the structure of the molecule.

Example 5.1 Overdamped System In order to "feel" the magnitude of thermal fluctuations for biomolecules, let us compare the magnitude of viscous drag and inertia on a biomolecular scale. Consider a spherical protein with a mass of 100 [kDa] ($m = 1.66 \times 10^{-22}$ [kg]) and radius $a = 3$ [nm]. If the viscosity is set to γ, Stokes' law in hydrodynamics yields the following:

$$\gamma = 6\pi a\eta \tag{5.2}$$

and the viscosity of water is $\eta = 10^{-3}$ [kg/m·s] $= 1$ [pN·ns/nm^2]. Thus, $\gamma \approx 60$ [pN·ns/nm]. The correlation time of the velocity of the protein molecule is τ [s], which is determined by the relationship between the inertial force and viscous drag. From the equation of motion,

$$m\frac{dv}{dt} = -\gamma v, \tag{5.3}$$

and thus we get

$$v(t) \propto e^{-\frac{\gamma}{m}t} = e^{-\frac{t}{\tau}} \tag{5.4}$$

and $\tau = m/\gamma$ gives a rough estimate for the velocity correlation time. Within this correlation time, the protein molecule has a velocity in a certain direction due to inertial forces, but thereafter has a random velocity in an unrelated direction owing to thermal fluctuations. As a result, we can see that the correlation time for this protein molecule is quite short as

$$\tau = \frac{1.66 \times 10^{-22} \text{ [kg]}}{6.0 \times 10^{-11} \text{ [kg/s]}} \approx 3 \text{ [ps]}. \tag{5.5}$$

At 25°C, the average velocity of this protein molecule is based on the law of equipartition of energy:

$$\sqrt{\langle v^2 \rangle} = \sqrt{\frac{3k_\mathrm{B}T}{m}} = \sqrt{\frac{3 \times 1.38 \times 10^{-23} \times 298}{1.66 \times 10^{-22}}} \approx 8.6 \text{ [m/s]}, \tag{5.6}$$

and the distance that this protein molecule travels in a constant direction during the correlation time of 3 [ps] is only a small distance,

$$\tau \times \sqrt{\langle v^2 \rangle} \approx 0.03 \text{ [nm]}, \tag{5.7}$$

which is only 1/100 of the molecular radius. That is, for biomolecule-scale motion, the effect of inertia is negligible compared to that of viscous drag [14, 32]. Such a system is called an overdamped system.

Example 5.2   Reynolds Number Let us examine the hydrodynamic environment in which biomolecules exist. The Reynolds number is a dimensionless number that characterizes the behavior of a viscous fluid in hydrodynamics. Using the typical flow velocity of the fluid v[m/s], the size of the object in the fluid L[m], the viscosity of the fluid η[N·s/m^2], and the density of the fluid ρ[kg/m^3], the Reynolds number Re is defined as the ratio of the inertial force to the viscous force:

$$Re = \frac{inertial\ force}{viscous\ drag} = \frac{\rho L^3 v^2/L}{\eta v L} = \frac{\rho v L}{\eta}. \tag{5.8}$$

The smaller the size of the object, the smaller the Reynolds number, even in the same fluid. For example, in the case of fish such as the crucian carp swimming in water, $v \approx 1\,[\text{m/s}]$, $L \approx 0.1\,[\text{m}]$, $\rho \approx 10^3\,[\text{kg/m}^3]$, and $\eta \approx 10^{-3}\,[\text{kg/m·s}]$, so $Re \approx 10^5$ and inertia dominates the system (underdamped system), whereas in the case of bacteria, $v \approx 10\,[\mu\text{m/s}]$, $L \approx 1\,[\mu\text{m}]$, ρ, and η are the same as previously, and $Re \approx 10^{-5}$. In other words, biomolecules and microorganisms are in a world with low Reynolds numbers (see Section 5.10).[1]

Example 5.3   Chemical Reactions Let us turn our attention to chemical reactions between molecules. This is realized by the existence of thermal fluctuations. In reaction kinetics, the van't Hoff–Arrhenius equation for the reaction rate constant is known as

$$k \propto e^{-\frac{E_a}{k_B T}}. \tag{5.9}$$

Since this equation can be rewritten as

$$e^{-\frac{E_a}{k_B T}} \propto \int_{E_a}^{\infty} e^{-\frac{E}{k_B T}}\,dE, \tag{5.10}$$

it can be interpreted that the reaction rate is the sum of the probability of crossing the potential barrier of free energy (E_a: activation energy) by a large thermal fluctuation, which rarely occurs. Because enzymatic catalysis is common in living organisms, in which E_a is not too large compared to thermal fluctuations, the reaction can proceed by the repeated formation or destruction of the enzyme-substrate complex due to thermal fluctuations.

Furthermore, let us consider molecular motors in which chemical reactions and mechanical processes are coupled (chemomechanical coupling). There are two types of molecular motors, linear motors and rotary motors. Here we take linear motors as examples. Linear motors contain proteins, such as myosin, kinesin, and dynein. They use the free energy of ATP hydrolysis, which is the "energy currency" of the organism, to perform directional movement on actin, microtubules, and other "line"-shaped proteins, respectively. They perform functions such as muscle contraction and active transport of materials. A series of single-molecule measurements have shown that Brownian motion, due to thermal fluctuations, is essentially incorporated in the mechanism of "walking" [46]. For example, in the case of a bipedal "walking" motor such as myosin V, there is a single-leg bonded state in the

---

[1] The Reynolds number is a measure that gives a criterion for whether a fluid is laminar or turbulent. In the case of a low Reynolds number, the viscosity is dominant, and the flow becomes sticky, which makes it difficult to generate turbulence and generates laminar flow. In the case of a high Reynolds number, inertia dominates, and the flow is smooth and tends to generate turbulence.

middle of the reaction, and the dissociated leg "searches" for the next bonded site in Brownian motion. Such a mechanism is discussed to enable the molecular motor to operate with a high energy efficiency.

Example 5.4 Protein Folding To examine the "softness" of a protein molecule more closely, we check its thermodynamic stability. As exemplified by the rock and key model of enzymatic reactions, protein molecules perform specific functions by folding into specific conformations. However, the free energy difference ΔG between the folded and unfolded states in water is only about 5~12 [kcal/mol], and it is marginally stable to the extent that it is denatured by a small temperature or pH change. A funnel-type energy landscape has been theoretically proposed, which facilitates not only the thermodynamic stability but also the folding kinetics [28]. This means that in the folding process thermal fluctuations play an important role.

In addition, protein molecules do not always have a unique three-dimensional structure (natural state) with the lowest free energy, but can also take multiple structures at the same level of free energy between the large entropy difference ΔS and the large enthalpy difference ΔH at the free energy difference $\Delta G = \Delta H - T\Delta S$. The former is derived from the difference in the number of possible structural states, and the entropy is larger in the unfolded state. The latter is derived from the energy of many noncovalent bonds, and the enthalpy is smaller in the unfolded state. The presence of such proteins has been reported in many cases. For example, actin and Ras molecules transition between multiple metastable structures in equilibrium with a time scale of more than several tens of milliseconds by thermal fluctuations (dynamical polymorphism), and this is shown to be important for the expression of their functions by single-molecule measurements of molecular structural change [3]. In addition, it has recently been clear that more than one-third of all eukaryotic proteins are "intrinsically disordered proteins" (IDPs) that do not form a specific conformation as a whole or in part [9]. These proteins exhibit interesting thermodynamic processes, such as simultaneous molecular binding and protein folding processes, suggesting that thermal fluctuations may play an important role.

Remark: As can be seen from the various examples just presented, the existence of thermal fluctuations is inevitable at the scale of biomolecules, and it is necessary to incorporate them. For biomolecules such as proteins, the ability to function normally under physiological temperature and pressure is a prerequisite. It is thought that evolutionary selection has been made to satisfy two conflicting properties: stability at physiological temperature and pressure and the ability to change state in response to environmental changes. As a result, the stability of the protein remains weak, and thermal fluctuations are not negligible [37]. At the same time, the other side of this is that at higher temperatures, thermal fluctuations dominate, and biomolecules do not maintain their shape and denature; at lower temperatures, binding forces dominate, and biomolecules are not able to have reversible interactions. The existence of molecular chaperones that assist in the formation of three-dimensional structures of various proteins may be an evolutionary reflection of this.

The physics that studies the properties of soft materials such as polymers, liquid crystals, colloids, and gels is called "soft matter physics." Nucleic acids (DNA and RNA), proteins, polysaccharides, and other biomacromolecules, which are the basic elements of living organisms, as well as lipid membranes, can also be classified as soft matter.

Column III: van't Hoff's Q10 Law

In enzymatic reactions such as metabolism, it is empirically known that the rate of reaction increases two to three times with an increase of environmental temperature by 10°C, which is van't Hoff's Q10 law. The ratio of the reaction rate constant at 25°C to the rate constant at a temperature 10 degrees higher is $\frac{k(T+10)}{k(T)} = e^{\frac{E_a}{k_B}\left(\frac{1}{T} - \frac{1}{T+10}\right)} = e^{7.90\times10^{18}E_a}$. Assuming that this equals 2, $E_a = 8.77 \times 10^{-20}$ [J] $= 21\, k_B T$ [J] $= 13$ [kcal/mol]. This value is approximately $20\, k_B T$, the free energy of ATP hydrolysis under physiological conditions, which is the energy source for the enzymatic reaction. If the activation energy of the metabolic reaction were much higher than the free energy of ATP hydrolysis, the reaction would rarely occur in the first place, and if it were much lower, the reaction would proceed with each thermal fluctuation and would be uncontrollable. Thus, it is highly likely that the similarity in magnitude between the activation energy and the free energy of ATP hydrolysis reflects the process of biological evolution.

5.3 The Size of the Cells, the Number of Molecules inside the Cells

Living cells exchange substances such as nutrients and waste, and exchange thermal energy with the environment. The ratio of their uptake into and out of the cell must be kept in a certain balance. Otherwise, the cell will not be able to maintain homeostasis and will eventually die. Therefore, there is a fundamental limitation in cell size.

For example, let us consider heat transfer in a cell. Heat production in a cell is approximately proportional to cell volume. The transfer of heat between the cell and the environment is carried out via the cell surface, and the total amount is proportional to the cell surface area. Therefore, as the cell size increases, the ratio of the cell surface area to cell volume decreases, and the cell is unable to exhaust sufficient heat to the environment, increasing the risk of heat death. The same is true for the transfer of nutrients and waste products. The larger the cell size, the less the cell is able to take up the nutrients it needs from the environment, and the less it is able to release waste products into the environment, which may accumulate in the cell and increase the risk of cell death. Thus, it is necessary for a cell to maintain a ratio of surface area to volume that is larger than a certain amount, and there is a functional significance in cell division from this point of view.[2]

The size (diameter) of most eukaryotic cells is on the order of 10 [μm] in size, and is approximately 1000 times larger than that of prokaryotic cells. Eukaryotic cells have developed a cell nucleus that contains a large genome and cytoskeleton to maintain the shape and strengthen the structure of a large cell. To solve the need for sufficient surface area to match with the volume increase, the cell surface is also created inside the cell by endocytosis (in which cells take up foreign substances into the cell). These intracellular membrane systems were not found in prokaryotes.

What is the typical number of biomolecules in a cell? In general, the number of components (DNA, RNA, protein, etc.) inside a cell is small compared to the macroscopic number ($1M = 6.02 \times 10^{23} \, [\ell^{-1}]$). Here, we estimated the number of molecules from the concentration of molecules in the cell. If the volume of a cell is approximated by a cube of 5 [μm] per side, it is $5^3 \, [\mu m^3] \approx 10^{-16} \, [m^3]$, and the mass of water in this volume is 10^{-13} [kg], which contains 10^{12} water molecules. If the concentration of a molecule is as low as 1 [nM] ($10^{-9} \, [\text{mol}/\ell]$), then there are only approximately 60 molecules in this volume. By using the same estimate for various intracellular concentrations of biomolecules, the number of molecules in the cell is shown in the following table.

concentration	1 nM	10 nM	100 nM	1 μM	10 μM	100 μM
number of molecules	60	600	6,000	60,000	600,000	6,000,000

Exercise 5.3.1 In the preceding example, we assumed that a cell is the size of a single yeast cell. Estimate the number of molecules in the cell, for a bacterium (approximately $1 \, [\mu m^3] = 10^{-18} \, [m^3]$) or a mammalian cell (approximately $10^3 \, [\mu m^3] = 10^{-15} \, [m^3]$). In addition, investigate how the order of the number of molecules changes along the central dogma flow of DNA, $\rightarrow$ mRNA, $\rightarrow$ protein.

The noncovalent nature of biomolecules, as discussed in the previous section, is an essential prerequisite not only for the contribution of biomolecules to structure

---

[2] Genome analysis has shown that the percentage of membrane proteins in all genes is constant across species, at almost 20% [35]. This suggests that there is a universal mechanism for the importance of cell membranes in living organisms.

formation, but also for interactions (i.e., bonding and dissociation, and chemical reactions) between biomolecules. In fact, many of the reactions that are important for biological functions are estimated in the standard state (25°C, 1 atm, 1 mol concentration), and the binding free energy is known: $\Delta G = -10\,[\text{kcal/mol}]$ [2, 45]. In the equilibrium state, according to the thermodynamics, it follows that

$$\Delta G = k_{\text{B}} T \ln\left(\frac{K_d}{1M}\right). \tag{5.11}$$

Thus, the energy conversion in Section 5.2 shows that the dissociation constant of the intermolecular interaction corresponding to the preceding binding free energy is approximately $K_d \approx 10^{-8}\,[\text{M}]$, where K_d is the concentration at which half of the molecules are bound. In biochemistry, the affinity of the binding reaction is generally interpreted as low if K_d is the order of mM, medium for µM, and high for nM or less.[3] Then, $K_d \approx 10^{-8}\,[\text{M}]$ implies a relatively strong affinity.

Now, let us also estimate the number of molecules in the cell, considering the two-body bond reaction $A + B \leftrightarrow AB$ of the two types of molecules, and let the dissociation constant be $10^{-8}\,[\text{M}]$. Then, we can also assume a state in which the concentration is $[A] \approx [B] \approx [AB]$ as an equilibrium state, and the concentration is $10^{-8}\,[\text{M}]$. If we approximate the size of a cell using a cube of $5\,[\text{µm}]$ per side, the number of molecules in the cell at this concentration is approximately 10^3, that is, the number of molecules is not large [39]. Thus, the affinity between molecules and the intracellular molecular concentration are linked through thermodynamics, and it can be seen that the thermal fluctuations of the molecules are related to the number fluctuations. In fact, the intercellular concentrations of many signaling molecules and transcription factors are known to be on the order of the nM, that is, the number of molecules ranging from a few to 100 [23, 41].

In the rate equations of chemical reactions, the average value is expressed as a continuous concentration quantity because the number of molecules is large. However, in cells, discreteness in the number of molecules for reactions is not negligible so that stochastic fluctuations cannot be ignored.

5.4 Continuous Equation and Stochastic Fluctuation

Let us look at the differences between the continuous and stochastic descriptions of chemical reactions. For example, consider the following chemical reaction:

---

[3] In order to increase the affinity of intermolecular interactions by weak noncovalent bonds, it is effective to increase the contact area between the molecules, which increases the binding energy by increasing the number of bonds. On the other hand, a trade-off can occur as the affinity of the interaction increases, and the specificity decreases. Therefore, the rock and key structure of enzymes and substrates is an example of how complementary surface structure between molecules, can achieve specificity.

$A \to X \to B$. For simplicity, let us assume that the reaction rate constants are $1[\mathrm{s}^{-1}]$ and $k[\mathrm{s}^{-1}]$, respectively, the volume of the chemical reaction tank is $V = 1[\ell]$, and the concentration of A is constant at $a[\mathrm{M}]$. Let us consider the change in the concentration of X as $x[\mathrm{M}]$ with the initial condition $x(0) = 0$. Because molecule X is generated at the rate $a[\mathrm{M}]$ per unit time and decomposed at the rate $kx[\mathrm{M}]$ stochastically, the concentration of X on the average follows the rate equation

$$\frac{\mathrm{d}x(t)}{\mathrm{d}t} = a - kx(t), \tag{5.12}$$

which is a continuous differential equation: The solution is $x(t) = a(1 - e^{-kt})/k$, and the stationary solution after sufficient time is $\bar{x} = a/k$.

Let us now examine the extent to which this continuous expression can maintain the accuracy of the description when the number of molecules is small. Specifically, we compare the results of the continuous expression with the results of computer simulations for the case where the average number of molecules X, $\bar{X} = \bar{x} \times V \times N_A$ (N_A is Avogadro's number) is 10000 at a maximum and 30 at a minimum, by changing $\bar{x}$. To simulate stochastic events, the method of simulating random numbers in a computer and making it react accordingly is generally called the Monte Carlo method (see also Section 8.3). To perform stochastic simulations, it is necessary to select particles randomly from a pool of particles and have them react according to the rate. Here, the Gillespie algorithm (see Section 5.4.1) was used to speed up the calculation. Figure 5.1 shows the results of stochastic simulations as a semi-log plot of the time series. Here, k, whose reciprocal is the relaxation time, is kept constant, and different steady-state solutions are realized by varying the value of a.

In the stochastic simulations, fluctuations started to become noticeable at around $\bar{X} = 1000$. For $\bar{X} = 30$, the fluctuations are so large that the width of the fluctuations may be more than half the steady-state value. In general, if the average of the total number of molecules is N, then the total number of molecules measured at each time is accompanied by variation of approximately $\pm\sqrt{N}$ as the sum of independent stochastic events (chemical reactions) ($\sqrt{N}$ law). The smaller N is, the larger the relative contribution of the error ($1/\sqrt{N}$). When the number of molecules is small, the discrete nature of the molecular number is not negligible, and a continuous description may be insufficient. Therefore, when considering chemical reactions in a cell, it is often important to describe them by incorporating stochasticity. We address this point in more detail in Chapter 6.

In addition, the intracellular environment is not an "ideal" reaction environment. In an artificial chemical reaction tank, the volume of the environment is sufficiently large, sufficiently stirred, in which certain chemical components exist uniformly and uncrowded. The environment of an intracellular chemical reaction, on the other hand, is quite different. The intracellular space is small and unstirred, with

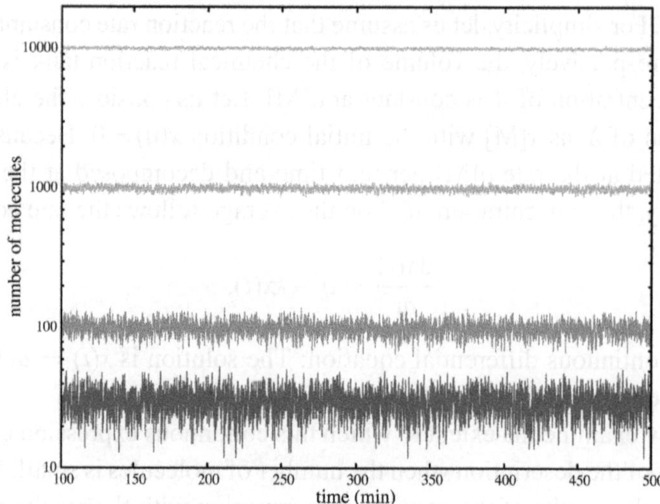

Figure 5.1 Temporal change of number of molecules in a stochastic simulation. From top to bottom, temporal changes corresponding to $\bar{X} = 10,000$, $1,000$, 100, and 30, respectively. A semilogarithmic plot is shown. $k = 0.1\,[\mathrm{s}^{-1}]$, and the parameter a is set to the value consistent with each $\bar{X}$.

a variety of intracellular organelles and a variety of molecules that do not contribute to specific reactions. Protein concentrations can exceed 100 [mg/ml] (molecular crowding) [2, 31]. In addition, a number of microstructures is formed by liquid–liquid phase separation in the cell. The importance of the intracellular crowding environment has been clearer recently [7]. Now, the cell can be considered to be a microsystem that operates robustly even under such fluctuating, crowded, and inhomogeneous conditions. Here it is worth considering the possibility that stochastic fluctuations of chemical reactions play an important role in cellular functions.

5.4.1 The Gillespie Algorithm

The Gillespie algorithm [12] is widely used as a method to speed up Monte Carlo simulations for the stochastic chemical reactions. The implementation is also simple and general, where discreteness in the number of molecules can be treated numerically against time.[4] As a simple example, we introduce the Gillespie algorithm using a chemical reaction system in which the chemical components C_1 and C_2 are reversible:

$$C_1 \underset{k_2}{\overset{k_1}{\rightleftarrows}} C_2.$$

---

[4] Even with this method, as the number of chemical reactions increases, the calculation requires a relatively large amount of computation time. Later, approximate computational methods such as the τ-leap method [12, 13] were proposed to further improve the speed.

Let the number of molecules and reaction rate constants of each component be c_i, and k_i ($i = 1, 2$), and let the volume of the system be V. Then, let us consider a probability density function $P(\tau, i)d\tau$, which represents the probability that for given time t, the next reaction to occur is reaction i and that it occurs in a short interval of time $(t + \tau, t + \tau + d\tau)$. Let us define

$$a_i \equiv k_i \frac{c_i}{V}, \tag{5.13}$$

$$a^t \equiv \sum_i k_i \frac{c_i}{V}. \tag{5.14}$$

Then

$$P(\tau, i)d\tau = P_0(\tau) \times a_i d\tau \tag{5.15}$$

holds, where $P_0(\tau)$ denotes the probability that none of the reactions will occur during time $(t, t + \tau)$. $P_0(\tau)$ satisfies the following equation:

$$P_0(\tau + d\tau) = P_0(\tau)(1 - a^t d\tau). \tag{5.16}$$

If time τ is expressed as a repetition of $d\tau$,

$$P_0(\tau) = \lim_{n \to \infty} \left(1 - \frac{a^t \tau}{n}\right)^n = e^{-a^t \tau} \tag{5.17}$$

is derived from the definition of the exponential function $e^x = \lim_{n \to \infty} \left(1 + \frac{x}{n}\right)^n$. Then,

$$P(\tau, i)d\tau = e^{-a^t \tau} \times a_i d\tau = \left(\frac{a_i}{a^t}\right) a^t e^{-a^t \tau} d\tau \tag{5.18}$$

holds. By denoting $P(i) = a_i/a^t$ and $P(\tau) = a^t e^{-a^t \tau}$,

$$P(\tau, i) = P(i)P(\tau) \tag{5.19}$$

is satisfied as the product of statistically independent probabilities. From the preceding, the following procedure can be introduced for a Monte Carlo simulation:

1. Determine the reaction rate constant $k_i(i = 1, 2)$, the volume of the system V, and the number of molecules $c_i(i = 1, 2)$ at time $t = 0$ (initial conditions).
2. Prepare two uniform random numbers r_1 and r_2 with $[0, 1]$ as the domain.
3. Using the transformation of the probability distribution, create a random number according to the exponential distribution as $\tau = - \ln(r_1)/a^t$ and determine the time of occurrence of the reaction $t + \tau$.
4. Suppose that reaction 1 occurs if $0 < r_2 \leq a_1/a^t$ and reaction 2 occurs if $a_1/a^t < r_2 < 1$.

5. After the reaction, time t is advanced to $t + \tau$ and the number of molecules is updated to $c_1 \to c_1 - 1$ and $c_2 \to c_2 + 1$ if reaction 1 occurs, and to $c_1 \to c_1 + 1$ and $c_2 \to c_2 - 1$ if reaction 2 occurs.

By repeating steps 2 to 5, we can achieve simulations that consider the stochasticity of chemical reactions. The case with more than two chemical components can be treated in the same way by counting up all the reactions, and the case where the reaction terms are more than second order can be treated in the same way by making the reaction probabilities based on the number of molecules in each component.

5.5 Extrinsic and Intrinsic Noise

Since around the year 2000, advances in measurement technology have revealed large fluctuations in living cells based on experimental evidence. In this section, we will examine fluctuations in gene expression and signaling systems. First, let us introduce some representative examples of fluctuations in gene expression systems. Elowitz et al. introduced genes encoding blue-green and yellow fluorescent proteins, called cyan fluorescent proteins (CFP) and yellow fluorescent proteins (YFP), into the *E. coli* genome by genetic engineering. Each gene is incorporated at opposite positions from the replication origin, and the lac (lactose) promoter, which is a common transcription start site, is located upstream of the gene.

When these *E. coli* are cultured, each cell showed various colors, even though the genotype and promoter were the same, and both genes received the same regulatory signals. The colors of these cells reflect the differences in the expression levels of the reporter proteins CFP and YFP within the cell, which is an example of the visualization of the stochasticity of gene expression at the single-cell level [11, 40].

Now, let us theoretically formulate fluctuations in gene expression. Here, "fluctuation" is quantitatively expressed by using the coefficient of variation, i.e., (standard deviation)/(mean). In the case of gene expression noise, two different types of noise sources exist in principle: direct noise due to the discreteness and stochasticity of the reaction process itself, such as transcription and translation, and indirect noise due to heterogeneity of cell states, such as cell size and cell cycle stages, or other concentrations of metabolites, ribosomes, and polymerases, which can affect each cell globally. Let us call them intrinsic and extrinsic noise, respectively. Then, the square sum of intrinsic and extrinsic noise will be equal to the square of the total noise. Note that extrinsic noise is applied commonly to each expression of genes, whereas intrinsic noise to each expression of genes is uncorrelated.

By using the fluorescence intensities of the fluorescent protein CFP (c) and YFP (y) described earlier, the noise intensities of each can be determined experimentally. After confirming that the distributions of c and y are identical, a correlation plot of the fluorescence intensities of CFP and YFP was obtained, and the components of the positive correlation corresponded to extrinsic noise, while the components orthogonal to the positive correlation components corresponded to intrinsic noise (Fig. 5.2). That is, it can be formulated as

$$\begin{cases} \eta_{\text{int}}^2 = \dfrac{\langle (c-y)^2 \rangle}{2 \langle c \rangle \langle y \rangle} \\[2mm] \eta_{\text{ext}}^2 = \dfrac{\langle cy \rangle - \langle c \rangle \langle y \rangle}{\langle c \rangle \langle y \rangle} \end{cases} \tag{5.20}$$

(see Section 5.5.1). In these two equations, the mean over the population of the cells is set to $\langle x \rangle$ (where x is either c or y). If the intrinsic noise in the gene expression process is assumed to obey a binomial or Poisson distribution (see also Section 6.3), then $\sigma_x^2 \propto \langle x \rangle$ is satisfied, and the following relationship holds:

$$\eta_{\text{int}}^2 = \frac{\sigma_x^2}{\langle x \rangle^2} \propto \frac{1}{\langle x \rangle}. \tag{5.21}$$

This indicates an inverse proportionality between the square of the intrinsic noise and the average expression level.

In recent years, noise in gene expression systems has been comprehensively measured in proteomic studies in *E. coli* and yeast, and it has become clear that the relationship between the mean expression level of each protein and the square of the noise is inversely proportional in regions where the mean expression level is low, as described earlier, while the noise strength is almost constant in regions where the mean expression level is high [41]. It can be concluded that the former region reflects the property of intrinsic noise, whereas the latter region mainly reflects the property of extrinsic noise. The intrinsic and extrinsic noise can be distinguished by characteristic time scales. For example, in *E. coli*, the correlation time of extrinsic noise is equal to or longer than the cell cycle in some cases (approximately 45 minutes), while the correlation time of intrinsic noise is as short as 10 minutes or less [33]. The longer correlation of extrinsic noise can be related to cellular memory and heterogeneity.

5.5.1 *The Formulation of the Intrinsic Noise and the Extrinsic Noise*

Let us theoretically formulate intrinsic and extrinsic noises [11]. If the expression level of a particular gene in the kth ($k = 1, \ldots, N$) cell is set to x_k, the gene expression noise can be expressed as $\eta = \sigma_x / \langle x \rangle$. If the variables related to intrinsic and extrinsic noise are grouped together in vector notation as I, E, then the mth-order moment of gene expression level is

$$\frac{1}{N}\sum_{k=1}^{N} x_k^m = \int\int \mathrm{d}E\mathrm{d}I x^m(E,I)\rho(EI)$$

$$= \int \mathrm{d}E\rho(E)\int \mathrm{d}I x^m(E,I)\rho(I|E)$$

$$= \int \mathrm{d}E\rho(E)\langle x(E)^m\rangle$$

$$= \overline{\langle x^m\rangle}, \tag{5.22}$$

where $\rho(EI)$ is the probability density function for intrinsic and extrinsic noise variables, and $x(E,I)$ is the measured gene expression for a specific value of E or I, then the preceding equation is obtained using the conditional probability equation. Note that the average value obtained by integrating I with fixed E is denoted as $\langle x(E)^m\rangle$, and the average value obtained by integrating E is denoted as $\overline{x^m}$. Using this notation, the square of the total noise of the gene expression η_{tot} is

$$\eta_{\mathrm{tot}}^2 = \frac{\overline{\langle x^2\rangle} - (\overline{\langle x\rangle})^2}{(\overline{\langle x\rangle})^2}$$

$$= \frac{\overline{\langle x^2\rangle - \langle x\rangle^2}}{(\overline{\langle x\rangle})^2} + \frac{\overline{\langle x\rangle^2} - (\overline{\langle x\rangle})^2}{(\overline{\langle x\rangle})^2} = \eta_{\mathrm{int}}^2 + \eta_{\mathrm{ext}}^2. \tag{5.23}$$

It can be expressed as the sum of squares of the intrinsic noise η_{int} and extrinsic noise η_{ext}. Now, suppose that there are two copy gene sequences in the kth cell, each of which produces proteins with expression levels of $x_k^{(1)}$ and $x_k^{(2)}$. Then, because they share extrinsic noise, we get

$$\frac{1}{N}\sum_{k=1}^{N} x_k^{(1)}x_k^{(2)} = \int\int\int \mathrm{d}E\mathrm{d}I_1\mathrm{d}I_2 x(E,I_1)x(E,I_2)\rho(EI_1I_2)$$

$$= \int \mathrm{d}E\rho(E)\left(\int \mathrm{d}I x(E,I)\rho(I|E)\right)^2$$

$$= \int \mathrm{d}E\rho(E)\langle x(E)\rangle^2$$

$$= \overline{\langle x\rangle^2}. \tag{5.24}$$

Then, the following relation is derived as

$$\frac{1}{N}\sum_{k=1}^{N}\left(x_k^{(1)} - x_k^{(2)}\right)^2 = \frac{1}{N}\sum_{k=1}^{N}\left(x_k^{(1)2} - 2x_k^{(1)}x_k^{(2)} + x_k^{(2)2}\right)$$

$$= \frac{2}{N} \left(\sum_{k=1}^{N} x_k^2 - \sum_{k=1}^{N} x_k^{(1)} x_k^{(2)} \right)$$

$$= 2 \left(\overline{\langle x^2 \rangle} - \overline{\langle x \rangle^2} \right)$$

$$= 2 \left(\overline{\langle x^2 \rangle - \langle x \rangle^2} \right)$$

$$= 2 \left(\langle x \rangle \right)^2 \cdot \frac{\left(\overline{\langle x^2 \rangle - \langle x \rangle^2} \right)}{\left(\langle x \rangle \right)^2}$$

$$= 2 \left(\langle x \rangle \right)^2 \cdot \eta_{\text{int}}^2$$

and the intrinsic noise is formulated as

$$\eta_{\text{int}}^2 = \frac{\dfrac{1}{N} \sum_{k=1}^{N} \left(x_k^{(1)} - x_k^{(2)} \right)^2}{2 \left(\langle x \rangle \right)^2}, \tag{5.25}$$

whereas, the extrinsic noise is formulated as

$$\eta_{\text{ext}}^2 = \frac{\overline{\langle x \rangle^2} - \left(\overline{\langle x \rangle} \right)^2}{\left(\overline{\langle x \rangle} \right)^2} = \frac{\dfrac{1}{N} \sum_{k=1}^{N} x_k^{(1)} x_k^{(2)} - \left(\overline{\langle x \rangle} \right)^2}{\left(\overline{\langle x \rangle} \right)^2}. \tag{5.26}$$

If the fluorescence intensities of the fluorescent protein CFP (c) and YFP (y) described earlier are set to $x_k^{(1)}$ and $x_k^{(2)}$, the respective noise intensities can be determined experimentally.

5.6 Noise in Signal Transduction System

The theoretical formulation of these noises was carried out by Paulsson in the gene expression system and by Shibata and Fujimoto in the signal transduction system [30, 38]. Quantitative relationships between intrinsic noise and the law of large numbers, and between extrinsic noise and propagation by the time constant of upstream fluctuations in the response have been revealed. The sum of both factors determines the total noise. At the same time, from the power spectrum analysis, it can be confirmed that the contribution of the extrinsic noise is large at the slow time scale and that of the intrinsic noise at the fast time scale. These are examples of linear response relationships, which will be described later, and show that information on the reaction process behind the system can be obtained from the characteristics of the system fluctuations and input/output variations.

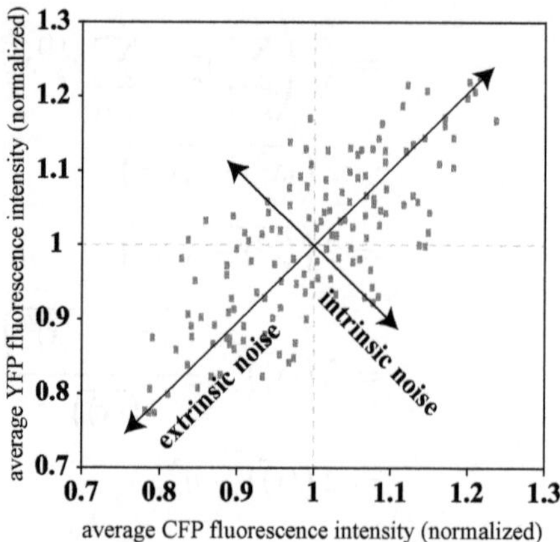

Figure 5.2 Schematic diagram of intrinsic and extrinsic noise. The individual points correspond to data from a single cell of *E. coli*, indicating that there is phenotypic variation even for the same genotype. The variability is expressed by decomposing it into two diagonal components (positive and negative correlation directions).

Next, we present an actual measurement of fluctuations in the signal transduction system. Fluctuations in the input–output relationship are also prominent in the signaling system. Let us consider the chemotaxis of the cellular slime mold, *Dictyostelium discoideum*, as an example. *Dictyostelium* cells are known to detect the concentration gradient of cAMP in the environment and exhibit motility in the direction of high concentration (chemotaxis) (see also Section 8.2). Chemotaxis signaling is mediated by the G protein-coupled receptor (GPCR)-type cAMP receptors on the cell membrane and is transmitted to the cell. Although *Dictyostelium* cells are capable of detecting spatial differences in cAMP concentrations at both ends of the cell, even in the absence of cell motility (called "spatial sensing") [29], the number of cAMP-bound receptors on the cell surface fluctuates around the temporal and spatial averages because the binding and dissociation reactions between cAMP and its receptors occur stochastically [43, 44].

Now, assuming that there is a constant cAMP concentration gradient in the external environment, we determine the anterior and posterior sides of the cell along the gradient and compare the number of cAMP-bound receptors in a region with equal area on each side. If the average number of binds per location encodes information on the gradient of external cAMP concentration, the cell performs spatial sensing by reading the difference of the number of binds between each side. Therefore, the fluctuation in the number of binds is considered to be due to the noise

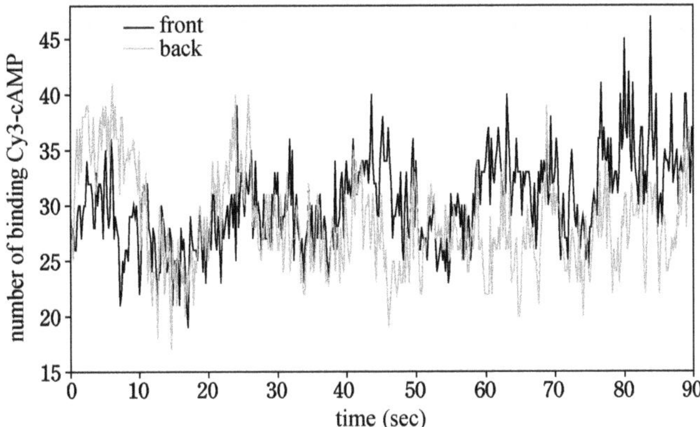

Figure 5.3 Single-molecule measurements of cAMP binding to cAMP-binding receptors in a *Dictyostelium* cell. In the presence of a gradient of cAMP concentration in the environment, the number of cAMPs bound to equal areas in the anterior and posterior portions of the cell base varied with time (by courtesy of Dr. Masahiro Ueda (Osaka University, RIKEN BDR)).

of the concentration. The sensitivity of *Dictyostelium* cells to detect concentration gradients is high, and it is known that chemotactic response is achieved even when the gradient is only approximately 2% [43, 44]. This means that under such conditions, as shown in Fig. 5.3, the difference between the mean number of binds between each side is equal to or smaller than the noise. In addition, in the signaling system downstream of the receptor, there are many chemical reactions with fluctuations, which are accompanied by various propagation noises.

As described earlier, cells are subject to fluctuations in the networks of gene expression and signal transduction systems. Therefore, a cell should have mechanisms that enable it to function robustly even under noisy conditions. In addition, the complexity of the internal degrees of freedom in the protein reaction dynamics should be considered when considering the source of the fluctuations. In fact, there are various examples that span multiple spatiotemporal scales, such as the existence of slow relaxation processes in protein molecules, large difference in the rate constants of individual molecules within the same protein species (static disorder), fluctuation in the rate constants of individual molecules and a memory of the reaction (dynamic disorder), and the partial "breaking" of ergodicity (the coincidence between the time and population averages of the system) [22, 25]. In addition to the dynamics at the reaction network level, the relationship between protein fluctuations and reaction dynamics is also important when discussing cellular fluctuations.

5.7   The Law of Large Numbers and the Central Limit Theorem

So far, we have confirmed the universality of fluctuations in intracellular dynamics. In this section, we introduce the basic theoretical framework to deal with "fluctuations." The law of large numbers and the central limit theorem, as well as random walks and diffusion processes, are examined, and a bridge to the theory of Brownian motion is provided. Suppose independent stochastic variables exist as $X_i (i = 1, 2, \ldots, N)$ following the same probability distribution with mean μ and variance σ^2. Then, the sample mean $\overline{X} = \frac{1}{N} \sum_{i=1}^{N} X_i$ is also a stochastic variable, and its expected value and variance are denoted as

$$E(\overline{X}) = E\left(\frac{1}{N}\sum_{i=1}^{N} X_i\right) = \frac{1}{N} \cdot N \cdot E\left(X_i\right) = \mu, \tag{5.27}$$

$$V(\overline{X}) = V\left(\frac{1}{N}\sum_{i=1}^{N} X_i\right) = \frac{1}{N^2} \cdot N \cdot V\left(X_i\right) = \frac{\sigma^2}{N}, \tag{5.28}$$

respectively, from the additivity. That is, the expected value of the random variable in the sample mean is consistent with that of the original random variable, and the standard deviation is $1/\sqrt{N}$ compared to that of the original random variable. Furthermore, if N is sufficiently large, it is shown that $\overline{X}$ stochastically converges to μ. This is called the "law of large numbers". This law implies that by increasing the number of independent trials, the theoretical probability can be determined from the probability of an event's occurrence (empirical probability).

Example 5.5   The Law of Large Numbers by Coin Tosses (Bernoulli Trial)
Given a coin with probability p as head (1) and probability $q(=1-p)$ as tails (-1), a single coin toss can be represented by the following random variables:

$$X_i = \begin{cases} 1 \ (probability \ p) \\ -1 \ (probability \ q = 1 - p). \end{cases} \tag{5.29}$$

Now, suppose that the coin tosses are tried N times independently and the number of times the head appears is k. The sample mean of that random variable is

$$\overline{X} = \frac{1}{N}(k \cdot 1 + (N - k) \cdot (-1)) = \frac{k}{N} - \left(1 - \frac{k}{N}\right). \tag{5.30}$$

The law of large numbers states that if N is sufficiently large, $k = Np$ and X converge to $\overline{X} = p - (1 - p) = p - q$. If we actually perform a computer simulation of a coin toss with N that is not so large, there remains a large fluctuation around $\bar{x}$. N must be at least several thousand times to reach sufficient accuracy (Fig. 5.4).

Next, what will be the shape of the distribution that the sample mean $\overline{X}$ follows? The "central limit theorem" helps to explain this.

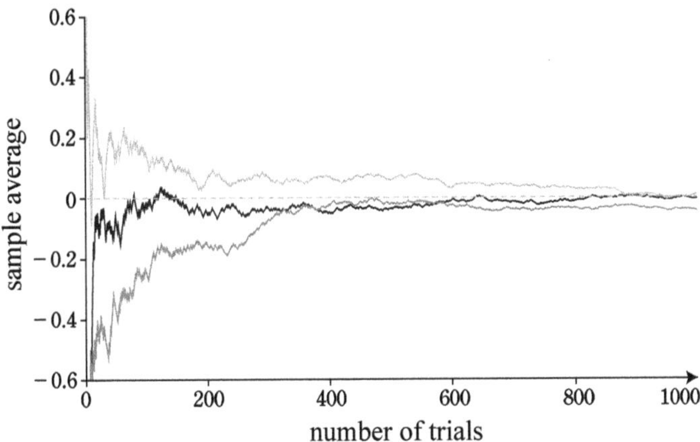

Figure 5.4 How the law of large numbers is established in a coin toss. In the case of $p = q = 0.5$, three independent trial examples are shown.

Theorem: If N random variables are independent of each other and follow the identical probability distribution with mean μ and variance σ^2, the probability distribution of their means converges to a normal (Gaussian) distribution with mean μ and variance σ^2/N in the large N limit.

This theorem was presented by de Moivre and Laplace as a limit theorem for the binomial distribution and was later generalized as a universal theorem that is not limited to the binomial distribution, and it literally has become the "central" limit theorem in probability and statistics.

Example 5.6   Central Limit Theorem by Coin Tosses (Bernoulli Trial) If the previous example is extended upon, the distribution of coin tosses (Bernoulli trial) is binomial, but the central limit theorem states that the random variable $\overline{X}$ of the sample mean of the coin toss is $\mu = p - q$ and the variance is

$$\frac{\sigma^2}{N} = \frac{E((X_i - \mu)^2)}{N} = \frac{p(1 - (p - q))^2 + q(-1 - (p - q))^2}{N} = \frac{4pq}{N}, \quad (5.31)$$

and as N becomes larger, the distribution approaches the normal distribution

$$\mathcal{N}\left(p - q, \frac{4pq}{N}\right) = \frac{1}{\sqrt{2\pi\left(\frac{4pq}{N}\right)}} e^{-\frac{(X-(p-q))^2}{2\left(\frac{4pq}{N}\right)}}.$$

If we treat the sample sum as a random variable rather than the sample mean, then the probability distribution that the random variable follows in the limit of large N is $\mathcal{N}(N\mu, N\sigma^2) = \mathcal{N}(N(p - q), 4Npq)$ (Fig. 5.5). The process of deriving the normal distribution from the binomial distribution is essentially the same as that of the random walk, which will be explained in the next section.

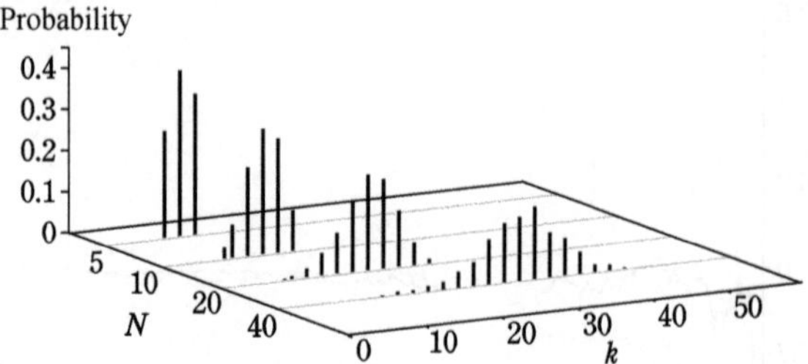

Figure 5.5 How the central limit theorem is established in a coin toss. Each probability distribution for the summed variables (denoted as k) of $N = 5, 10, 20, 40$, is shown in the case of $p = 0.8$.

For the central limit theorem to hold, it is necessary for the mean and variance to have finite values and for each trial to be independent, but there are some distributions that do not fit this criterion. For example, if the tail of a probability density function according to a random variable decays with a power of $|x|^{-\alpha-1}(0 < \alpha < 2)$, its tail is fat, so that the variance diverges. In this case, the limiting distribution converges not to the normal distribution but to the stable distribution with the characteristic index α. Here, stable distributions are the distributions in which the sum of independent random variables becomes the same distribution as the original random variable by an appropriate first-order transformation. A stable distribution can be thought of as an extension of the normal distribution, and can express the fat-tail property. For example, the Levy distribution is a type of stable distribution, and due to its characteristic fat-tail property and scale invariance, it is used as a probability distribution suitable for modeling extreme events.

5.8 Random Walk and Diffusion Process

5.8.1 Random Walk

We introduce the random walk theory as a basic framework for dealing with the diffusion processes of particles. For simplicity, we consider a one-dimensional system. Initially, a particle was placed at the origin and a coin was tossed at each time, and the particle was moved by one step in the positive direction if it was head, and by one step in the negative direction if it was tail. When this process is repeated, the position of the particle changes randomly over time because the trials at each time are independent. This process is called a one-dimensional "random walk" (the name was proposed by the statistician Pearson in 1905). Here, "walk" refers to a process with discrete steps, and the unit step length Δx is moved at the unit step time Δt. Without losing generality, we can set $\Delta t = 1$ and $\Delta x = 1$ by suitably choosing the unit for space and time. As for the random walk in a higher-dimensional system, where each dimension can be treated independently, it can be described as a natural extension of the results of the one-dimensional system.

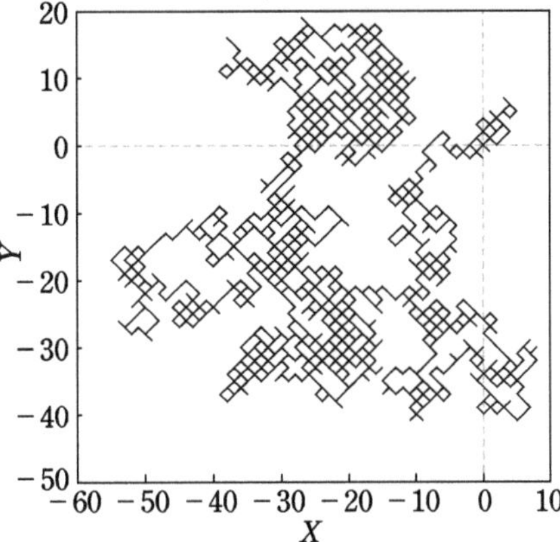

Figure 5.6 An example trajectory of a two-dimensional random walk.

Considering $\Delta x = 1$, we can introduce the stochastic variable of coin so that the head of the coin is set to 1 and the tail of the coin is set to -1. Then, as in the previous section, its expected value and variance are given by

$$\begin{cases} E(X_i) = p - q \\ V(X_i) = 4pq, \end{cases} \tag{5.32}$$

respectively, where p (q) is the probability for head (tail). If $p = q = 0.5$, it is a random walk with no bias. If there is a difference between p and q, there is a bias toward positive or negative. Figure 5.6 shows an example of a two-dimensional random walk with no bias.

The position of the particle after moving N times is determined by the integration $S_N = X_1 + X_2 + \cdots + X_N$ of the random variable, and because each trial is independent, its expected value and variance are respectively

$$\begin{cases} E(S_N) = N(p - q) \\ V(S_N) = 4Npq, \end{cases} \tag{5.33}$$

and both are proportional to the number of steps (note that if $p = q$, the mean is 0 and the variance is N).

Now, we consider the nonbiased coin, that is, a symmetric random walk ($p = q = \frac{1}{2}$) in a one-dimensional system, and obtain the probability density $P(S, N)$ of a particle at position S in the Nth step. Within N steps, we assume N_+ steps in the positive direction and N_- steps in the negative direction, as

$$\begin{cases} N = N_+ + N_- \\ S = N_+ - N_-, \end{cases} \tag{5.34}$$

then

$$\begin{cases} N_+ = \frac{N+S}{2} \\ N_- = \frac{N-S}{2} \end{cases} \tag{5.35}$$

hold. The number of cases in which the particle is at position S in the Nth step is $\binom{N}{N_+} = \frac{N!}{N_+!N_-!}$, and the number of all cases up to the Nth step is 2^N, which is a binomial distribution.

$$P(S,N) = \frac{1}{2^N} \frac{N!}{N_+!N_-!} = \frac{1}{2^N} \frac{N!}{(\frac{N+S}{2})!(\frac{N-S}{2})!}. \tag{5.36}$$

Now, assuming that N is sufficiently large, and by using Stirling's formula $N! \approx \sqrt{2\pi N}(\frac{N}{e})^N$, we get

$$\begin{aligned}
P(S,N) &\approx \frac{\sqrt{2\pi N}(\frac{N}{e})^N}{2^N \sqrt{\pi(N+S)}(\frac{N+S}{2e})^{\frac{N+S}{2}} \sqrt{\pi(N-S)}(\frac{N-S}{2e})^{\frac{N-S}{2}}} \\
&= \sqrt{\frac{2N}{\pi}} \frac{N^N}{(N+S)^{\frac{N+S+1}{2}}(N-S)^{\frac{N-S+1}{2}}} \\
&= \sqrt{\frac{2}{\pi N}} \left(1+\frac{S}{N}\right)^{-\frac{N+S+1}{2}} \left(1-\frac{S}{N}\right)^{-\frac{N-S+1}{2}}.
\end{aligned} \tag{5.37}$$

Taking the logarithm of both sides,

$$\begin{aligned}
\log P(S,N) = \log\left(\sqrt{\frac{2}{\pi N}}\right) &- \left(\frac{N+S+1}{2}\right) \log\left(1+\frac{S}{N}\right) \\
&- \left(\frac{N-S+1}{2}\right) \log\left(1-\frac{S}{N}\right),
\end{aligned} \tag{5.38}$$

and then performing Taylor expansion $\log(1 \pm x) = \pm x - \frac{x^2}{2} \pm \frac{x^3}{3} + \cdots$, we get

$$\begin{aligned}
\log P(S,N) = \log\left(\sqrt{\frac{2}{\pi N}}\right) &- \left(\frac{N+S+1}{2}\right)\left(\frac{S}{N}-\frac{1}{2}\left(\frac{S}{N}\right)^2 + o\left(\left(\frac{S}{N}\right)^3\right)\right) \\
&- \left(\frac{N-S+1}{2}\right)\left(-\frac{S}{N}-\frac{1}{2}\left(\frac{S}{N}\right)^2 + o\left(\left(\frac{S}{N}\right)^3\right)\right).
\end{aligned} \tag{5.39}$$

Considering $N \gg 1$,

$$\begin{aligned}
\log P(S,N) &\approx \log\left(\sqrt{\frac{2}{\pi N}}\right) - \left(\frac{N+S}{2}\right)\left(\frac{S}{N}-\frac{1}{2}\left(\frac{S}{N}\right)^2 + o\left(\left(\frac{S}{N}\right)^3\right)\right) \\
&\quad - \left(\frac{N-S}{2}\right)\left(-\frac{S}{N}-\frac{1}{2}\left(\frac{S}{N}\right)^2 + o\left(\left(\frac{S}{N}\right)^3\right)\right) \\
&= \log\left(\sqrt{\frac{2}{\pi N}}\right) - \frac{N}{2}\left(1+\frac{S}{N}\right)\left(\frac{S}{N}-\frac{1}{2}\left(\frac{S}{N}\right)^2 + o\left(\left(\frac{S}{N}\right)^3\right)\right)
\end{aligned}$$

$$-\frac{N}{2}\left(1-\frac{S}{N}\right)\left(-\frac{S}{N}-\frac{1}{2}\left(\frac{S}{N}\right)^2+O\left(\left(\frac{S}{N}\right)^3\right)\right)$$

$$=\log\left(\sqrt{\frac{2}{\pi N}}\right)-\frac{N}{2}\left(\frac{S}{N}-\frac{1}{2}\left(\frac{S}{N}\right)^2+\left(\frac{S}{N}\right)^2+O\left(\left(\frac{S}{N}\right)^3\right)\right)$$

$$-\frac{N}{2}\left(-\frac{S}{N}-\frac{1}{2}\left(\frac{S}{N}\right)^2+\left(\frac{S}{N}\right)^2+O\left(\left(\frac{S}{N}\right)^3\right)\right)$$

$$=\log\left(\sqrt{\frac{2}{\pi N}}\right)-\frac{N}{2}\left(\left(\frac{S}{N}\right)^2+O\left(\left(\frac{S}{N}\right)^3\right)\right). \tag{5.40}$$

If we ignore the last higher-order term, assuming that it is small, then $\log P(S,N)\approx$ $\log\left(\sqrt{\frac{2}{\pi N}}\right)-\frac{N}{2}(\frac{S}{N})^2$, which means that it is $P(S,N)\approx\sqrt{\frac{2}{\pi N}}e^{-\frac{S^2}{2N}}$. Since the evenness of S is equal to the evenness of N, we need to further divide this equation by two, and the final result is

$$P(S,N)\approx\frac{1}{\sqrt{2\pi N}}e^{-\frac{S^2}{2N}}. \tag{5.41}$$

This is a Gaussian distribution with a mean of 0 and a variance of N. This result can be regarded as the derivation of the probability density function from the microscopic description of the position of the particle.

Furthermore, the case of a biased random walk with anisotropic transition probabilities can also be discussed in a similar way, as,

$$P(S,N)=\frac{N!}{N_+!N_-!}p^{N_+}q^{N_-}=\frac{N!}{\left(\frac{N+S}{2}\right)!\left(\frac{N-S}{2}\right)!}p^{\frac{N+S}{2}}q^{\frac{N-S}{2}}, \tag{5.42}$$

and it can be shown that there is a Gaussian distribution with an average of $N(p-q)$ and variance of $4Npq$.

5.8.2 Derivation of the Diffusion Equation from Random Walk

Next, let us consider the macroscopic description of the probability density function $P(S,N)$ of the particle position under a random walk. Assuming that the initial position of the particle is the origin, $P(0,0)=1, P(S,0)=0$ $(S\neq0)$. We can consider a continuous model of the random walk by treating $P(S,N)$ as a probability distribution function $P(x,t)$ with position $S\to x$ and time $N\to t$.

Then, we introduce a "flow" of the distribution function of particles. Here, the flow refers to the probability flow of a particle moving in the positive direction through a unit area perpendicular to the x-axis at time t. Considering that the probability density function of a particle at each position is $P(x,t)$, the flow can be defined as

$$J(x,t)\Delta t = \frac{1}{2}(P(x - \Delta x, t) - P(x,t))\Delta x \tag{5.43}$$

if we consider flow in the positive direction. The reason for the 1/2 coefficient is that the contribution of the probability density function $P(x,t)$ to the flow is half in the positive direction and half in the negative direction. Here, if we perform a Taylor expansion and ignore the higher-order terms,

$$P(x - \Delta x, t) = P(x,t) - \frac{\partial P(x,t)}{\partial x}\Delta x + \frac{\partial^2 P(x,t)}{\partial x^2}\frac{(\Delta x)^2}{2} - \cdots \tag{5.44}$$

Assuming that Δx is small, and by using Eq. (5.43),

$$J(x,t) = -\frac{(\Delta x)^2}{2\Delta t}\frac{\partial P(x,t)}{\partial x} = -D\frac{\partial P(x,t)}{\partial x} \tag{5.45}$$

is obtained ($D = \frac{(\Delta x)^2}{2\Delta t}$, diffusion coefficient). This is "Fick's first law," which indicates that the rate at which a substance diffuses in the direction of resolving a concentration gradient is proportional to its concentration gradient. Furthermore, considering the conservation of the probability, its temporal change over Δx and the flow over Δt should be balanced, so that

$$(J(x,t) - J(x + \Delta x, t))\Delta t = (P(x, t + \Delta t) - P(x,t))\Delta x. \tag{5.46}$$

Accordingly,

$$\frac{\partial P(x,t)}{\partial t} = -\frac{\partial J(x,t)}{\partial x}, \tag{5.47}$$

which is a continuous equation for the conservation of probability. Substituting Fick's first law into this equation yields the following:

$$\frac{\partial P(x,t)}{\partial t} = D\frac{\partial^2 P(x,t)}{\partial x^2}, \tag{5.48}$$

which is "Fick's second law," that is, the diffusion equation. The solution of this equation starting from $P(x,0) = \delta(x)$ is given by the Gaussian distribution

$$P(x,t) = \frac{1}{\sqrt{4\pi Dt}}e^{-\frac{x^2}{4Dt}} \tag{5.49}$$

(see Section 4.1.2 for the specific derivation of the solution).

Now, let us compare the result of the random walk (5.41) with that of the diffusion process (5.49). Here, the unit step time $\Delta t = 1$ and the unit step length $\Delta x = 1$, which are used for the discrete random walk notation. Then for the N steps corresponding to $t = N\Delta t$, $V(S_N) = N(\Delta x)^2$. Accordingly by eliminating N, we obtain $V(S_N) = \frac{(\Delta x)^2 t}{\Delta t}$. If we set $D = \frac{(\Delta x)^2}{2\Delta t}$, we find $V(S_N) = 2Dt = V(x)$. From this, we can see a consistent relationship between the microscopic descriptions of the random walk and the macroscopic diffusion process. The ratio of Δt to $(\Delta x)^2$ is kept constant in both the micro and macroscopic discussions, in taking the zero limit, of Δt and Δx, which is important in considering the continuous description of Brownian motion to be discussed in the next section. The diffusion equation itself was also derived in Section 4.1.1, but here it should be noted that we have derived it from the standpoint of the probability distribution of a random walk particle.[5] Note that the same description as the one given earlier can be obtained for the case where N particles randomly walk simultaneously, assuming that each particle does not interact and is independent.

5.8.3 Random Walk in Higher Dimensions

We have treated a one-dimensional random walk here. In contrast, a three-dimensional random walk has qualitatively different recurrence probabilities from those of a one-dimensional or two-dimensional system. In the case of a symmetric random walk, the event of returning to the original location (recurrence) occurs with probability 1 in one-dimensional and two-dimensional systems if one waits for an infinitely long period of time, whereas there is a finite probability that one cannot return to the same location in a three-dimensional system or more (Polya's theorem).

For example, consider the density of the step points using a random walk. If the number of steps of a random walk is N and the average distance from the starting point is r, then $r^2 = N$, regardless of the dimension. Considering this, in the one-dimensional case, there are N points in the linear distance $r \propto \sqrt{N}$, so that the density of points is dependent on $\sqrt{N}$. In the two-dimensional case, because there are N points in the circle area $\pi r^2 \propto N$, the density of points is constant regardless of N. In the three-dimensional case, there are N points in the sphere volume $\frac{4}{3}\pi r^3 \propto N^{\frac{3}{2}}$, so that the density of points is proportional to $\frac{1}{\sqrt{N}}$. In other words, as the number of steps N increases, the number of returns to any point increases in the one-dimensional case, but in the three-dimensional case, the density of points decreases and the number of points that cannot be returned increases. On the other hand, in the two-dimensional case, we can keep the "just right" arrangement, and we can always return to the point in some time,[6] although the expected value of the recurrence time is infinite. It is also known that the rate of diffusion-limiting chemical reactions is dimension dependent. These findings provide a suggestive perspective for considering the structure of the cell: the one-dimensional nature of molecular motor systems, the two-dimensional nature of the cell membrane, and the three-dimensional nature of the cytoplasm.

---

[5] Here, the treatment of higher-order terms in Taylor expansion is made more explicit, and the diffusion coefficients and Fick's two laws are organized into the derivation of the diffusion equation.

[6] For this reason, there is a joke that "two-dimensional romance is preferred because on average there is always an encounter in the two-dimensional world."

5.9 Brownian Motion and Einstein Relation

In a macroscopically homogeneous solvent environment at temperature $T[K]$, consider a situation in which particles with a size larger than that of the solvent molecules exist. For example, the size of the solvent molecule is of the order of 1 nm, and the size of the particle is on the order of 1 µm. Both sizes are approximately 1,000 times different from each other, and the large number of solvent molecules around the particle continuously impacts the particle from all directions due to thermal motion. Although the particle is large, it is not on a macroscopic scale, and the position of the particle fluctuates owing to the collision of solvent molecules, resulting in random motion. This is "Brownian motion".

When a particle is in motion, the impact of the solvent molecules from the front outweighs the impact from the back in both the number and intensity of the particle, just as it does when a person runs in the rain. Therefore, the impact of a solvent molecule always gives a force to hold down a particle by subtracting it from the front and back. This is the viscous resistance, which prevents the particle from continuing to accelerate, resulting in the achievement of a steady random motion.

As an example, the simulation results of Brownian motion in a two-dimensional system are shown in Fig. 5.7. Brownian motion is a universal physical phenomenon that appears in particles in a medium in thermal equilibrium regardless of whether they are inorganic or organic, and because biomolecules also exhibit Brownian motion, it is also of fundamental importance in biology.[7]

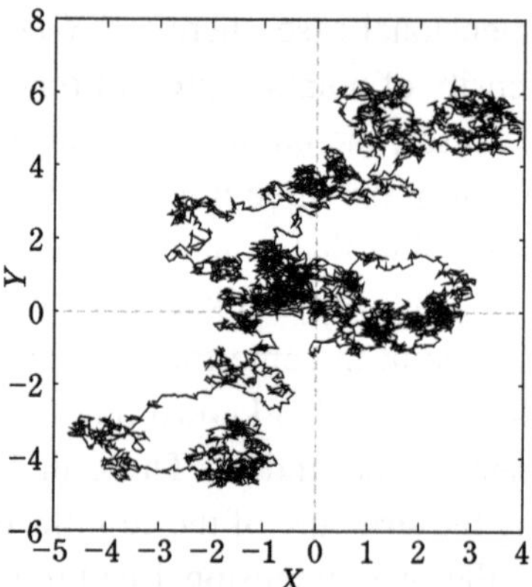

Figure 5.7 An example trajectory of two-dimensional Brownian motion.

[7] The Scottish botanist Robert Brown used a microscope to observe pollen in an aqueous solution and found that the small particles from the pollen were in intense motion (1827). At first he thought the particles were living "atoms of life," but he remained cautious as a scientist and conducted a series of experiments with inorganic and nonliving powders, confirming that similar motions occurred in all of them.

Example 5.7 Effect of Gravitational and Buoyant Forces on Brownian Particles Let us try to quantitatively evaluate the effects of gravity and buoyancy on Brownian particles (particles in Brownian motion). For example, consider a polystyrene bead with a density equal to water, $\rho = 1$ [g/cm^3], and a particle diameter of $r = 100$ [nm]. The mass is $m = \frac{4}{3}\pi\rho r^3 \approx 4 \times 10^{-18}$ [kg], and the change in gravitational potential energy due to falling a distance equal to the particle size is $mgr \approx 4 \times 10^{-24}$ [J]. On the other hand, the average energy of thermal fluctuations is $1k_{\mathrm{B}}T$ [J] $\approx 4 \times 10^{-21}$ [J], confirming that thermal fluctuations contribute more to Brownian particles than gravity or buoyancy. For example, particles do not settle in colloids such as mud because of the Brownian motion of the particles. (It is also affected by the presence of repulsive force due to electric charge.)

Now, let us theoretically formulate Brownian motion. What constitutes Brownian motion? Although there are several methods,[8] the one that is intuitively easier to understand is the method being constructed as the limit of the random walk. The random walk is constructed as discrete time increments (Δt) and spatial increments (Δx); then, if we take the limits of $\Delta t \to 0$ and $\Delta x \to 0$ while keeping $\frac{(\Delta x)^2}{2\Delta t}$ finite, we make a continuous description of the Brownian motion, considering that there are many particles in the solvent, and their concentration is $c(x, t)$. If the solvent is subjected to an external force f [N] and a drift of constant velocity v_d [m/s] exists in the flow,

$$J(x, t) = -D\frac{\partial c(x, t)}{\partial x} + v_d c(x, t) \tag{5.50}$$

is obtained from Fick's first law. Here, if the external force is derived from a potential, it is given by

$$f(x) = -\frac{\partial U(x)}{\partial x}. \tag{5.51}$$

On the other hand, the viscous drag is given by $-\gamma v_d$, where γ is the viscous drag coefficient. As the external force and the viscous drag have to be balanced, it follows as

$$v_d = -\frac{1}{\gamma}\frac{\partial U(x)}{\partial x}. \tag{5.52}$$

Now, in equilibrium, the flow of the solute is zero where the flow of the solvent by the external force and that by the diffusion process are balanced, leading to

$$0 = -D\frac{\partial c(x, t)}{\partial x} - \frac{c(x, t)}{\gamma}\frac{\partial U(x)}{\partial x}. \tag{5.53}$$

---

[8] The first mathematically rigorous formulation of Brownian motion as a stochastic process was given by Wiener (1923), and this formulation is called the Wiener process.

Accordingly,

$$\frac{\partial c(x,t)}{c(x,t)} = -\frac{\partial U(x)}{D\gamma},\tag{5.54}$$

and we get

$$c(x,t) \propto e^{-\frac{U(x)}{D\gamma}}.\tag{5.55}$$

On the other hand, in the equilibrium $c(x,t)$ follows $e^{-\frac{U(x)}{k_B T}}$. To satisfy the agreement, the relationship

$$D = \frac{k_B T}{\gamma}\tag{5.56}$$

has to be satisfied. This equation is called the "Einstein relation" (or the "Sutherland–Einstein relation"), and it shows that the higher the temperature of the solvent and the smaller the viscosity, the larger the diffusion rate due to Brownian motion.

Column IV: Einstein and the Theory of Brownian Motion

The theoretical formulation of Brownian motion was accomplished by Einstein in 1905. His objective was to demonstrate the existence of atoms and molecules by proposing an experimental method for determining Avogadro's number based on measurable long-term behavior, such as diffusion and distribution of Brownian particles. Several years later, Perrin's series of precise experiments beautifully verified the predictions of the theory, confirming the physical reality of atoms and molecules and revealing that thermodynamic laws have a statistical foundation.

Einstein's theory of Brownian motion took principles from the microscale theory of dilute solutions consisting of atomic and molecular constituents, extrapolating the application of energy equipartition and the formula for osmotic pressure (van't Hoff's formula) to the motion of a particle with a size of $1\,[\mu m]$. Simultaneously, it also took principles from macroscopic fluid theory, extrapolating the application of Stokes' law of viscosity and Fick's law of flow conservation. Assuming mechanical equilibrium, he equated the terms involving forces appearing in both approaches to derive the "Einstein relation" that connects the diffusion coefficient with temperature and viscosity.

Furthermore, Einstein derived the Gaussian distribution of fluctuating Brownian particle positions and the time dependence of their variance using probability theory. With the Einstein relation, by equating the diffusion coefficient obtained from the mechanical description to the diffusion coefficient derived from the stochastic description, he established a comprehensive theoretical framework that allowed direct determination of Avogadro's number from experimental measurements of Brownian motion dispersion ($\langle x^2 \rangle = 2Dt$; solving for D, $D = k_B T/\gamma$, yields the Boltzmann constant $k_B = R/N_A$, where R is the gas constant. This in turn allows experimental determination of Avogadro's number N_A.)

This bold and innovative approach, combined with relatively accessible mathematical methods, led to profound theoretical insights, making it one of the gold standards of physics. For further details, see [42, 21]. This theory of Brownian motion served as a precursor to later nonequilibrium statistical physics and also embraced the concept of fractals. The micro-macro consistency principle is also applied to the total path level in Kawasaki's mode-coupling theory, and to the Jarzynski equality and the fluctuation theorem [16, 19]. Moreover, the principle of consistency between genotypes and phenotypes in evolution has led to the proposal of relations similar to those of the Einstein relation concerning the evolution rate of phenotypes and fluctuations [18].

5.10 Examples in Cell Biology

The average travel distance of diffusion due to Brownian motion in an n-dimensional space is given by

$$\sqrt{\langle r(t)^2 \rangle} = \sqrt{2nDt}. \tag{5.57}$$

Since it will spread only in proportion to the square root of time, it is expected that it will take a large amount of time to transport a substance over a long distance by diffusion. However, on a spatial scale smaller than that of a cell, it may be possible to transport materials over a wider range by diffusion than by straightforward transport. In fact, the trajectory of Brownian motion in a two-dimensional system (Fig. 5.7) shows that the trajectory locally fills the plane, but it takes a long time to move on a larger scale.

Example 5.8 Diffusion of Proteins inside of the Cells and Péclet Number As a concrete example, let us compare the distance traveled between the three-dimensional diffusion of proteins in the cell and the one-dimensional active transport by molecular motors. Because the diffusion coefficient in the cytoplasm is approximately $D = 5\,[\mu m^2/s]$ [23, 24] and the typical velocity of a molecular motor is approximately $v = 1\,[\mu m/s]$ [14, 46], the average travel distance of the two turns out to be equal at

$$\sqrt{6Dt} = vt. \tag{5.58}$$

By introducing the preceding D and v values, the two coincide at $t = 30\,[s]$, and the distance traveled at this time is $L = 30\,[\mu m]$. This is slightly larger than the size of eukaryotic cells. In the world below this spatio-temporal scale, diffusion may be more effective than active transport, and the size of the cell is close to the size that can take advantage of such diffusion properties (see Fig. 5.8). Moreover,

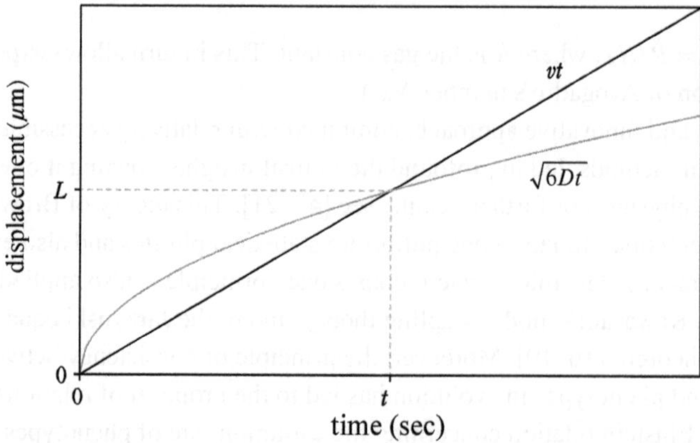

Figure 5.8 Comparison of the time dependence of spatial spread in active transport and diffusion.

because material transport by diffusion is possible under thermal equilibrium, it is possible without additional energy cost, and it is also possible to transport materials to unspecified sites in the cell. On the other hand, material transport by diffusion is insufficient in neural axons that may be a few *micrometers* thick but are more than *millimeters* long. In fact, molecular motor systems such as microtubules and kinesins are used for active substance transport at the cost of the free energy of ATP hydrolysis.

Now, the Péclet number is introduced to quantify the ratio between the contributions of diffusion and active transport. The Péclet number is a dimensionless number that characterizes the transport phenomena in a continuous system. In the case of material transport, the Péclet number is obtained from the ratio of the transport rate by advection to that by diffusion by using the typical advection rate v[m/s], the characteristic advection distance L[m], and the diffusion coefficient D[m^2/s] as

$$Pe = \frac{transport\ rate\ due\ to\ advection}{transport\ rate\ due\ to\ diffusion} = \frac{Lv}{D}. \tag{5.59}$$

For $Pe \ll 1$, diffusion is dominant, whereas advection is dominant for $Pe \gg 1$. In the preceding example with $v = 1$ [μm/s] and $D = 5$ [μm^2/s], and the cell size $L = 10$ [μm], $Pe = 2$, so that the contributions of advection and diffusion are nearly comparable for $Pe = 2$, indicating that both are important in cell systems. It should be noted that actual diffusion in cytoplasm is even more complex: in contrast to normal diffusion, some molecules have been found to exhibit "anomalous diffusion," in which the mean square displacement depends on the power of time ($\langle r(t)^2 \rangle \propto t^\alpha$ ($0 < \alpha < 2$, $\alpha \neq 1$)). This can be realized as a random walk according to the Levy distribution (also see Section 5.7, and the examples later in this section). It can also appear when the particle motion has the memory and constraints which may be due to intracellular structures and molecular crowding [17].

Example 5.9 Diffusion-Limited Reaction In chemical reactions, molecules "meet" by colliding with each other through diffusion, and then a reaction occurs. There are two extreme situations in such chemical reactions; diffusion-limited or reaction-limited. The former is defined as the reaction in which the "encounter" of molecules becomes a rate-limiting step. In case of solution reaction, the probability of the "encounter" depends on environmental conditions such as the viscosity of the solvent.

Now, let us consider that the binding of a receptor and a substrate molecule on the cell membrane is a diffusion-limiting reaction due to the Brownian motion of the substrate molecule [31]. Suppose that a cell is a sphere of radius $R = 5\,[\mu\mathrm{m}]$, and that the total number of receptors is N, distributed uniformly on the cell membrane. The diffusion coefficient of the receptor on the membrane is known to be around $0.1\,[\mu\mathrm{m}^2/\mathrm{s}]$ based on single-molecule measurements [34, 36], which is much smaller than the diffusion coefficient of the substrate molecules in the extracellular solution, $D \approx 100\,[\mu\mathrm{m}^2/\mathrm{s}]$. Thus it is reasonable to assume that the receptor configuration is stationary on the cell membrane when comparing the two.

Here we consider the situation in which substrate molecules collide with the receptor due to diffusion. Let the concentration of the substrate molecule be $c(x, y, z, t)$. Then the first and second of Fick's laws are

$$\begin{cases} J(x, y, z, t) = -D\nabla c(x, y, z, t) \\ \frac{\partial c(x,y,z,t)}{\partial t} = D\nabla^2 c(x, y, z, t). \end{cases} \tag{5.60}$$

By transforming this into a three-dimensional polar coordinate representation and imposing a spherical symmetry from the isotropy of space, the preceding equations can be reduced to a differential equation for the distance $r[\mathrm{m}](> R)$ from the center of the cell, as

$$\begin{cases} J(r, t) = -D\frac{\partial c(r,t)}{\partial r} \\ \frac{\partial c(r,t)}{\partial t} = \frac{D}{r^2}\left(\frac{\partial}{\partial r}\left(r^2 \frac{\partial c(r,t)}{\partial r}\right)\right). \end{cases} \tag{5.61}$$

Because $\frac{\partial c(r,t)}{\partial t} = 0$ for the steady state, implying constant $(c(r, t) = c(r))$, $\frac{dc(r)}{dr} = \frac{A}{r^2}$ follows, where A is a constant. Then $J(r, t) = J(r) = -\frac{DA}{r^2}$. Accordingly, the total flow toward the center of a sphere of radius R (inflow onto the cell membrane) is given by

$$J_{\mathrm{tot}} = -J(R) \cdot 4\pi R^2 = 4\pi DA. \tag{5.62}$$

Because influx on the cell membrane results in collision with the receptor of the substrate molecule and binding at a constant rate, the reaction rate constant of the binding reaction is k_{on} and $c(R, t) = c(R)$ in the steady state, $J_{tot} = Nk_{on}c(R)$, then $A = \frac{Nk_{on}c(R)}{4\pi D}$, which results in

$$\frac{dc(r)}{dr} = \frac{Nk_{on}c(R)}{4\pi Dr^2}.$$ (5.63)

This differential equation is solved under the conditions $R \leq r < \infty$ and $c(\infty) = c_0$, as

$$c_0 - c(R) = \frac{Nk_{on}c(R)}{4\pi DR}.$$ (5.64)

Thus we get, $c(R) = \frac{c_0}{1+\left(\frac{Nk_{on}}{4\pi DR}\right)}$. Therefore, the binding rate of the substrate to the receptor is given by

$$v = Nk_{on}c(R) = \frac{Nk_{on}c_0}{1 + \left(\frac{Nk_{on}}{4\pi DR}\right)} = \frac{4\pi DRNk_{on}c_0}{4\pi DR + Nk_{on}}.$$ (5.65)

From Eq. (5.65), as in the case of the Michaelis–Menten-type reaction rate equation (see Section 2.2.1), we get (i) $v \approx 4\pi DRc_0$ (maximum velocity at diffusion rate-limit) for $Nk_{on} \gg 4\pi DR$ (complete adsorption), (ii) $v = Nk_{on}c_0 \propto Nk_{on}$ for $Nk_{on} \ll 4\pi DR$ (weak adsorption), and (iii) $v = 2\pi DRc_0$ (half of the maximum velocity at a diffusion rate-limit) for $Nk_{on} \approx 4\pi DR$.

Now, we examine which case is valid for cells. It is experimentally known that in many cases, the binding reactions of substrates by receptors are fast ($k_{on} \approx \mu M^{-1} \cdot s^{-1}$) with moderate affinity ($K_d \approx \mu M$) [45]. If $k_{on} = 1\ [\mu M^{-1} \cdot s^{-1}]$ and the cell size and the diffusion coefficients of substrates are the values described earlier, the number of receptors can be estimated to be approximately $N \approx 10^5 \sim 10^6$ in order for condition (iii) to hold. In fact, the number of receptors is approximately $10^5 \sim 5 \times 10^6$ per cell, so that condition (iii) is satisfied, from the order estimation.

Let us discuss why case (iii) is relevant to cells. In the first place, with complete adsorption condition (i), $c(R) \approx 0$ and all the substrates in the vicinity of the cell bind to the receptor, so the ability to detect environmental information is high. However, if k_{on} is of the order described earlier, the cell must express a huge number of receptors. On the other hand, for weak adsorption condition (ii), $c(R) \approx c_0$ and the number of receptors that bind to the substrate is nearly 0, so the detection of the external input itself may not be possible. Given that cells have been evolutionarily selected to express a number of receptors that can sensitively detect environmental information at the lowest possible energy cost, under the surface area of the cell membrane, it makes sense that condition (iii) would hold.

The same framework can be applied to estimate the binding rate constant in a two-body binding reaction between molecules. Assume that the reaction is diffusion-limited, a spherical molecule of radius $R[m]$ is fixed in a three-dimensional space, and a substrate molecule collides with the molecule due to diffusion. To estimate the maximum value of the binding rate constant, all collisions are assumed to be accompanied by binding with a probability of 1.

In the steady state, as described previously, the differential equation by Fick's law becomes $\partial c(r, t)/\partial r = A/r^2$. Under the boundary conditions $c(R, t) = 0$ and $c(\infty, t) = c_0$, the solution of the equation can be obtained as $c(r, t) = c_0(1 - R/r)$ with $A = c_0 R$. Then, the total flow from the sphere of radius R to the center of the sphere (inflow to the solid molecule) is given by

$$J_{\text{tot}} = -J(R) \cdot 4\pi R^2 = 4\pi D c_0 R. \tag{5.66}$$

Since the maximum rate constant in the binding reaction multiplied by the concentration of substrate molecules in the environment equals this total flow rate, it can be expressed as

$$k_{\text{on}} = 4\pi DR. \tag{5.67}$$

This is called the "Smoluchowski kinetic equation." Because $D = \frac{k_B T}{\gamma}$ from the Einstein relation and $\gamma = 6\pi R\eta$ from Stokes' law, this leads to

$$k_{\text{on}} = \frac{2k_B T}{3\eta}. \tag{5.68}$$

In this equation, the binding rate constant turns out to be independent of the size of the particle. This is because small (large) molecules diffuse faster (slower), but the reaction surface area is smaller (larger), and consequently, the two cancel each other out.

In the preceding discussion, we formulated the relationship between particles with fixed positions and colliding substrates. If we consider the collision of molecules moving differently, we can formulate the diffusion process of a general two-body chemical reaction. In this case, we set the radii of the respective molecular species as R_A and R_B and the diffusion coefficients as D_A and D_B, and by adding the contributions of both, we can formulate the diffusion rate process of a general two-body chemical reaction, as

$$k_{\text{on}} = 4\pi (D_A + D_B)(R_A + R_B). \tag{5.69}$$

Now, assuming that $D_A = D_B$, $R_A = R_B$, then

$$k_{\text{on}} = 16\pi D_A R_A = \frac{8k_B T}{3\eta}. \tag{5.70}$$

Noting that $\eta = 10^{-3}$ [Pa·s] for water and $k_B T = 4.1 \times 10^{-21}$ [J] at room temperature, $k_{\text{on}} = 7 \times 10^9$ [M$^{-1} \cdot$ s^{-1}] is obtained. This is the maximum reaction rate in the diffusion-limited process.

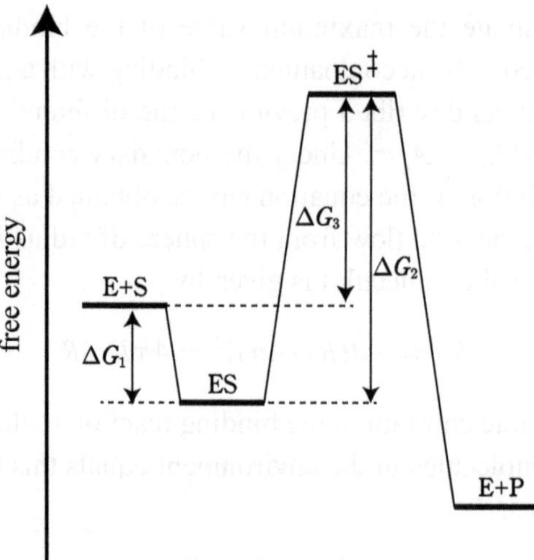

Figure 5.9 A schematic diagram of enzyme reactions. Enzymes, substrates, complexes, their activated states, and products are denoted as E, S, ES, ES‡, and P, respectively. $\Delta G_1 = -RT \ln(K_{\mathrm{M}})$, $\Delta G_2 = -RT \ln(k_{\mathrm{cat}})$, and $\Delta G_3 = \Delta G_2 - \Delta G_1$ correspond to $\frac{k_{\mathrm{cat}}}{K_{\mathrm{M}}}$.

Example 5.10   Perfect Enzyme Using the previously mentioned theory, we study the diffusion-limited enzyme reaction. The reaction scheme for the enzymatic reaction is shown in Fig. 5.9, where the catalytic efficiency (specificity constant) can be expressed as $\frac{k_{\mathrm{cat}}}{K_{\mathrm{M}}}$ from the Michaelis–Menten rate equation (see Section 2.2.1). When the amount of substrate is sufficiently small, the catalytic efficiency is given by the relationship between the strength of substrate binding and the reaction rate of the enzyme. If K_{M} is small (high affinity) and the reaction speed is high, the catalytic efficiency is high.

Here, we consider the maximum value of the catalytic efficiency by adopting the diffusion limit case of the enzyme reaction with $k_{\mathrm{off}} \ll k_{\mathrm{cat}}$. Then,

$$\frac{k_{\mathrm{cat}}}{K_{\mathrm{M}}} = \frac{k_{\mathrm{cat}}}{\left(\frac{k_{\mathrm{off}} + k_{\mathrm{cat}}}{k_{\mathrm{on}}}\right)} = k_{\mathrm{on}}, \tag{5.71}$$

showing that the maximum value of the catalytic efficiency is equal to that of k_{on}. Indeed, for enzymes such as acetylcholinesterase and fumarase, called "perfect enzymes," the $\frac{k_{\mathrm{cat}}}{K_{\mathrm{M}}} \approx 10^8 \sim 10^9 \, [\mathrm{M}^{-1} \cdot \mathrm{s}^{-1}]$ is reached, which is almost the same order as $k_{\mathrm{on}} = 7 \times 10^9 \, [\mathrm{M}^{-1} \cdot \mathrm{s}^{-1}]$, the maximum k_{on} in the diffusion-limited process (see Example 5.9). This indicates that the catalytic activity has evolved to near the physical upper limit as a diffusion-limited reaction [2, 45].

Next, let us look at an example of the relationship between cell motility and random walk. Although Brownian motion can be realized by thermal fluctuations,

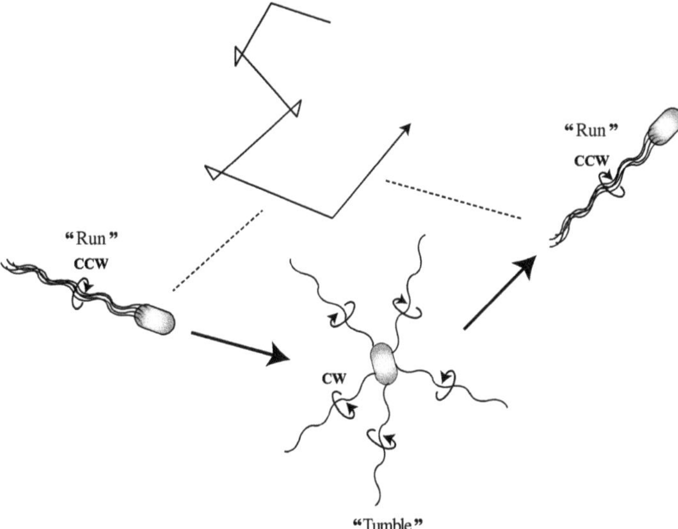

Figure 5.10 Motility of *E. coli*. It swims by switching between two states: straight (run) and random change of direction (tumble).

cell movement is not passively driven by thermal fluctuations; rather, it makes full use of its own locomotor apparatus to perform active motion. The active nature of cell is linked to spontaneous movement and fluctuation of the cell.

Example 5.11 *E. coli* **in motion** Although *E. coli* is a unicellular organism with a small size of approximately 1 [μm], it can swim at an average speed of approximately 20 [μm/s] in solution. It should be noted that the swimming of microorganisms such as *E. coli* is fundamentally different from the swimming of large organisms because the inertia of water does not work under low Reynolds number conditions (see Section 5.2). They cannot be propelled by reciprocating motions with time-reversal symmetry using inertia, such as the opening and closing of a scallop (scallop theorem [32]). Therefore, *E. coli* migrate using the screw-like rotational motion of the flagella with time-reversal asymmetry.[9]

E. coli are known to swim by switching between two states: a straight-line motion (called "Run") and a random reorientation motion (called "Tumble"), as shown in Fig. 5.10. *E. coli* have multiple long left-handed flagella that grow from the cell membrane, and in the Run state, the flagella are rotated counterclockwise (CCW) and are bundled in parallel to each other to achieve synchronized rotational motion, resulting in smooth movement. In contrast, in the tumble state, the flagella are rotated clockwise (CW) to separate the flagella in various directions, and the

---

[9] The discovery of the rotary motor of the flagellum was the first discovery of the existence of rotary motors in the natural world. Note that eukaryotic and bacterial flagella are completely different in nature and origin.

smooth movement of *E. coli* is inhibited by rotating them arbitrarily, resulting in a significant change in the direction of motion. The mean duration of the tumble state is approximately 0.1 [s], whereas the mean duration of the run state is $1 \sim 10$ [s]. The presence of the characteristic time of each state and the distribution of the distance traveled by Run reflect the stochasticity of the *E. coli* locomotor system itself. Because the overdamped condition is satisfied for a small size of *E. coli*, and the correlation time of motion is much shorter than the duration of tumble, 0.1 [s], the memory of motion direction is lost after tumble, and the run starts again in a random direction. This is an example of a three-dimensional random walk in a broad sense.[10]

In addition, *E. coli* can detect the concentration gradient of nutrients such as sugars and amino acids in the outside world and aggregate at high concentrations, while certain toxic substances such as metal ions and indole can be avoided and moved away (chemotaxis; also see Section 2.4). *E. coli* ascends the concentration gradient by repeating the following processes: the frequency of tumble declines and the duration of the run increases when the concentration gradient of the trigger substance is raised, whereas the run motion that goes down the concentration gradient is the same as the motion without the concentration gradient. By repeating this process, *E. coli* moves up the concentration gradient. In the case of repellants, the opposite is true, and *E. coli* acts by the strategy of "not stopping immediately when the situation is getting better." This can be said to be a biased random walk, because the local speed does not change with or without concentration gradients. This mechanism is called "temporal sensing", because *E. coli* has a signaling system related to chemotaxis and switches its movement by sensing the temporal change of the external concentration with its molecular system. Therefore, *E. coli* can detect the concentration gradient in the external world by moving on its own. This is a different information processing mechanism from the "spatial sensing" described in the chemotaxis of *Dictyostelium* cells. A series of studies on motility and chemotaxis responses in *E. coli* were pioneered by Berg et al. [4, 5].

Exercise 5.10.1 Computer simulations were used to confirm whether the bacteria could actually move up the gradient of the chemoattractant or not in the process just described. For simplicity, consider a one-dimensional system with a linear chemoattractant gradient that does not change over time, and assume that the bacterial velocity is constant. Start the simulation with a large number of bacteria in a spatially random initial arrangement and study the spatial distribution of the bacteria after a sufficient time. Comparing the results in

---

[10] With bacterial size, the effect of Brownian motion cannot be ignored. It should be noted that the Brownian motion of *E. coli* is approximately 1 [μm^2/s], which is relatively small compared to the speed of *E. coli*, but this does not allow *E. coli* to make a complete straight forward motion or a complete stop.

the following three cases: (1) the bacteria are unable to detect changes in concentration over time and change the frequency of change in the direction of movement based only on the concentration information of the current location; (2) the bacteria vary the frequency of change in the direction of movement based on the average value of the most recent past and current concentration information; and (3) the bacteria can detect changes in the concentration gradient over time. In these cases, can bacteria assemble at sites with high concentrations of chemoattractants?

Remark: If the tumbling frequency is only a function of the external concentration of chemoattractant, the directional movement for chemotaxis is not achieved. If the adaptation dynamics discussed in Section 2 are present, this is possible. Consider the following cellular response: Upon concentration change, the internal state of the cell first responds with the time scale t_s and then returns to the original state with the scale t_r, and the tumbling frequency (t_f) changes with the internal state. Then, when the condition $t_s < t_f < t_r$ is satisfied, the chemotaxis of cells is achieved [27, 15].

Column V: The Scaling Relationship between the Body Length and Locomotion Speed of Organisms

There is a scaling relationship between body length and locomotion speed of organisms with a relatively wide range of scales: proportional relationship between the body length and motion speed. This is valid in the region of μm $\sim$ mm [6]. The dimensional analysis shows that this relationship corresponds to the condition for overdamped motion in the fluid [1]. Now, let the length of an organism moving in the fluid be a[m]. In the case of overdamped motion, the motion of an individual is governed by the viscous drag γv, and because the coefficient of viscosity is proportional to the body length from Stokes' law $\gamma = 6\pi a \eta$, the work required for the individual to move a distance of its own body length is $W \propto a\gamma v \propto a^2 v$. On the other hand, the energy an individual expends for movement is proportional to the volume of the individual, so $E \propto a^3$. Because these are equal ($W = E$), we can derive $v \propto a$. If the body length of the organism is so large that it cannot be in a low Reynolds number regime, this proportional relationship deviates, and either v must be less than a, or there is no dependence on it. In the case of underdamped motion, the motion of an individual is governed by inertial forces, and the kinetic energy is $K = \frac{1}{2}mv^2 \propto a^3 v^2$. Because $K = E \propto a^3$ from the conservation law of energy, v does not depend on the body length a.

Example 5.12 Amoeboid Movements of Eukaryotic Cells

Amoeboid motility is a mode of movement found in amoeboid cells. This is a crawling motion in which the cell sticks to the environmental substrate, produces

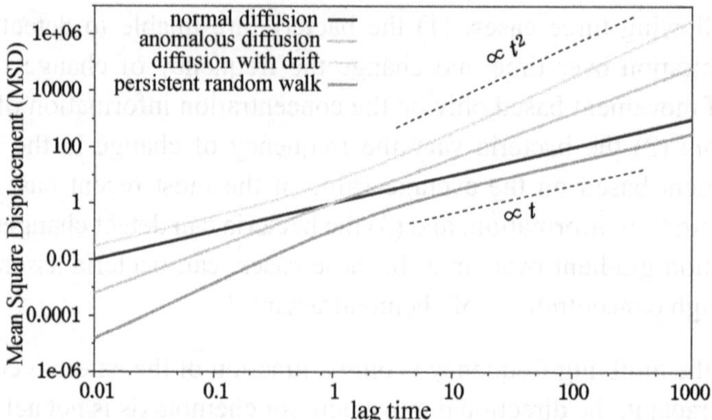

Figure 5.11 Various modes of motility are shown in MSD. Normal diffusion, anomalous diffusion, diffusion with drift, and persistent random walk are illustrated.

a pseudopod (general term including filopod, lamellipod, etc.), and deforms itself by projecting its cytoplasm into the pseudopod. In addition to amoeboid cells such as *Dictyostelium* cells, this movement is widely observed in many migratory cells, such as nerve cells, leukocytes, and cancer cells. In the case of amoeboid motility, it is basically unnecessary to consider the effect of Brownian motion from the solvent, because the amoeboid motility does not "swim" in the fluid like *E. coli*. It is the motility of eukaryotic cells, which are much larger than *E. coli*. The biological process of amoeboid movement is generally understood as a cycle; extension and adhesion of the pseudopod from the anterior part of the cell, forward movement of the cell body including the cell nucleus, and dissociation of the adhesion at the posterior part of the cell [8]. The basic method to statistically analyze such cell motions is the mean square displacement (MSD) of the position of the cell's mass center. The typical modes of motion that can be classified based on the mean square displacement are as follows:

$$\text{MSD}(t) = \begin{cases} At^\alpha \ (0 \le \alpha \le 2) \\ At + Bt^2 \\ \frac{B}{A}(t + \frac{1}{A}(e^{-At} - 1)). \end{cases} \tag{5.72}$$

Here the motion with $\text{MSD}(t) \propto t^\alpha$ with $\alpha = 1$ is a random walk (or Brownian motion or diffusion), and that with $\alpha = 2$ is ballistic motion, and that with $\alpha \neq 1$ gives anomalous diffusion (see Fig. 5.11). For example, the random walk following the Levy distribution (see Example 5.8 for an example of protein diffusion in this section) is one of the mechanisms to generate such anomalous diffusion.

If $\text{MSD}(t) \propto At + Bt^2$, it is a random walk with mean displacement drift (bias); if $\text{MSD}(t) \propto \frac{B}{A}(t + \frac{1}{A}(e^{-At} - 1))$, it is a ballistic motion on a fast time scale ($t \ll 1/A$)

as $\text{MSD}(t) \propto \frac{Bt^2}{2}$, but a persistent random walk on a time scale slower than the relaxation time $(1/A)$. In the case of amoeboid movement, the direction of motion is random on time scales slower than the correlation time (see Eqs. (5.3) to (5.5)), which corresponds to the time for a cell to move directionally through its own size. The mathematical model of persistent random walk is discussed in Section 6.1.

Exercise 5.10.2 Create a number of time series of one-dimensional symmetric random walk using simulations, and check if the law of large numbers holds, by computing the change of the mean position over time. In addition, confirm that the mean square displacement is proportional to time and that the central limit theorem is valid.

References

[1] J. A. Adam. *Mathematics in Nature: Modeling Patterns in the Natural World.* Princeton, Princeton University Press, 2003.

[2] B. Alberts *et al. Molecular Biology of the Cell.* New York, Garland Science, 2014.

[3] Y. Arai *et al.* Dynamic polymorphism of Ras observed by single molecule FRET is the basis for molecular recognition. *Biochemical and Biophysical Research Communications*, **343**(3): 809–815, 2006.

[4] H. C. Berg. *Random Walks in Biology.* Princeton, Princeton University Press, 1993.

[5] H. C. Berg. *E. coli in Motion.* Springer, 2003; related movies www.rowland.harvard.edu/labs/bacteria/movies/index.php.

[6] J. T. Bonner. *Why Size Matters: From Bacteria to Blue Whales.* Princeton, Princeton University Press, 2006.

[7] C. P. Brangwynne *et al.* Germline P granules are liquid droplets that localize by controlled dissolution/condensation. *Science*, **324**: 1729–1732, 2009.

[8] D. Bray. *Cell Movements: From Molecules to Motility.* New York, Garland Science, 2001.

[9] A. K. Dunker *et al.* Intrinsically disordered protein, *Journal of Molecular Graphics and Modelling*, **19** (1), 26–59, 2001.

[10] H. J. Dyson and P. E. Wright. Intrinsically unstructured proteins and their functions, *Nature Reviews Molecular Cell Biology*, **6**: 197–208, 2005.

[11] M. B. Elowitz *et al.* Stochastic gene expression in a single cell, *Science*, **297** (5584): 1183–1186, 2002.

[12] D. T. Gillespie. Exact stochastic simulation of coupled chemical reactions, *The Journal of Physical Chemistry*, **81** (25): 2340–2361, 1977.

[13] D. T. Gillespie. Approximate accelerated stochastic simulation of chemically reacting systems, *Journal of Chemical Physics*, **115** (4): 1716–1733, 2001.

[14] J. Howard. *Mechanics of Motor Proteins and the Cytoskeleton.* Sunderland, Sinauer Associates, 2001.

[15] M. Inoue and K. Kaneko. Condition for intracellular adaptive dynamics for chemotaxis, *Physical Review E*, **74**: 011903, 2006.

[16] C. Jarzynski. Nonequilibrium equality for free energy differences, *Physical Review Letters*, **78** (14): 2690, 1997.

[17] J.-H. Jeon *et al.* In vivo anomalous diffusion and weak ergodicity breaking of lipid granules, *Physical Review Letters*, **106**: 048103, 2011.

[18] K. Kaneko. *Life: An Introduction to Complex Systems Biology: An Introduction to Complex Systems Biology*. Springer, 2010.

[19] K. Kawasaki. Kinetic equations and time correlation functions of critical fluctuations. *Annals of Physics (N. Y.)*, **61**(1): 1–56, 1970.

[20] J. Kozuka *et al.* Dynamic polymorphism of single actin molecules in the actin filament, *Nature Chemical Biology*, **2**, 83–86, 2006.

[21] R. Kubo, M. Toda, N. Hashitsume. *Statistical Physics II: Nonequilibrium Statistical Mechanics*. London, Springer, 2013.

[22] H. P. Lu, L. Xun and S. Xie. Single-molecule enzymatic dynamics. *Science*, **282** (5395), 1877–1882, 1998.

[23] R. Milo and R. Phillips. *Cell Biology by the Numbers*. New York, Garland Science, 2016.

[24] R. Milo and R. Phillips. *BioNumbers: The Database of Useful Biological Numbers*. http://bionumbers.hms.harvard.edu.

[25] M. Morimatsu *et al.* Multiple-state reactions between the epidermal growth factor receptor and Grb2 as observed by using single-molecule analysis. *PNAS*, **104**(46): 18013–18018, 2007.

[26] K. Ohki and H. Miyata. *Physical Principles of Biomembranes and Cells*. Springer, 2018.

[27] F. Oosawa and Y. Nakaoka. Behavior of micro-organisms as particles with internal state variables, *Journal of Theoretical Biology*, **66**: 747–761, 1977.

[28] R. H. Pain (editor). *Mechanisms of Protein Folding*. Oxford University Press, 2001.

[29] C. A. Parent and P. N. Devreotes. A cell's sense of direction, *Science*, **284** (5415):765–770, 1999.

[30] J. Paulsson. Summing up the noise in gene networks, *Nature*, **427**, 415–418, 2004;

[31] R. Phillips *et al. Physical Biology of the Cell*. New York, Garland Science, 2012.

[32] E. M. Purcell. Life at low Reynolds number, *American Journal of Physics*, **45**(1), 1977.

[33] N. Rosenfeld *et al.* Gene regulation at the single-cell level, *Science*, **307** (5717): 1962–1965, 2005.

[34] Y. Sako and A. Kusumi. Compartmentalized structure of the plasma membrane for receptor movements as revealed by a nanometer-level motion analysis, *Journal of Cell Biology*, **125**(6): 1251–1264, 1994.

[35] R. Sawada and S. Mitaku. Biological Meaning of DNA compositional biases evaluated by ratio of membrane proteins. *Journal of Biochemistry*, **151** (2):189–196, 2012.

[36] M. J. Saxton and K. Jacobson. Single-particle tracking: Applications to membrane dynamics, *Annual Review of Biophysics and Biomolecular Structure*, **26**: 373–399, 1997.

[37] K. Sekimoto. *Stochastic Energetics*. Springer, 2010.

[38] T. Shibata and K. Fujimoto. Noisy signal amplification in ultrasensitive signal transduction, *PNAS*, **102** (2): 331–336, 2005.

[39] K. Sneppen and G. Zocchi. *Physics in Molecular Biology*. Cambridge University Press, 2005.

[40] P. S. Swain, M. B. Elowitz and E. D. Siggia. Intrinsic and extrinsic contributions to stochasticity in gene expression, *PNAS*, **99** (20): 12795–12800, 2002.

[41] Y. Taniguchi *et al.* Quantifying *E. coli* proteome and transcriptome with single-molecule sensitivity in single cells. *Science*, **329**(5991): 533–538, 2010.

[42] M. Toda, R. Kubo and N. Saito. *Statistical Physics I: Equilibrium Statistical Mechanics*. London, Springer, 2013.

[43] M. Ueda *et al.* Single-molecule analysis of chemotactic signaling in *Dictyostelium* cells. *Science*, **294**(5543): 864–867, 2001.

[44] M. Ueda and T. Shibata. Stochastic signal processing and transduction in chemotactic response of eukaryotic cell. *Biophysical Journal*. **93**(1): 11–20, 2007.

[45] D. Voet and J. G. Voet. *Biochemistry*. Hoboken, Wiley, 2010.

[46] T. Yanagida and Y. Ishii (Editors). *Single Molecule Dynamics in Life Science*. Weinheim, Wiley-VCH, 2008.

6

Langevin Equation and Fokker–Planck Equation

6.1 Microscopic Description of Brownian Motion (Langevin Equation)

The universality of Brownian motion on molecular scales and its significance were described in Chapter 5. In this chapter, by referring to the mathematical description of the random walk described in Chapter 5, we introduce the micro- and macroscopic descriptions of Brownian motion (Langevin equation and Fokker–Planck equation, respectively), and confirm that they are consistent with each other. In addition, the chemical Langevin equation and the corresponding Fokker–Planck equation, which are applied to the formulation of molecular number fluctuations in chemical reactions, are introduced, and the mechanism of state transitions based on fluctuations in dynamical systems and their functional significance in cells are discussed.

In Brownian motion, each time a solvent molecule collides with a particle, it is subjected to a force that changes the velocity of the particle. Here, the environment is a thermal equilibrium system at temperature $T[K]$ in which the number of solvent molecules around the particle is very large, and individual collisions occur in any direction for a very short period of time, smaller than the time scale of the particle's motion. Because the fluctuating force generated in this particle is the result of many simultaneous collisions, it is irregular and does not have a specific direction. As a result, the particle performs a random walk. As explained in the previous chapter, the random walk is discrete in both time and space. Can we express this in the form of differential equations that are continuous in both time and space? To accomplish this, first, let us consider Brownian motion in one dimension. By using the random force $\xi(t)$, the equation of motion is

$$m\frac{\mathrm{d}v(t)}{\mathrm{d}t} = -\gamma v(t) + \xi(t). \tag{6.1}$$

Then, the equation corresponding to the overdamped condition where the inertia can be neglected is $0 = -\gamma v + \xi(t)$, that is,

176

$$\frac{\mathrm{d}x(t)}{\mathrm{d}t} = \frac{\xi(t)}{\gamma} \equiv \eta(t). \tag{6.2}$$

Now, let us consider this motion not at the molecular level, but at the macroscopic level. Let us assume that during a short time Δt at the macroscopic level, collisions of microscopic solvent molecules occur many times, and as a result, random walks occur during Δt. Then, it is discontinuous at the microscopic level but can be described continuously at the macroscopic level. Since the change Δx between Δt is the result of the random walk, the statistical mean (of the random force) $\langle \Delta x \rangle$ must be 0, and its variance $\langle (\Delta x)^2 \rangle$ must be $2D\Delta t$. In fact, the solution of the preceding equation yields

$$\Delta x = \int_{t}^{t+\Delta t} \eta(t')\mathrm{d}t', \tag{6.3}$$

and $\langle \Delta x \rangle = 0$ because of $\langle \eta(t) \rangle = 0$. Also,

$$\begin{aligned}
\langle (\Delta x)^2 \rangle &= \int_{t}^{t+\Delta t} \int_{t}^{t+\Delta t} \langle \eta(t')\eta(t'') \rangle \mathrm{d}t' \mathrm{d}t'' \\
&= \int_{t-t'}^{t-t'+\Delta t} \int_{t}^{t+\Delta t} \langle \eta(t')\eta(t'+\tau) \rangle \mathrm{d}t' \mathrm{d}\tau \\
&= 2D\Delta t
\end{aligned} \tag{6.4}$$

needs to be satisfied. If the properties of this solvent do not change with time, that is, if the properties of the random force do not change with time (stationary), then $\langle \eta(t')\eta(t'+\tau) \rangle$ does not depend on t'. Then the upper integral is

$$\langle (\Delta x)^2 \rangle = \Delta t \int_{t-t'}^{t-t'+\Delta t} \langle \eta(t')\eta(t'+\tau) \rangle \mathrm{d}\tau. \tag{6.5}$$

By setting $\tau = t - t' - \tau'$, we obtain

$$\Delta t \int_{0}^{\Delta t} \langle \eta(t')\eta(t'+\tau') \rangle \mathrm{d}\tau' = 2D\Delta t. \tag{6.6}$$

Here, the collision of solvent molecules is random in macroscopic time, and there is no time correlation between them: if they hit right, next they hit right or left equally. In other words, this random force has no past memory and is $\langle \eta(t')\eta(t'+\tau') \rangle = 0$ ($\tau' \neq 0$). Then $\langle \eta(t')\eta(t'+\tau') \rangle$ needs to be a strange function of τ' as it is zero for $\tau' \neq 0$, but if we integrate over τ', it should satisfy $\int_{0}^{\Delta t} \langle \eta(t')\eta(t'+\tau') \rangle \mathrm{d}\tau' = 2D$. This is represented by using the delta function, that is, $\delta(x) = 0$ for $x \neq 0$ but $\int_{-\infty}^{\infty} \delta(x)dx = 1$, as

$$\langle \eta(t')\eta(t'+\tau') \rangle = 2D\delta(\tau'). \tag{6.7}$$

This is called white noise because the delta function in time contains all components uniformly in the frequency space when transformed by the Fourier transform, and has no special frequency component ("color"). Now this random force $\eta(t)$ (or $\xi(t)$) is a result of random microscopic collisions of molecules. For macroscopic time integral Δt, there are many collisions, leading to random walk. Hence for this time span, we can use the central limit theorem as in the previous chapter. Here the force $\eta(t)$ (or $\xi(t)$) follows Gaussian distribution, then this random force is also called Gaussian white noise.

Equation (6.1) appears to be an ordinary differential equation, but it is different from the usual one because the force $\eta(t)$ is expressed as a delta function, as shown in the preceding discussion. The value of $\langle \eta(t')\eta(t' + \tau')\rangle$ is zero for any short time τ', but if we integrate it over $\tau' = 0$, we get a finite magnitude, that is, a force of infinite magnitude can be added at any instant in time. Therefore, $x(t)$ itself is an undifferentiable function, even though it is continuous. This is because there are many random walks taking place at any given moment. The original reason why such a strange differential equation had to be introduced is that in a random walk or diffusion process, Δx is not proportional to time Δt, as in the ordinary differential equation, but $\langle (\Delta x)^2 \rangle$ is proportional to time Δt. This strange dependence on Δt had to be expressed by a continuous differential equation. Such equations were originally formulated in physics as the Langevin equation and in mathematics as the stochastic differential equation.

Next, let us deal with the Langevin equation (Eq. (6.1)) with inertial force and underdamped conditions. This is called the Ornstein–Uhlenbeck process, which is one of the simplest stochastic processes following a stationary distribution. For simplicity, let the mass $m = 1$. Then $\frac{dv(t)}{dt} = -\gamma v(t) + \xi(t)$ and the solution is

$$v(t) = v_0 e^{-\gamma t} + \int_{-\infty}^{t} e^{-\gamma(t-t_1)}\xi(t_1)dt_1, \tag{6.8}$$

with v_0 as the initial condition. After a sufficiently long time, the first term of the right-hand side of the preceding equation decays to 0, and

$$v(t) = \int_{-\infty}^{t} e^{-\gamma(t-t_1)}\xi(t_1)dt_1 \tag{6.9}$$

is derived. Here, $\langle v(t)\rangle = 0$ since $\langle \xi(t)\rangle = 0$. As $\xi(t)(= \gamma\eta(t))$ is white noise, it satisfies

$$\langle \xi(t_1)\xi(t_2)\rangle = 2I\delta(t_1 - t_2), \tag{6.10}$$

then

$$\langle v(t)v(t')\rangle = \int_{-\infty}^{t}\int_{-\infty}^{t'} e^{-\gamma(t-t_1)}e^{-\gamma(t'-t_2)}\langle \xi(t_1)\xi(t_2)\rangle dt_1 dt_2$$

$$= \int_{-\infty}^{t}\int_{-\infty}^{t'} e^{-\gamma(t-t_1)}e^{-\gamma(t'-t_2)}2I\delta(t_1 - t_2)dt_1 dt_2 \tag{6.11}$$

is derived. Denoting $u = t - t_1$ and $u' = t' - t_2$, and by assuming $\tau = t' - t \geq 0$,

$$\langle v(t)v(t+\tau)\rangle = \int_0^\infty \int_0^\infty e^{-\gamma u} e^{-\gamma u'} 2I\delta(u' - (u+\tau))\,du\,du'$$

$$= \int_0^\infty 2Ie^{-2\gamma u} e^{-\gamma \tau}\,du = \frac{I}{\gamma}e^{-\gamma\tau} \tag{6.12}$$

is obtained. That is, the correlation function of the velocity of the particle shows an exponential decay. As mentioned previously, since the system is in equilibrium for a sufficiently long time, the fluctuation behavior does not depend on how the time origin is taken. Therefore, by rewriting the time origin appropriately and letting its velocity be $v(0)$, the correlation function satisfies $\langle v(t)v(t+\tau)\rangle = \langle v(0)v(\tau)\rangle$. Then,

$$\begin{cases} \langle (v(t))^2 \rangle = \frac{I}{\gamma} \\ \int_0^\infty \langle v(0)v(\tau)\rangle\,d\tau = \frac{I}{\gamma^2} \end{cases} \tag{6.13}$$

is derived. Since the kinetic energy of a particle is calculated from the law of equipartition of energy in thermal equilibrium,

$$\frac{1}{2}\langle v(t)^2\rangle = \frac{1}{2}k_B T \tag{6.14}$$

has to be satisfied. (This is for one-dimensional systems, but the same is true for higher dimensions.) Hence,

$$I = \gamma k_B T \tag{6.15}$$

is derived from the upper equation in Eq. (6.13). By using Eq. (6.10) and the stationarity of the noise, the following relation is obtained:

$$\gamma = \frac{1}{2k_B T} \int_0^\infty \langle \xi(0)\xi(\tau)\rangle\,d\tau. \tag{6.16}$$

This relation shows that the time correlation function of the random force (fluctuation force) by the solvent molecules and the size of the viscous resistance of the particles (energy dissipation to the solvent) are connected with the temperature as a coefficient. This is called the "fluctuation-dissipation theorem (FDT)" (more precisely, the second fluctuation-dissipation theorem). Since the original source of the random force comes from solvent molecules, the connection between ξ and γ is natural and consistent, and the higher the temperature of the solvent, the greater the fluctuation force.

Furthermore, consider the case where the particles are subjected to a constant external force F. Then, the Langevin equation is given by

$$\frac{dv}{dt} = -\gamma v + F + \xi(t), \tag{6.17}$$

and the average speed in the steady state is

$$\langle v \rangle_{st} = \frac{F}{\gamma} = \mu F, \tag{6.18}$$

where μ indicates the degree of response of the particle to an external force (the coefficient of proportionality between the external force and velocity is called mobility) and is generally called the transport coefficient. Following the lower Eq. (6.13), Eq. (6.15), and Eq. (6.18), one gets

$$\int_0^\infty \langle v(0)v(\tau) \rangle d\tau = \frac{I}{\gamma^2} = \frac{k_B T}{\gamma} = \mu k_B T. \tag{6.19}$$

Thus,

$$\mu = \frac{1}{k_B T} \int_0^\infty \langle v(0)v(\tau) \rangle d\tau \tag{6.20}$$

is derived. This equation is also called "fluctuation-dissipation theorem" (more precisely, the first fluctuation-dissipation theorem) and it shows the relationship between the correlation function of the velocity of a particle in the equilibrium state without external force and the response rate (transport coefficient) of the system in the nonequilibrium state with external force. It is a pioneering example of the framework generally understood as a "linear-response theory" [24].

In addition, the Langevin equation without external force can be expressed in terms of position rather than velocity:

$$\frac{d^2 x}{dt^2} = -\gamma \frac{dx}{dt} + \xi. \tag{6.21}$$

By multiplying both sides of the equation by x and arranging it,

$$\frac{d^2}{dt^2} \left(\frac{x^2}{2} \right) - \left(\frac{dx}{dt} \right)^2 = -\gamma \frac{d}{dt} \left(\frac{x^2}{2} \right) + \xi x \tag{6.22}$$

is derived. By taking the time average of both sides of the equation,

$$\frac{d^2}{dt^2} \langle x^2 \rangle + \gamma \frac{d}{dt} \langle x^2 \rangle = 2 \langle v^2 \rangle = 2 k_B T \tag{6.23}$$

is derived, by noting that $\langle \xi x \rangle = 0$ since ξ and x are statistically independent and the cross-correlation between them is 0, and $\langle \xi \rangle = 0$, and by applying Eq. (6.14). By defining $\frac{d}{dt} \langle x^2 \rangle \equiv y$, the preceding equation becomes $\frac{dy}{dt} = -\gamma y + 2 k_B T$ and is solved as

$$y(t) = y(0)e^{-\gamma t} + \frac{2 k_B T}{\gamma} (1 - e^{-\gamma t}). \tag{6.24}$$

Note that the initial condition of $\frac{d}{dt} \langle (x(0))^2 \rangle$ is set to be 0, that is, $y(0) = 0$ due to the steady-state condition; then

$$y(t) = \frac{2 k_B T}{\gamma} (1 - e^{-\gamma t}) \tag{6.25}$$

is derived. By returning to the original variable x,

$$\frac{\mathrm{d}\langle x(t)^2\rangle}{\mathrm{d}t} = \frac{2k_{\mathrm{B}}T}{\gamma}(1 - e^{-\gamma t}).\tag{6.26}$$

Accordingly we get

$$\langle x(t)^2\rangle = \frac{2k_{\mathrm{B}}T}{\gamma}\left(t + \frac{1}{\gamma}(e^{-\gamma t} - 1)\right),\tag{6.27}$$

which is known as Fürth's formula. This is the formula for the mean square displacement of persistent random walks discussed in Section 5.10. By comparing it with the known equation $\langle x(t)^2\rangle = 2Dt$ under the condition that $t \gg 1/\gamma$ is satisfied after a long enough time, the Einstein relation is again derived:

$$D = \frac{k_{\mathrm{B}}T}{\gamma}\tag{6.28}$$

This can also be confirmed by combining $I = D\gamma^2$ from Eq. (6.2) and the two Eqs. (6.7) and (6.10) for noise and $I = \gamma k_{\mathrm{B}}T$ in Eq. (6.15).

Column VI: Anomalous Diffusion in Cell Movement

The amoeboid movement mentioned at the end of the previous chapter is similar to the persistent random walk because there is a velocity memory and the direction of motion is random when viewed from a time scale slower than the correlation time. In fact, it has been experimentally shown that the cell movements of various organisms have the characteristics of persistent random walk in the first approximation. As a typical mathematical model, the Ornstein–Uhlenbeck process has been employed to derive Eq. (6.27), as described earlier (note that the noise in cell movements is not solvent-derived thermal fluctuations, but nonthermal fluctuations derived from the dynamics of the signaling system in the cell). In recent years, however, more quantitative kinetic analysis of *Dictyostelium* and other amoeboid movement cells has revealed that the nature of the fluctuations is not a Gaussian distribution, but rather a more complex and rich motional mode with multiple time scales, including anomalous diffusion [16, 22], and their relationship with the aforementioned dynamic processes in cell biology and their physiological and functional significance are now being studied extensively.

6.2 Macroscopic Description of Brownian Motion (Fokker–Planck Equation)

In the previous section, we used the Langevin equation to describe the motion of the Brownian particle in a microscopic way. Now, let us describe the time evolution of the Brownian particles in a macroscopic way, as in the case of the random walk. If we integrate the Langevin equation without external forces $\frac{\mathrm{d}v}{\mathrm{d}t} = -\gamma v + \xi(t)$ for a short time Δt, we find that it is

$$\Delta v = -\gamma v \Delta t + \int_{t}^{t+\Delta t} \xi(t')\mathrm{d}t',\tag{6.29}$$

and by taking the time average and the variance,

$$
\begin{cases}
\langle \Delta v \rangle = -\gamma v \Delta t \\
\langle (\Delta v)^2 \rangle = \gamma^2 v^2 (\Delta t)^2 + \int_t^{t+\Delta t} dt' \int_t^{t+\Delta t} dt'' \langle \xi(t')\xi(t'') \rangle \\
\quad\quad\quad = 2I\Delta t + \gamma^2 v^2 (\Delta t)^2
\end{cases}
\tag{6.30}
$$

are derived. The $\langle (\Delta v)^3 \rangle$ and the higher-order moments are likewise of the order $(\Delta t)^2$ or higher, and are neglected. First, let us consider the probability density distribution of velocity, $P(v, t)$. In the Brownian motion, the velocity of a particle at $t = 0$ is $v = v_0$ and is a delta function, but the distribution expands with time. If the probability that v increases by Δv between Δt is $\phi(\Delta v; \Delta t)$,

$$
P(v, t + \Delta t) = \int_{-\infty}^{\infty} P(v - \Delta v, t)\phi(\Delta v; \Delta t)\mathrm{d}(\Delta v)
\tag{6.31}
$$

is required. Δt and Δv are assumed to be small, and by using Taylor expansion, we can write

$$
P(v, t) + \frac{\partial P(v, t)}{\partial t}\Delta t + \cdots
\tag{6.32}
$$

$$
= \int_{-\infty}^{\infty} \left(1 - \Delta v \frac{\partial}{\partial v} + \frac{(\Delta v)^2}{2}\frac{\partial^2}{\partial v^2} + \cdots \right) P(v, t)\phi(\Delta v; \Delta t)\mathrm{d}(\Delta v).
$$

From Eq. (6.30),

$$
\begin{cases}
\int_{-\infty}^{\infty} \phi(\Delta v; \Delta t)\mathrm{d}(\Delta v) = 1 \\
\int_{-\infty}^{\infty} \Delta v\phi(\Delta v; \Delta t)\mathrm{d}(\Delta v) = \langle \Delta v \rangle = -\gamma v \Delta t \\
\int_{-\infty}^{\infty} (\Delta v)^2 \phi(\Delta v; \Delta t)\mathrm{d}(\Delta v) = \langle (\Delta v)^2 \rangle = 2I\Delta t + \gamma^2 v^2 (\Delta t)^2
\end{cases}
\tag{6.33}
$$

are derived. The second term of the right-hand side of Eq. (6.32) can be arranged by the first order of Δt,

$$
\int_{-\infty}^{\infty} \Delta v \frac{\partial}{\partial v}(P(v, t)\phi(\Delta v; \Delta t))\mathrm{d}(\Delta v)
$$

$$
= \frac{\partial}{\partial v}\left(P(v, t)\int_{-\infty}^{\infty}(\Delta v\phi(\Delta v; \Delta t))\mathrm{d}(\Delta v) \right)
$$

$$
= -\gamma \frac{\partial}{\partial v}(vP(v, t))\Delta t.
\tag{6.34}
$$

As a result, $\frac{\partial P(v,t)}{\partial t}\Delta t = \gamma \frac{\partial}{\partial v}(vP(v, t))\Delta t + I\frac{\partial^2 P(v,t)}{\partial v^2}\Delta t$, that is,

$$
\frac{\partial P(v, t)}{\partial t} = \gamma \frac{\partial}{\partial v}(vP(v, t)) + I\frac{\partial^2 P(v, t)}{\partial v^2}.
\tag{6.35}
$$

This is the time-evolving equation of the probability density distribution of particles in velocity space, which is called the Fokker–Planck equation [18]. The steady-state solution of this equation is obtained by solving $\gamma v P = -I \frac{\partial P}{\partial v}$, leading to $P(v) \propto \exp(-\frac{\gamma v^2}{2I})$ (see Example 5.7). This is consistent with the Maxwell distribution that the velocity of the particle in equilibrium satisfies,

$$P(v) = \frac{1}{\sqrt{2\pi k_B T}} e^{-\frac{v^2}{2k_B T}}, \tag{6.36}$$

recalling that $I = \gamma k_B T$. This indicates that the Ornstein–Uhlenbeck process has a Gaussian distribution (Maxwell distribution is also a kind of it) as the equilibrium distribution of the velocity.

Next, let us consider the positional probability density distribution $P(x, t)$ rather than the velocity of the Brownian particle. If we can coarse-grain the time scale in which the inertial force works by focusing on the slow time scale behavior, then the overdamped Langevin equation

$$\gamma \frac{dx}{dt} = \xi(t) \tag{6.37}$$

is integrated for the time Δt,

$$\Delta x = \frac{1}{\gamma} \int_t^{t+\Delta t} \xi(t') dt' \tag{6.38}$$

and

$$\begin{cases} \langle \Delta x \rangle = 0 \\ \langle (\Delta x)^2 \rangle = \frac{1}{\gamma^2} \int_t^{t+\Delta t} dt' \int_t^{t+\Delta t} dt'' \langle \xi(t')\xi(t'') \rangle = \frac{2I}{\gamma^2} \Delta t \end{cases} \tag{6.39}$$

is derived. Thereafter, the same procedure as for derivation of $P(v, t)$ can be followed to obtain

$$\frac{\partial P(x, t)}{\partial t} = \frac{I}{\gamma^2} \frac{\partial^2 P(x, t)}{\partial x^2} = \frac{k_B T}{\gamma} \frac{\partial^2 P(x, t)}{\partial x^2}. \tag{6.40}$$

Since this is the Fokker–Planck equation for displacement and the form of the equation indicates that it is a diffusion equation, the Einstein relation of

$$D = \frac{k_B T}{\gamma} \tag{6.41}$$

is confirmed here as well. As discussed in the Chapter 5, the steady-state solution of this equation is a Gaussian distribution:

$$P(x, t) = \frac{1}{\sqrt{4\pi D t}} e^{-\frac{x^2}{4Dt}}. \tag{6.42}$$

When the drift with the velocity exists, $\langle \Delta x \rangle = v \Delta t$. Then, the Fokker–Planck equation becomes

$$\frac{\partial P(x, t)}{\partial t} = -\frac{\partial}{\partial x}(v P(x, t)) + D \frac{\partial^2 P(x, t)}{\partial x^2} \tag{6.43}$$

(this equation is called the advection-diffusion equation, especially the Smoluchowski equation), even when the velocity v drifts in the $+x$ direction due to a constant external force F. The first term on the right-hand side is the drift term of the mean displacement due to external force, and the second term is the dispersion term due to the diffusion. These results can also be confirmed by using Fick's law, which is a macroscopic law. Since Fick's first law plus the drift term of velocity v results in

$$J(x,t) = vP(x,t) - D\frac{\partial P(x,t)}{\partial x} \tag{6.44}$$

for diffusion under an external force, Fick's second law, together with the equation of continuity,

$$\frac{\partial P(x,t)}{\partial t} = -\frac{\partial J(x,t)}{\partial x}, \tag{6.45}$$

results in

$$\frac{\partial P(x,t)}{\partial t} = -\frac{\partial}{\partial x}(vP(x,t)) + D\frac{\partial^2 P(x,t)}{\partial x^2}. \tag{6.46}$$

This agrees with the Fokker–Planck equation (Eq. (6.43)).[1]

6.3 Poisson Process

The Poisson process is a steady-state stochastic process that satisfies the following conditions:

- The event occurs repeatedly with a fixed probability in time.
- Each event occurs independently.

Many steady-state chemical reactions can be described as Poisson processes. In the Poisson process, the latency period between events follows an exponential distribution, and the probability of n events occurring during time t follows the Poisson distribution. In the following, we will derive these properties.

Let $N(t)$ be the number of times that an event occurs during time t in the Poisson process ($N \geq 0$). Considering the short unit time Δt, the probability of two or more events occurring during the time is negligible, so either no event occurs or only one event occurs. Let k be the probability of an event per unit time (rate constant) and each probability will be

$$P(N(\Delta t) = 0) = 1 - k\Delta t, \quad P(N(\Delta t) = 1) = k\Delta t \tag{6.47}$$

---

[1] The Langevin equation, which is a molecular kinetic description of Brownian motion, was formulated by Langevin in 1908, based on Einstein's theory in 1905. Later, a method of projection operators to construct the Langevin equation from the microscopic dynamics with many degrees of freedom was developed by Mori and Zwanzig et al. [24, 15]. In addition, the "energy theory" for the Langevin equation as a "kinetic theory" has recently been formulated by Sekimoto et al. as a "stochastic energetics" [20].

From this, the probability of an event occurring n times during time t, $P_n(t) \equiv P(N(t) = n)$, satisfies the following master equation by the independence of each event:

$$P_{n+1}(t + \Delta t) = P_0(\Delta t)P_{n+1}(t) + P_1(\Delta t)P_n(t), \tag{6.48}$$

where the left side of the equation is the probability that the event occurs $n + 1$ times during time $t + \Delta t$, the first term on the right side of the equation is the probability that the event occurs $n + 1$ times during time t and not one time during the next short time Δt, and the second term on the right side of the equation is the probability that the event occurs n times during time t and one time during the next short time Δt. Substituting Eq. (6.47) into Eq. (6.48), we find the probability that

$$P_{n+1}(t + \Delta t) = (1 - k\Delta t)P_{n+1}(t) + k\Delta t P_n(t),$$
$$\frac{P_{n+1}(t + \Delta t) - P_{n+1}(t)}{\Delta t} = -k(P_{n+1}(t) - P_n(t)).$$

This equation is rewritten to the corresponding differential equation as $\Delta t \to 0$,

$$\frac{dP_{n+1}(t)}{dt} = -k(P_{n+1}(t) - P_n(t)) \tag{6.49}$$

($n \geq 0$). Note that in the case of $P_0(t)$, the master equation becomes

$$P_0(t + \Delta t) = P_0(\Delta t)P_0(t) = (1 - k\Delta t)P_0(t), \tag{6.50}$$

and the differential equation that shifts at $\Delta t \to 0$ is

$$\frac{dP_0(t)}{dt} = -kP_0(t). \tag{6.51}$$

In this case, the steady-state solution of Eq. (6.51) that satisfies the initial condition $P_0(0) = 1$ is

$$P_0(t) = e^{-kt}.$$

If the probability of the first event occurring during time $[t, t + \Delta t]$ is $Q_1(t)\Delta t$, it is equal to the probability that no event occurs before time t and only one event occurs during the next short time Δt, that is, $Q_1(t)\Delta t = P_0(t)k\Delta t$; then

$$Q_1(t) = ke^{-kt}, \tag{6.52}$$

which is the exponential distribution. Since each event is an independent and stationary stochastic process, $Q_1(t)$ gives the probability that the latency period between each event is t. In other words, in the Poisson process, the distribution of the latency period between each events follows the exponential distribution.

Returning to Eq. (6.49), the stationary solution that satisfies the initial condition $P_n(0) = 0$ ($n \geq 1$) can be confirmed by substitution,

$$P_n(t) = \frac{(kt)^n e^{-kt}}{n!}, \tag{6.53}$$

which is the Poisson distribution.[2] In other words, the probability of events occurring n (≥ 1) times during time t in a Poisson process follows the Poisson distribution. Let $Q_n(t)\Delta t$ be the probability that the nth (≥ 1) event occurs during time $[t, t + \Delta t]$, which is equal to the probability that the event occurs $n - 1$ times by time t and only once during the next short time Δt; then $Q_n(t)\Delta t = P_{n-1}(t)k\Delta t$, and

$$Q_n(t) = \frac{k^n t^{n-1} e^{-kt}}{(n-1)!},\qquad(6.54)$$

which is the Gamma distribution. In other words, the latency period distribution until the event occurs n (≥ 1) times in the Poisson process is the Gamma distribution (Allan distribution: a Gamma distribution in which n is an integer.)

6.4 Chemical Langevin Equation

As we have already mentioned, the amount of each component in the cell fluctuates in time, and also from cell to cell. On the other hand, in the usual dynamical systems, variables change in time deterministically in a unique manner for a given initial value. When a fixed point is reached, the amount remains constant and there is no fluctuation. This is not true in the cells, as molecular reactions are stochastic in nature. Then, how should we incorporate the fluctuations into the dynamical system? Now, let us look back at the differential equations of concentration. For example, for the reaction $X_2 + X_3 \rightarrow X_1$ with the reaction rate constant r, the change in the concentration x_1 of X_1 is given by the concentrations x_2, x_3 of X_2 and X_3 and rate constants, and is written as $\frac{\mathrm{d}x_1}{\mathrm{d}t} = rx_2x_3$ in the dynamical system. However, a chemical reaction is originated from collisions of X_2 and X_3 molecules with random motion to produce a reaction product with a certain probability. The rx_2x_3 is the average amount of reaction events per unit time. If many samples of these concentrations are prepared, the actual number of reactions will fluctuate around this average value by samples. Therefore, it is necessary to have a formulation in which the variation around the mean is incorporated. If we consider the change over a short period of time, we can consider x_1, x_2, x_3 to be constant within that time. Then, the actual reaction events are distributed around this mean rx_2x_3 without any bias, that is, the change of concentration and fluctuations can be considered to be separated at each instant. On the other hand, if there are many molecules and sufficient reactions occur in a short time, we can treat this fluctuation as a statistical distribution. That is to say, the change in concentration over time can be treated as a

---

[2] From the setting of Eq. (6.47), the probability of the event occurring n times at time $t = m\Delta t$, $P_n(m\Delta t)$, is a binomial distribution. By taking the limit $m \rightarrow \infty$ with $\Delta t = t/m$, then $\Delta t \rightarrow 0$, $k\Delta t \rightarrow 0$, $mk\Delta t = kt$, and $P_n(t)$ become the Poisson distribution.

stochastic differential equation, in which the change of the mean quantity is added to the surrounding fluctuation term.

The preceding consideration is what Turing discussed in his posthumous paper (see collected works [27]), which is probably the first treatment of chemical reaction systems by stochastic differential equations. Although it is somewhat different from the current standard form of stochastic differential equations and is written in a similar manner as the linear fluctuation, the essence of stochastic differential equations is pioneered in this paper. Since then, the treatment of such fluctuations has been established in a promising form by the expansion by the inverse of the system size [10, 14], and the chemical reaction Langevin equation by the time-scale separation [7]. In this section, we will explain the stochastic differential equation of the chemical reaction system based on the Langevin equation, which seems to be conceptually easy to understand. In fact, this is almost the same form as that described by Turing in his manuscript.

First, consider the case where the molecular number X_i of each component $i = 1, \ldots, k$ changes in the reaction of M kinds. We consider the change in each reaction at a certain short time Δt, which is short enough that the concentration does not change in the interval, but long enough that the reaction occurs sufficiently at the molecular level and can be treated statistically. Let us assume that the time scales of micro molecular collisions and macro concentration changes are separated and can be treated in this way (see Fig. 6.1). Then, in this time scale, the change in each reaction can be considered independently. Therefore, the change in one reaction can be divided into its mean and its surrounding fluctuations, and then all the reactions can be added together.

Now, consider the case where a single reaction causes a change of mean $\Delta X_i = F_i(\{X_j\})(j = 1, \ldots, k)$ during Δt. For example, in the case of the reaction described earlier, $F_1 = rX_2X_3$. Since this is an average, there is a fluctuation around it. From the standpoint of each molecule, the stochastic reaction events between Δt can be considered to arise randomly according to a Poisson process with the mean F_i (see Section 6.3). Since it is a Poisson process, the variance of this probability distribution is the same as (the absolute value of) the mean, that is, the mean change between Δt is $F_i \Delta t$ and its variance is $|F_i|\Delta t$. Here, if many of these random events can occur during Δt, we can use the central limit theorem for the process between them. If the total number of molecules is sufficiently large, we may accept this assumption. Then, the fluctuations between Δt and the noise term around the mean can be expressed by the Gaussian distribution with the variance of $|F_i|\Delta t$. Now, in terms of the concentration change at the macroscopic side, the fluctuation terms between t and $t + \Delta t$, $t + 2\Delta t$ can be considered to be independent. If the limit of $\Delta t \to 0$ on the macro time scale is taken, the Langevin equation as $X = (X_1, X_2, \ldots, X_k)$,

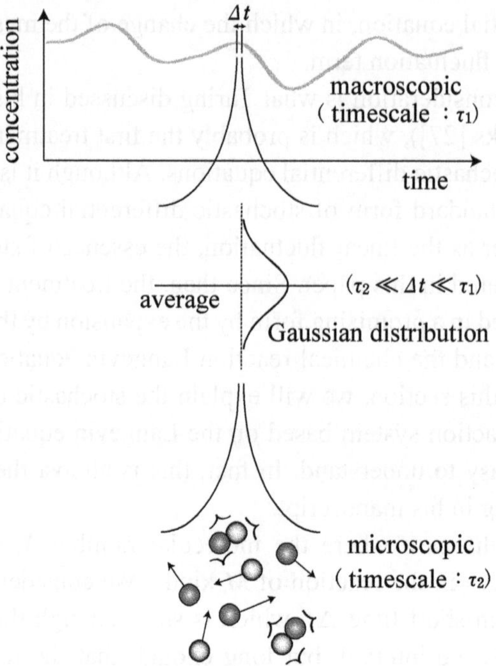

Figure 6.1 Micro–macro relationship in the condition where the Langevin equation holds

$$\frac{\mathrm{d}X_i}{\mathrm{d}t} = F_i(\{X_j\}) + \sqrt{|F_i(\{X_j\})|}\,\eta_i(t), \tag{6.55}$$

is obtained. Note that $\eta_i(t)$ is Gaussian white noise satisfying

$$\langle \eta_i(t)\eta_i(t')\rangle = \delta(t - t'). \tag{6.56}$$

By using the concentration $x_i = X_i/V$ and the reaction rate constant $f_i = F_i/V$ instead of the number of molecules X_i,

$$\frac{\mathrm{d}x_i}{\mathrm{d}t} = f_i(\{x_j\}) + \sqrt{\left|\frac{f_i(\{x_j\})}{V}\right|}\,\eta_i(t). \tag{6.57}$$

By summing up M kinds of chemical reactions,

$$\frac{\mathrm{d}x_i}{\mathrm{d}t} = \sum_{\mu=1}^{M} f_i^{\mu}(\{x_j\}) + \sum_{\mu=1}^{M} \sqrt{\left|\frac{f_i^{\mu}(\{x_j\})}{V}\right|}\,\eta_i^{\mu}(t), \tag{6.58}$$

$$\langle \eta_i^{\mu}(t)\eta_i^{\mu'}(t')\rangle = \delta_{\mu\mu'}\delta(t - t') \tag{6.59}$$

is derived. This is what is called the chemical Langevin equation ($\delta_{\mu\mu'}$ is based on the assumption that there is no correlation between each reaction).

The preceding derivation is based on the premise that the time range Δt can be such that the macroscopic concentration change within it is sufficiently small, whereas the microscopic reaction events occur sufficiently often between the macroscopic changes. Since macroscopic changes do not occur during Δt, the amplitude of the noise $\eta_i^{\mu}(t)$ is given by x of time t, while reaction events occur sufficiently many times on the microscopic view during this Δt. (A "safe" assumption to satisfy this premise would be that the system size is large and there are many molecules.) Then, it is possible to consider that many reactions are occurring during this Δt.

In the preceding equation, there are many noise terms $\eta_i^{\mu}(t)$, which may seem messy. In fact, they can be combined into one term. Let $\eta_i(t) = \sum_{\mu=1}^{M} \sqrt{|f_i^{\mu}(\{x_j\})/V|} \eta_i^{\mu}(t)$, which is the sum of the noise of the Gaussian distribution (see the central limit theorem and stable distribution in Chapter 5); then $\eta_i(t)$ is also Gaussian and its mean is 0, and the variance is as follows due to its additivity

$$\langle \eta_i(t)\eta_i(t') \rangle = \sum_{\mu=1}^{M} \left| \frac{f_i^{\mu}(\{x_j\})}{V} \right| \delta(t - t'). \tag{6.60}$$

That is, the noise term in Eq. (6.58) can be written as one term: $\left(\sqrt{\sum_{\mu=1}^{M} |f_i^{\mu}(\{x_j\})|} / \sqrt{V} \right) \eta_i(t)$ Then Eq. (6.58) is reduced to

$$\frac{\mathrm{d}x_i}{\mathrm{d}t} = f(\{x_j\}) + \sqrt{\epsilon} g(\{x_j\})\eta_i(t) \tag{6.61}$$

by using $f(\{x_j\}) = \sum_{\mu=1}^{M} f_i^{\mu}(\{x_j\})$, $g(\{x_j\}) = \sqrt{\sum_{\mu=1}^{M} |f_i^{\mu}(\{x_j\})|}$, and $\epsilon = 1/V$.[3]

As already mentioned, the Langevin equation represents the time evolution for each sample, while the Fokker–Planck equation represents the time evolution of the distribution $P(\{x_j\})$. In the following, we consider the probability distribution for a single component x, and abbreviate i of x_i.[4] The Fokker–Planck equation corresponding to Eq. (6.61) is given by

$$\frac{\partial P}{\partial t} = -\frac{\partial}{\partial x}(f(x)P(x, t)) + \frac{\epsilon}{2}\frac{\partial^2}{\partial x^2}(g(x)^2 P(x, t)). \tag{6.62}$$

Note that in the chemical Langevin equation, the diffusion coefficient corresponding to a certain x is $g(x)^2$. This term in the chemical Langevin equation is derived based on the assumption that macroscopic changes in x are paused during Δt. In stochastic differential equations, this interpretation is called the Ito type. On the other hand, if we take into account the effect that x changes little by little during Δt, it seems natural to include this effect in the noise term as well. An

---

[3] The noise $\eta_i(t)$ and $\eta_j(t)$ between the different components is not necessarily uncorrelated. If component i and component j are involved in the same reaction, they are attributed to a common $\eta^{\mu}(t)$. Then we need to go back to the definition of $\eta_i(t)$ and consider the correlation.

[4] Similar equations can be obtained for multicomponent systems. However, the steady-state distribution is not always easy to obtain as follows.

interpretation of this form (the Stratonovich type) would lead to including the x-dependence of $g(x)$ during the partial derivative. The Fokker–Planck equation in this case has a slightly different order between derivatives and $g(x)$, which gives

$$\frac{\partial P}{\partial t} = -\frac{\partial}{\partial x}(f(x)P(x,t)) + \frac{\epsilon}{2}\frac{\partial}{\partial x}(g(x)\frac{\partial}{\partial x})(g(x)P(x,t)). \tag{6.63}$$

In the ordinary chemical Langevin equations, the Ito-type interpretation is used, but since this latter form is also natural as a physical interpretation, it is also often used (see Column VII for details). In fact, replacing f with $f + \frac{\epsilon g(x)}{2}\frac{\mathrm{d}g(x)}{\mathrm{d}x}$ in the Stratonovich type gives the Ito-type formula, so this difference is not so important as long as the last noise term $\epsilon = 1/V$ is small (or if the volume V or the number of particles is large).

In the steady state of Eq. (6.62),

$$-f(x)P(x) + \frac{\epsilon}{2}\frac{\mathrm{d}}{\mathrm{d}x}(g(x)^2 P(x)) = 0 \tag{6.64}$$

is satisfied and it follows $-\frac{f(x)\phi}{g^2} + \frac{\epsilon}{2}\frac{\mathrm{d}\phi}{\mathrm{d}x} = 0$ by substituting $g^2 P = \phi$. Thus, $\phi = \exp\left(\frac{2}{\epsilon}\int \frac{f(x)}{g(x)^2}\mathrm{d}x\right)$ is derived. Then, the steady-state distribution of Eq. (6.62) is expressed as

$$P(x) = \frac{1}{g(x)^2}\exp\left(\frac{2}{\epsilon}\int \frac{f(x)}{g(x)^2}\mathrm{d}x\right). \tag{6.65}$$

In the Stratonovich-type interpretation, it is only necessary to replace f with $f + \frac{\epsilon g}{2}\frac{\mathrm{d}g}{\mathrm{d}x}$, and the solution is given by replacing the prefactor $\frac{1}{g(x)^2}$ in Eq. (6.65) with $\frac{1}{g(x)}$.

The preceding equation for the time evolution of $P(\{x_1, x_2, \ldots, x_k; t\})$ can also be obtained by starting from an equation representing a change in the probability distribution and expanding it with $1/V$. Specifically, consider the probability $P(\{N_1, N_2, \ldots, N_k; t\})$ that there are N_i particles for species i ($i = 1, \ldots, k$). By each reaction, the number of associated particles increases or decreases. The probability of the reaction depends on the rule of the reaction and the distribution of the number of particles $P(\{N_1, N_2, \ldots, N_k; t\})$ at that time. For example, if the reaction occurs with a rate constant k of $X_2 + X_3 \rightarrow X_1$, then

$$\frac{\mathrm{d}P(\{N_1, N_2, \ldots, N_k; t\})}{\mathrm{d}t}$$
$$= k(P(\{N_1 - 1, N_2 + 1, N_3 + 1, \ldots, N_k; t\}) - P(\{N_1, N_2, \ldots, N_k; t\})).$$

If there are many reactions, you can add more terms of such a time evolution for each type of reaction. The time evolution of $P(\{N_1, N_2, \ldots, N_k; t\})$ thus derived is called the master equation. Since each N_i can take a value from 0 to the maximum number of N_i, this equation involves a huge number of terms. By assuming that N_i is large and that the variance around the mean value of N_i is small compared to it, one can simplify the equation by expanding it around the mean. This expansion is called the Kramers–Moyal expansion, and if we take this expansion to the second order of $1/N_i \propto 1/V$, the Fokker–Planck equation is obtained, which agrees with that derived from the chemical Langevin equation.

Column VII: The Ito Type and the Stratonovich Type of Stochastic Differential Equation

When considering the noise term, one might think that it is a little uncomfortable to fix $x(t)$ as the value of t during Δt, and that it would be better to use the value between t and $t + \Delta t$. In fact, the stochastic differential equation using the former form is called the Ito type, and the interpretation using $t + \Delta t/2$ is called the Stratonovich type. Since x can vary slightly during Δt, the Stratonovich-type interpretation could be natural from a physical point of view, while the Ito type is easier to handle from a mathematical analysis. Of course, there is no difference between the two types if the term of the noise strength $g(x)^2$ does not depend on the variable x. Also, the two are convertible with each other, even if they depend on x. In fact, if we consider Eq. (6.61)), the difference between the two comes from the change in $g(x)$ within Δt, so we only need to add $\frac{\epsilon g(x)}{2} \frac{dg(x)}{dx}$ to $f(x)$ to convert the Stratonovich type to the Ito type. This can be also confirmed by looking at the derivation of the Fokker–Planck equation. The interpretation of the Ito type is that the diffusion coefficient does not change while x diffuses in a random walk, resulting in Eq. (6.62). On the other hand, the Stratonovich type takes in the effect that the diffusion coefficient changes, so Eq. (6.63) is obtained. However, in the current derivation of the chemical Langevin equation, there will be no precision to decide which of the two should be adopted. In any case, the difference is negligible where V is large (ϵ is small).

On the other hand, the preceding derivation of the Langevin equation may not be so rigorous: The magnitude of the fluctuation is assumed not to change in Δt, but the magnitude of the fluctuation in the Langevin equation depends on the concentration variable itself, as $\sqrt{|f_i(\{x_j\})|}$. In the 1970s, van Kampen [10] and Kubo, Matsuo, Kitahara [14] formulated the expansion by the inverse of the system size V instead of imposing the existence of an appropriate Δt. In this case, it is formulated so that the mean value of $\overline{x_i}$ first evolves in time as a dynamical system at the $V \to \infty$ limit, and then fluctuations change around it. Suppose that there are $N_i \propto V$ particles of each component i in the system of volume V, and the concentration $x_i = N_i/V$ is divided into its ensemble mean value $\overline{x_i}$ and its fluctuations $\delta x_i = x_i - \overline{x_i}$, which are treated as separate variables. In this case, the fluctuation is assumed to be small, that is, the order of $1/\sqrt{V}$. Then, Eq. (6.57) turns out to be divided into

$$\frac{d\bar{x}_i}{dt} = f_i(\{\bar{x}_j\}); \quad \frac{d\delta x_i}{dt} = \sqrt{\left|\frac{f_i\{\overline{x_j(t)}\}}{V}\right|}\,\eta_i(t).$$

In this way, the temporal evolution of the fluctuation is determined by $\overline{x_i(t)}$. Then, there is no question of interpreting it as Ito or Stratonovich. However, any of the preceding forms of the Langevin equations, and the two chemical Langevin equations of Ito or Stratonovich type agree asymptotically as V (or the number of particles) is large.

For simplicity, we will explain this derivation by taking an example of a reaction in which the number of particles in a single component increases or decreases by one (an extension to the general equation is easy). By representing the probability that the number of particles is n as $P(n, t)$, and representing the reaction rate constant to increase the component by one as $r(n)$ and to decrease by one as $d(n)$, the master equation is given by

$$\frac{dP(n, t)}{dt} = r(n-1)P(n-1, t) - r(n)P(n, t) + d(n+1)P(n+1, t) - d(n)P(n, t).$$

$$(6.66)$$

Now assuming that n is greater than 1, we expand the function as

$$F(n \pm 1) = F(n) \pm \frac{dF(n)}{dn} + \frac{1}{2}\frac{d^2 F(n)}{dn^2} + \cdots .$$

By taking the derivative to the second-order term, Eq. (6.66) turns out to be

$$\frac{dP(n, t)}{dt} = -\frac{\partial}{\partial n}((r(n) - d(n))P(n, t)) + \frac{1}{2}\frac{\partial^2}{\partial n^2}((r(n) + d(n))P(n, t)). \quad (6.67)$$

If we consider $x = n/V$ as a concentration, the preceding Taylor expansion is an expansion by $1/V$ against the change in the distribution $P(x, t)$ of concentration x, so that $f(x) = r(x) - d(x)$, $g(x) = \sqrt{|r(x)| + |d(x)|}$. Then, Eq. (6.67) agrees with the Fokker–Planck equation (Eq. 6.62) derived from the chemical Langevin equation.

Example 6.1   Let us consider the reaction $A \rightarrow X \rightarrow$ (decomposition) as discussed in Section 5.4. By denoting the rate constant of each reaction as k' and k, temporal change of the concentration $x = X/V$ is given by

$$\frac{dx}{dt} = r - kx + \sqrt{\frac{r + kx}{V}}\,\eta(t), \quad (6.68)$$

where $r = ak'$, with a as the concentration of A. The corresponding Fokker–Planck equation is given by

$$\frac{\partial P(x, t)}{\partial t} = \frac{\partial}{\partial x}((kx - r)P(x, t)) + \frac{1}{2} + \frac{\partial^2}{\partial x^2}\left(\frac{r + kx}{V}P(x, t)\right). \quad (6.69)$$

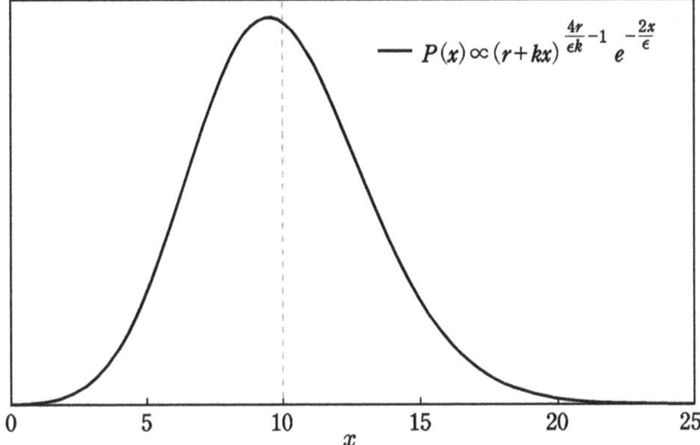

Figure 6.2 An example of Gamma distribution (the ordinate is in the normal scale).

As the steady-state solution of $dx/dt = r - kx$ is given by $x^* = r/k$, so that the variance of fluctuations around the steady-state solution is given by $2r/V$. Also, by using the formula (6.65) with $f = r - kx$ and $g = \sqrt{r + kx}$ the stationary distribution of (6.69) is given by

$$P(x) \propto (r + kx)^{\frac{4r}{\epsilon k}-1} \exp\left(-\frac{2x}{\epsilon}\right). \tag{6.70}$$

By shifting the origin of x to $x \to x + r/k$, this distribution turns out to be the Gamma distribution (Fig. 6.2).

Exercise 6.4.1 Verify that the Gamma distribution of Eq. (6.70) converges to the Gaussian distribution around the mean $x^* = r/k$ in the limit of $\epsilon \to 0$, and derive the variance of the distribution.

6.5 Transition due to Minority

Naturally, as the volume or total number of molecules is smaller, the fluctuation term is larger and the width of the distribution is broader. In the previous example, the variance of the distribution around a fixed point $x^* = r/k$ is given by $2r/V$, which increases as V is decreased. Then, is there a case where the nature of the distribution changes, not only the magnitude of fluctuation (which increases with the decrease of the volume or the total number of molecules)? This was discussed as a noise-induced transition in the 1970s [9]. Since this is not limited to chemical reaction systems, we first discuss it generally in the Langevin equation, including the chemical Langevin equation. For simplicity, let us again consider the Langevin equation of one variable x:

$$\frac{dx}{dt} = f(x) + \sqrt{\epsilon} g(x) \eta(t). \tag{6.71}$$

In this case, the stationary distribution of the corresponding Fokker–Planck equation is

$$P(x) = \frac{1}{g(x)^2} \exp\left(\frac{2}{\epsilon} \int \frac{f(x)}{g(x)^2} dx\right). \tag{6.72}$$

The question here is whether the shape of the distribution changes with increasing ϵ. (Note that in the study of noise-induced transitions, it is better to use the form in physics (Stratonovich type) (6.63), which can be obtained by replacing the first term (prefactor) $\frac{1}{g(x)^2}$ with $\frac{1}{g(x)}$ in Eq. (6.72).) Here, if $g(x) = $ constant, the distribution is proportional to $\exp(\frac{2}{\epsilon} \int f(x) dx)$. That is, by using the potential $U(x)$ with $f(x) = -\frac{dU}{dx}$, $P(x) \propto \exp(-\frac{2}{\epsilon} U(x))$. Then, as ϵ is larger, $P(x)$ only spreads around the potential bottom $x = x^*$ that satisfies $f(x^*) = \frac{dU(x^*)}{dx} = 0$. (For example, if $U(x) = \frac{1}{2}kx^2$, the steady-state distribution becomes Gaussian around $x^* = 0$, and the variance of the Gaussian distribution is larger as ϵ increases.) On the other hand, if the noise intensity $g(x)$ depends on x, the shape of the distribution may change as ϵ increases. Let us consider the following example.

Example 6.2   Consider the cases $f(x) = ax - cx^m$, $g(x) = x$ [19, 21]. For example, x yields x with a reaction rate constant a, which fluctuates. This alone would cause x to diverge. We assume that when m $(m > 1)$ molecules of x are assembled, they are decomposed. Then, the term $-cx^m$ is introduced and the divergence is suppressed. The steady-state distribution in this case is given by Eq. (6.65) as

$$P(x) \propto x^{\frac{2a}{\epsilon}-2} \exp\left(\frac{-2cx^{m-1}}{\epsilon(m-1)}\right). \tag{6.73}$$

This distribution has a peak at a finite value of x when $a > \epsilon$, but has a peak at $x = 0$ when $\epsilon > a$ (see Fig. 6.3). In other words, the position of the peak is shifted to zero with the magnitude of the noise. In this system, in the absence of noise, the equation of the dynamical system $dx/dt = ax - cx^m$ has two fixed points, $x = 0$ and $x = (a/c)^{1/(m-1)}$, and at $a > 0$, the latter is the stable fixed point (attractor). If the noise is not so large, x is distributed around this fixed point, However, the larger x is, the relatively larger noise intensity is applied in the Langevin equation (Eq. 6.71). Therefore, when the noise increases beyond a certain level, the distribution "shifts" so that it has a peak toward $x = 0$.

Now, in a chemical reaction system, can such a transition occur with the decrease in molecular numbers? An example using four components is given by [25], but let us simplify this to a reaction with two components [4, 17]. Specifically, consider the reaction $X + Y \rightarrow 2X$, then $Y + X \rightarrow 2Y$ (respective reaction speed a), and then $X \rightarrow Y$, $Y \rightarrow X$ (respective reaction speed b). Then, the change of X amount in each reaction is proportional to aXY, $-aYX$, $-bX$, and bY. Since the sum of X

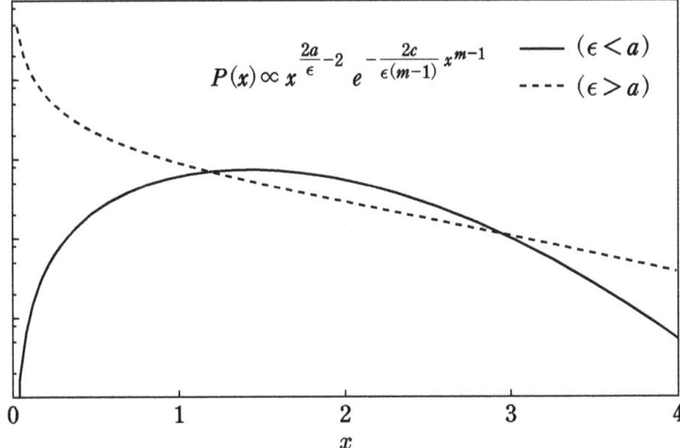

Figure 6.3 An example of noise-induced transition (the ordinate is in the logarithmic scale).

and Y is conserved in this model, if the total concentration of X and Y is set to 1, the concentration of Y is given by $1 - x$, and the rate constants for each reaction are $\pm ax(1 - x)$, $-bx$, and $b(1 - x)$. Therefore, it follows that

$$\frac{dx}{dt} = b(1 - 2x) + \sqrt{\frac{2ax(1 - x) + b(x + 1 - x)}{V}}\,\eta(t). \tag{6.74}$$

Let $z = 2(x - 1/2)$; then $-1 \le z \le 1$ becomes, because $0 \le x \le 1$ and $0 \le y \le 1$. We then obtain

$$\frac{dz}{dt} = -2bz + \sqrt{\frac{2a(1 - z^2) + 4b}{V}}\,\eta(t). \tag{6.75}$$

By using $f(z) = -2bz$, $g(z) = \sqrt{2a(1 - z^2) + 4b}$ in Eq. (6.65), the stationary distribution is given by

$$P(z) \propto g(z)^{\frac{2bV}{a} - 2}. \tag{6.76}$$

If the noise term can be ignored, the stationary solution of equation $\frac{dz}{dt} = -2bz$ is given by $z = 0$. That is, $x = 1/2$, implying that there are equal quantities of both X and Y. In Eq. (6.75), the right side is only $-2bz$, so even if noise enters, as long as it is small, i.e., total molecular number is large, the distribution in Eq. (6.76) will show a single-peaked distribution around $z = 0$ (Fig. 6.4). In fact, in the preceding equation, if $bV > a$, then the distribution has a peak at $z = 0$.

However, when V becomes smaller and $bV < a$, this distribution transitions to that with peaks at $z \sim -1$ and $z \sim 1$, as shown in Fig. 6.4. In other words, the distribution makes the transitions to a state where only X exists for $x = 1$ or Y exists for $x = 0$ in time. This is an example of a noise-induced transition, in which the peak position of the distribution shifts due to the increase of noise, occurring due

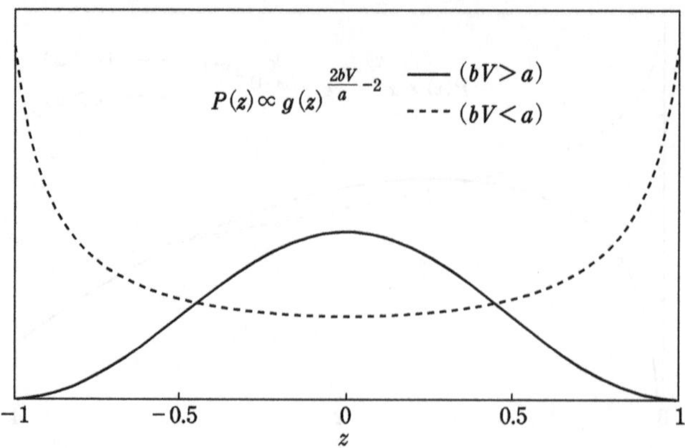

Figure 6.4 An example of noise-induced transition (the ordinate is in the normal scale).

to the decrease of size V. Of course, if V (or the number of particles) is smaller, the expansion at $1/V$ itself may be dangerous, and the approximation expressed in the Langevin equation itself may be worse. However, according to numerical calculations, the Langevin equation can almost quantitatively predict such transition that will occur in the chemical reaction (although the transition point of V may deviate slightly depending on the interpretation of the stochastic differential equation). When V is much smaller, phenomena beyond the scope of this approximation occur. In fact, if the total number of particles goes down to a few, or if the number of kinds of reacting molecules increases, the collision of relevant reacting molecules does not occur easily. In this case, the steady state continues for a long period of time, and when a reaction occurs, the reaction continues like an avalanche [1]. Although there are many kinds of molecules in a cell, there are many components with a small number of molecules each, as seen in Chapter 5, so this phenomenon derived from the small number of molecules is important in considering biological phenomena.

6.6   Intracellular Fluctuations and Their Functional Significance

In fact, the distribution of the amount of protein in a cell has been extensively studied. The distribution of protein concentrations in cells is found to be not a Gaussian (normal) distribution, but often follows approximately a log-normal distribution in which the logarithm of the amount is a Gaussian distribution, that is,

$$P(x) \propto \frac{1}{x} \exp\left(-\frac{(\log(x) - \log(c))^2}{\sigma^2}\right), \tag{6.77}$$

or a Gamma distribution,

$$P(x) \propto x^{K-1} \exp\left(-\frac{x}{c}\right) \tag{6.78}$$

[5, 23]. The origins of such distributions have been discussed in various ways.

According to the central limit theorem, the Gaussian distribution appears by the addition of random variables (additive mean). In contrast, let us consider the case where the statistic is determined by the multiplication of random variables (multiplicative mean). In this case, since multiplication is transformed to addition by taking the logarithm, one can apply the central limit theorem for the variables after the logarithm is taken. This is what can be easily considered as the origin of log-normal distribution [5], as the concentration change by catalytic reactions (as seen in the term $x_i x_j$ in dynamics) is multiplicative in nature. On the other hand, the Gamma distribution is a Poisson distribution with x^{K-1}, which tends to appear as the result of a random process in K steps (see Section 6.3). As seen in Eq. (6.70), it also appears as a result of the chemical Langevin equation of a simple synthesis-decomposition process. When there is a multiplicative process, the distributions are skewed to a larger side of variables, as seen in log-normal and Gamma distributions.

As another example of multiplicative process, let us consider the concentration distribution due to fluctuations in growth rate. Suppose that the cell volume, V, is expressed as $\frac{dV}{dt} = \mu V$ using the growth rate μ. Then, the concentration is diluted by the ratio. As we have already discussed, let us assume that a protein is produced by a reaction rate constant r and decomposed by k. The change in concentration includes the dilution term due to volume growth, and is given by

$$\frac{dx}{dt} = r - kx - \mu x, \tag{6.79}$$

which leads to the stationary value $x = r/(k + \mu)$. Let us then assume that the fluctuations in the growth rate follow a Gaussian distribution around the mean value of μ with the variance ϵ_g. Then, not only are the concentration fluctuations based on the reaction, but also the growth rate fluctuations are included, and the Langevin equation is given by

$$\frac{dx}{dt} = r - kx - \mu x + x\eta_g(t) + \sqrt{\frac{r + kx}{V}}\eta(t), \tag{6.80}$$

where $\langle \eta_g(t)\eta_g(t')\rangle = \epsilon_g\delta(t - t')$. Let us consider only the effect of $\eta_g(t)$, assuming that V is large. Then the Fokker–Planck equation for the preceding equation is given by

$$\frac{\partial P}{\partial t} = -\frac{\partial}{\partial x}((r - (k + \mu)x)P(x, t)) + \frac{\epsilon_g}{2}\frac{\partial}{\partial x}(x\frac{\partial}{\partial x})(xP(x, t)). \tag{6.81}$$

The steady-state distribution is given by[5]

$$P(x) \propto \frac{1}{x} \exp\left(\frac{2}{\epsilon_g} \int \frac{r - (k + \mu)x}{x^2} \, \mathrm{d}x \right) = x^{-\frac{2(k+\mu)}{\epsilon_g} - 1} \exp\left(-\frac{2r}{\epsilon_g x} \right). \tag{6.82}$$

Indeed, this distribution agrees well with the experimental results [26]. On the other hand, when we estimate the magnitude of the noise $\eta_g(t)$ and $\sqrt{(r + kx)/V}\,\eta(t)$ from the experiment, the former is several tens of times larger for the cells with sufficiently fast growth. These two noises correspond to the extrinsic and intrinsic noises mentioned in Chapter 5. Unlike fluctuations in the expression of each gene, fluctuations in the growth rate are exogenous and confer a common noise on all genes. Hence, growth fluctuations produce a positive correlation between all expression fluctuations This has also been observed in experiments [26].

Exercise 6.6.1 Assume that the previously mentioned proteins synthesize a factor for cell division. Suppose that the factor is produced in proportion to the protein concentration x and the cell-division time is inversely proportional to x. Show that the division time has a Gamma distribution.

Attractor selection and adaptation: As described in Section 1.14, from the standpoint of the dynamical system, different cell types were considered to fall into different attractors. If there is no noise, once the cell has fallen into an attractor, it does not leave the attractor, and the cell remains in that state. On the other hand, if there is a noise term, then switching to another attractor can occur, and if sampled over a long period of time, the probability of its existence at a certain value x is given by the stationary distribution $P(x)$ of the Fokker–Planck equation. Therefore, the distributions of the cells to different states are obtained.

If $\frac{\mathrm{d}x}{\mathrm{d}t} = f(x) + \sqrt{\epsilon}g(x)\eta(t)$ and $g(x)$ of the noise term $\sqrt{\epsilon}g(x)\eta(t)$ is constant, then the steady-state distribution $P(x)$ in Eq. (6.72) is given by $P(x) \propto \exp(-\frac{2}{\epsilon}U(x))$ by the potential $U(x)$ satisfying $f(x) = -\frac{\mathrm{d}U(x)}{\mathrm{d}x}$ as described in Section 6.4. Since $\frac{\mathrm{d}f}{\mathrm{d}x} = -\frac{\mathrm{d}^2U}{\mathrm{d}x^2} < 0$ at the stable fixed point x^*, $P(x)$ has a peak at x^* regardless of the magnitude of ϵ. In other words, the distribution is concentrated around the attractor. On the other hand, if $g(x)$ depends on x, the steady-state distribution is replaced by $U_{\mathrm{eff}} = -\int f(x)/g(x)^2 \mathrm{d}x - \frac{\epsilon}{2} \log(g(x))$, resulting in $P(x) \propto \exp(-\frac{2}{\epsilon}U_{\mathrm{eff}}(x))$, according to Eq. (6.72). Thus, where $g(x)$ is large, the probability decreases along with ϵ. As a result, as ϵ increases, the distribution may not peak around an attractor with a large $g(x^*)$. An example of this can be seen in Section 6.5. In other words, the noise causes the state to flow out of the attractor. In the case of a cell, it is possible to select a state depending on the intensity of the noise. However, it may be difficult to

---

[5] Here, the Stratonovich type is used. If we use the Ito type, then $\frac{1}{x}$ in Eq. (6.82) is replaced by $\frac{1}{x^2}$.

control the intensity of the noise in a cell. Instead, Kashiwagi et al. [13] introduced "activity A" (activity) and considered the following equation

$$\frac{dx}{dt} = A(x)f(x) + \sqrt{\epsilon}g(x)\eta(t) \tag{6.83}$$

and discussed the selection of attractors by this A. In this case, under

$$U_{\text{eff}} = -\int \frac{A(x)f(x)}{g(x)^2}dx - \frac{\epsilon}{2}\log(g(x)), \tag{6.84}$$

the distribution is $P(x) = \exp(-\frac{2}{\epsilon}U_{\text{eff}})$, where the noise intensity is constant. Therefore, the attractor with large $A(x^*)$ is chosen even if the noise term is constant ($g(x) = 1$). Thus, the noise-induced selection of the adapted state with high activity A is expected.

Example 6.3 For example, consider the second-order synthesis and dilution (decomposition) reaction as described in Example 1.18 in Section 1.15. This is represented by

$$\frac{dx}{dt} = \frac{rx^2}{1+x^2} - kx + \sqrt{\epsilon}g(x)\eta(t). \tag{6.85}$$

This system has fixed-point attractors $x_1^* = \frac{(r/k)+\sqrt{(r/k)^2-4}}{2}(>0)$ and $x_2^* = 0$ for $r > 2k$. Now, considering the dilution due to the growth rate μ, the term $-\mu x$ is added into the preceding equation. If the dilution is more important than decomposition, we can ignore kx and put in μx instead. In addition, this growth rate generally depends on the state of the cell. For example, if the amount of this protein, x, contributes to the growth of this cell, then μ is an increasing function of x. On the other hand, growth induces a dilution, and in order to balance this dilution, the synthesis needs to increase with the growth rate μ. In other words, the synthesis term also increases with μ. To achieve a perfect balance, it should be proportional to μ. Therefore, by replacing r with $\mu(x)r$, the equation should be $dx/dt = f(x) = \mu(x)(rx^2/(1+x^2) - x)$. That is, this equation is of the form in which the growth rate μ is the activity A. In this case, the fixed points x_1^* and x_2^* do not depend on the growth rate μ. However, because $\mu(x_1^*) > \mu(x_2^*)$, the probability of x_1^* is high in the presence of noise. As a result, a state with a high growth rate is chosen (for simplicity, the x-dependence of $g(x)$ is assumed to be small).

For this attractor selection to work, f does not necessarily have to be proportional to $A(x)$. If it is proportional, the properties of the dynamical system of $dx/dt = f(x)$ are kept preserved, and the activity A represents only the speed of change, in order to ease the analysis. However, in general, if $f(x, A(x))$ is an increasing function of

A, then the discussion here is valid. On the other hand, if the noise term $g(x)$ is proportional to $\sqrt{A(x)}$ and is not large (i.e., ϵ is small), the effect of selecting an attractor with high A is canceled, as can be seen from the form of U_{eff} in Eq. (6.84). In particular, if all the reaction processes are proportional to A and the noise is derived from the chemical Langevin equation, this effect is canceled out. However, there are various sources of noise, including the cell growth-rate fluctuations, and these will remain to exist even when A (or the growth rate) is zero. If this is the case, then the selection of high-activity attractors by this noise will work.

Example 6.4 In *E. coli*, an artificial gene network of toggle switches described in Sections 1.4 and 1.13 has been incorporated (with plasmids) and its expression state has been investigated in response to changes in the environment [13]. As already mentioned, this network consists of two gene expression systems: expression of gene 2 is suppressed when gene 1 is expressed, and that of gene 1 is suppressed when gene 2 is expressed, and as analyzed in Example 1.5, there are two attractors, $p_1 \gg p_2$ and $p_2 \gg p_1$, with p_i as the concentration of corresponding protein. In the experiment, protein A required for environment A was embedded into the gene network on the side of 1 and protein B required for environment B on the side of 2 (see the original paper for specific protein species). In many cells, attractor 1 was selected for environment A and attractor 2 was selected for environment B. In environment A, the growth rate μ of the attractor with $p_1 \gg p_2$ is much higher than that of the attractor with $p_2 \gg p_1$. Therefore, as the growth rate increases, both the protein synthesis rate and the dilution rate are expected to increase, and the discussion in this section can be applied. In this case, the following equations need to be considered based on Eqs. (1.3) and (1.4):

$$\frac{dp_1}{dt} = \mu(p_1, p_2)\left(\frac{\alpha}{1 + p_2^2} - p_1\right) + \eta_1(t), \tag{6.86}$$

$$\frac{dp_2}{dt} = \mu(p_1, p_2)\left(\frac{\alpha}{1 + p_1^2} - p_2\right) + \eta_2(t). \tag{6.87}$$

Here, the growth rate μ takes a large value in environment A for $p_1 \gg p_2$ and is close to 0 for $p_2 \gg p_1$. If we compute this model numerically, we can select the attractor with the higher growth rate within a range of moderate noise level. From these results, it is expected that the cell has a general adaptive process in which a state with a high growth rate is selected due to noise.

Although these are theoretical predictions that fluctuations are effective for adaptation, the significance of fluctuations to other problems has been discussed elsewhere. This includes the proportionality between fluctuations and evolutionary speed, and the correlation between robustness to noise and robustness to mutation [12].

6.7 Adiabatic Elimination in Stochastic System

In Section 1.12, we discussed how fast element(s) are adiabatically eliminated to obtain the closed equation for dynamics only for slow variable(s). The method presented therein is to obtain the fixed-point solution for the fast variable y for a given slow variable x [8]. This was achieved for deterministic dynamical systems for x and y. Now, can we extend this method to stochastic differential equations? For this purpose, we consider a coupled Langevin equation for a slow variable x and a fast variable y, as

$$\frac{dx}{dt} = f(x,y) + \eta_x(t); \quad \frac{dy}{dt} = \gamma g(x,y) + \sqrt{\gamma}\eta_y(t), \tag{6.88}$$

where η_x and η_y are Gaussian white noise satisfying $<\eta_x(t)\eta_x(t')> = 2\epsilon_x\delta(t-t')$ and $<\eta_y(t)\eta_y(t')> = 2\epsilon_y\delta(t-t')$, whereas mutual correlation in the two noise terms is assumed to be zero: $<\eta_x(t)\eta_y(t')> = 0$. As y is a fast variable, $\gamma \gg 1$ is assumed. Without the noise term, we eliminate the variable y by solving $\frac{dy}{dt} = 0$, to get $y^* = a(x)$, where $a(x)$ is given by solving $g(x, a(x)) = 0$. This is the adiabatic elimination method in Section 1.12. Now, in the Langevin equation, there exists a noise term around this fixed-point state. Then how this noise term influences the dynamics of slow variable is a question to be addressed here. For the moment, we assume that the noise magnitude is small, and furthermore, we treat the noise term as if it were just a small constant term. Then by considering the perturbation as $y = y^* + \delta y$, and by Taylor series expansion, we get $g(x,y) = g(x,y^*) + c(x)\delta y + O((\delta y)^2)$, where $c(x) = \frac{\partial g(x,y)}{\partial y}|_{y=y^*}$. By solving the equation for $dy/dt = 0$ with $g(x,y^*) = 0$, we can get

$$c(x)\delta y + \frac{\eta_y(t)}{\sqrt{\gamma}} = 0. \tag{6.89}$$

Then, by inserting this $y = a(x) + \delta y$ into the equation for x, we *formally obtain*

$$\frac{dx}{dt} = f(x, a(x) + \frac{\eta_y(t)}{\sqrt{\gamma}c(x)}) + \eta_x(t). \tag{6.90}$$

(As the noise distribution is symmetric between plus and minus, one can change $\eta_y(t)$ to $-\eta_y(t)$.) In fact, this *derivation* has some problem in principle and in practice, even though it contains some truth therein: First, for the elimination procedure, we set $dy/dt = 0$ by pretending as if the noise term were constant, but actually the noise term fluctuates. Second, if $f(x,y)$ is nonlinear against y, the preceding equation includes a nonlinear term for $\eta_y(t)$ (such as $\eta_y(t)^2$), whose meaning is not validated, as one knows that the white noise is an unusual variable (as one can see in the delta function in the definition).

For simplicity we take the following case and examine the validity of the preceding *naive derivation*:

$$\frac{\mathrm{d}x}{\mathrm{d}t} = f_0(x) + f_1(x)y + f_2(x)y^2 + \eta_x(t); \qquad \frac{\mathrm{d}y}{\mathrm{d}t} = -\gamma c(x)(y - a(x)) + \sqrt{\gamma}\,\eta_y(t) \quad (6.91)$$

For further simplicity, we neglect the term of the order ϵ_x/γ, assuming that the noise to the slow variable is not so large (recall γ is large). The higher-order term in y for $\frac{\mathrm{d}x}{\mathrm{d}t}$ also is neglected. Recalling that γ is large, y can be solved by fixing x as

$$y(t) = a(x) + \int_0^t \exp(-\gamma c(x)(t - t'))\sqrt{\gamma}\,\eta_y(t')\mathrm{d}t'. \qquad (6.92)$$

If γ is large enough, $y(t) \approx a(x) + \frac{\eta_y(t)}{\sqrt{\gamma}c(x)}$. The insertion of this form into $\mathrm{d}x/\mathrm{d}t$ corresponds to the preceding naive derivation, whereas it remains elusive whether the noise term is represented by the Ito or the Stratonovich type, and also how the term $\eta_y(t)^2$ is interpreted. Indeed, to answer this question and get a general form for the elimination of a fast variable requires intricate calculation with somewhat advanced technique in mathematical physics.

Here we give only a result derived from the complicated calculation [11, 18]. First, we consider the linear case to y, that is, the case with $f_2(x) = 0$. In this case, the elimination procedure of the fast variable leads to the Langevin equation for x, as

$$\frac{\mathrm{d}x}{\mathrm{d}t} = (f_0(x) + f_1(x)a(x))(1 - \frac{a'(x)f_1(x)}{\gamma c(x)}) + \eta_x(t) + \frac{f_1(x)}{\sqrt{\gamma}c(x)}\eta_y(t), \qquad (6.93)$$

where the noise should be interpreted to be of Stratonovich type. Equation (6.93), aside from the correction in the drift term $(1 - \frac{a'(x)f_1(x)}{\gamma c(x)})$ agrees with the previous naive derivation, by using the Stratonovich interpretation.[6]

Next, we include the higher-order term $f_2(x)y^2$ to the equation. In this case, we obtain

$$\frac{\mathrm{d}x}{\mathrm{d}t} = f_0(x) + f_1(x)a(x) + f_2(x)\frac{\epsilon_y a(x)^2}{c(x)} + \eta_x(t) + \frac{f_1(x) + 2a(x)f_2(x)}{\sqrt{\gamma}c(x)}\eta_y(t), \qquad (6.94)$$

where we took only the lowest order of $1/\gamma$ (i.e., by neglecting the correction term in Eq. (6.93)). Remarkably, by the term y^2 there appears an additional drift (flow) term given by $(a^2\frac{\epsilon_y}{c(x)})f_2(x)$ even in the zeroth order of $1/\gamma$. Due to the noise term in the fast variable, there appears a directional flow, even in the zeroth order of $1/\gamma$.

The preceding equation could be interpreted by the naive insertion of $y = a(x) + \frac{\eta_y(t)}{\sqrt{\gamma}c(x)}$, and interpreting as if $\eta_y(t)^2$ were $\gamma\epsilon_y$. Here this interpretation of $\eta_y(t)^2$ is rather tricky and nontrivial, because $<\eta(t)\eta(t')>$ is proportional to the delta function $\delta(t - t')$, and it is interpreted only after we get the preceding equation after a rather complicated calculation.

---

[6] An additional correction to the noise term for η_x in the order of ϵ_x/γ, which we neglected here.

Interestingly, the emergence of drift term induced by the noise in the fast variable, proportional to $\frac{\epsilon_y f_2(x)}{c(x)}$, can induce a nontrivial effect of noise. For instance, with this term, the bifurcation to a different attractor can be shifted. A critical example of this is the appearance of oscillatory behavior due to the noise, as noted in [11], whereas emergence of oscillatory behavior due to noise is generally called stochastic resonance. The significance of stochastic resonance to biology has also been explored extensively [2].

References

[1] A. Awazu and K. Kaneko. *Physical Review* E, **80** 010902 (R), 2009.

[2] R. Benzi, A. Sutera, and A. Vulpiani. The mechanism of stochastic resonance. *Journal of Physics A: Mathematical and General*, **14** (11), L453, 1981.

[3] R. Benzi, G. Parisi, A. Sutera, and A. Vulpiani. Stochastic resonance in climatic change. *Tellus*, **34** (1): 10–16, 1982.

[4] T. Biancalani, T. Rogers and A. J. McKane. Noise-induced metastability in biochemical networks. *Physical Review* E, **86** (1): 010106, 2012.

[5] C. Furusawa *et al*. Ubiquity of log-normal distributions in intra-cellular reaction dynamics. *BIOPHYSICS*, **1**: 25–31, 2005.

[6] L. Gammaitoni, P. Hänggi, P. Jung, and F. Marchesoni. Stochastic resonance. *Reviews of Modern Physics*, **70** (1): 223, 1998.

[7] D. T. Gillespie. The chemical Langevin equation. *The Journal of Chemical Physics*, **113** (1): 297–306, 2000.

[8] H. Haken. *Synergetics*. Berlin, Springer, 1979.

[9] W. Horsthemke and R. Lefever. *Noise-Induced Transitions*. Berlin, Springer, 1984.

[10] van Kampen. *Stochastic Processes in Physics and Chemistry*. Amsterdam, North Holland, 2007.

[11] K. Kaneko. Adiabatic elimination by the eigenfunction expansion method. *Progress of Theoretical Physics*, **66** (1), 129–142, 1981.

[12] K. Kaneko. *Life: An Introduction to Complex Systems Biology: An Introduction to Complex Systems Biology*. Berlin, Springer, 2010.

[13] A. Kashiwagi *et al*. Adaptive response of a gene network to environmental changes by fitness-induced attractor selection. *PLOS ONE*, **1** (1): e49, 2006.

[14] R. Kubo, K. Matsuo and K. Kitahara. Fluctuation and relaxation of macrovariables. *Journal of Statistical Physics*, **9** (1): 51–96, 1973.

[15] R. Kubo, M. Toda, N. Hashitsume. *Statistical Physics II: Nonequilibrium Statistical Mechanics*. Springer, 2013.

[16] L. Li, S.F. Nøfrelykke and E.C. Cox. Persistent cell motion in the absence of external signals: a search strategy for eukaryotic cells. *PLOS ONE*, **3** (5): e2093, 2008.

[17] J. Ohkubo, N. Shnerb and D. A. Kessler. Transition phenomena induced by internal noise and quasi-absorbing state. *Journal of the Physical Society of Japan*, **77**: 044002, 2008.

[18] H. Risken. Fokker–Planck equation. In *The Fokker–Planck Equation* (pp. 63–95), Berlin, Springer, 1996.

[19] A. Schenzle and H. Brand. Multiplicative stochastic processes in statistical physics. *Physical Review* A, **20** (4): 1628, 1979.

[20] K. Sekimoto. *Stochastic Energetics*. Berlin, Springer, 2010.

[21] M. Suzuki, K. Kaneko and F. Sasagawa. Phase transition and slowing down in non-equilibrium stochastic processes. *Progress of Theoretical Physics*, **65** (3): 828–849, 1981.

[22] H. Takagi *et al.* Functional analysis of spontaneous cell movement under different physiological conditions. *PLOS ONE*, **3** (7): e2648, 2008.

[23] Y. Taniguchi *et al.* Quantifying E. coli proteome and transcriptome with single-molecule sensitivity in single cells. *Science*, **329** (5991): 533–538, 2010.

[24] M. Toda, R. Kubo and N. Saito. *Statistical Physics I: Equilibrium Statistical Mechanics*. Springer, 2013.

[25] Y. Togashi and K. Kaneko. Transitions induced by the discreteness of molecules in a small autocatalytic system. *Physical Review Letters*, **86**: 2459–2462, 2001.

[26] S. Tsuru *et al.* Noisy cell growth rate leads to fluctuating protein concentration in bacteria, *Physical Biology*, **6**: 036015, 2009.

[27] A. M. Turing. *Collected Works of A.M. Turing*, Vol. 3. Morphogenesis.

7

Dynamical Systems Model of Cell Differentiation

7.1 Introduction

A multicellular organism is a cellular society composed of various cell types. These diverse types of cells are generated through proliferation and differentiation from a small number of cells, such as fertilized eggs, and ultimately form organisms with complex patterns of differentiated cell types. The proliferation and differentiation of cells are elaborately controlled during their development and tissue maintenance, and the dynamics to form a complex organism is a marvelous phenomenon in biology.

These differentiation dynamics are regulated by various intracellular and intercellular interactions. A simple example is the regulation of differentiation by an if-then type switch, whose input is extracellular signaling molecules, and the output is a cellular state. It has been shown that sophisticated patterns of differentiated cells can emerge by the if-then type switch, for example, based on the concentration gradient of signal molecules called morphogens. However, intuitively, such regulation of cell differentiation is not robust with respect to perturbations. For example, the number of such signal molecules is often small, and therefore stochastic fluctuations due to the small-number effect are inevitable (see Chapter 5). Accordingly, the regulation by signal molecules always involves large fluctuations, so that the on/off type control can be "wrong." Thus, it is not obvious whether the developmental process consisting of a sequence of such noisy regulations can be stably realized as a whole.

In this chapter, we discuss dynamical systems approaches for cellular differentiation. In particular, the following two problems are addressed.

1. How can cells with the same genome sequence, that is, the same intracellular network, undergo various different cellular states?
2. How are transitions between cell states (i.e., differentiation) regulated? In particular, how is stability at the cell-population level achieved under perturbations at various levels?

In Section 7.2, we will discuss how the intracellular reaction dynamics can generate various attractors using a simple discrete-time and discrete-state reaction model (Boolean network), and in Section 7.3, we will analyze how cell–cell interactions cause transitions between attractors and how stability at the cell-population level emerges by regulating the dynamic transitions.

7.2 Modeling Gene Regulation by Boolean Network

In a cell, mRNA is transcribed from genomic DNA and translated to synthesize proteins. Some of these proteins are responsible for regulating the transcription and translation of other genes. For example, a protein called a repressor binds to genomic DNA and suppresses the expression of genes downstream of the repressor binding site, while an activator protein promotes the expression of the target gene by, for example, altering the structure of the genomic DNA. As discussed in Section 1.12, in general, the synthesis and degradation of mRNA are faster than those of a protein. Thus, it would be better to adiabatically eliminate the change in the amount of mRNA and then consider the dynamics of the protein expression. Such protein expression is regulated by other proteins, resulting in a gene regulatory network in which genes regulate each other's expression. A pioneering study by Kauffman focused on the properties of gene regulatory networks and analyzed how the stable state (attractor) corresponding to various cell types emerged using a simple model [11, 12]. In the following section, we discuss the behavior of the simplified gene regulatory network.

Let us assume that a cell consists of N genes (proteins) and consider their regulatory interactions. First, let us assume that each gene has only two states of expression, on-state or off-state, which is represented by 1 or 0, respectively. This is a simplification, bearing in mind that if we follow Hill's formula as described in Section 2.1.2, protein abundance behaves as a threshold function of little or maximum expression. Next, we assume that the on/off of each gene is uniquely determined by the on/off of K genes that regulate it. Furthermore, assuming that gene expression has a characteristic time scale, let us consider the discrete-time evolution of how it changes in the next time step; if the expression state of each gene at a certain time t is determined, the state at time $t + 1$ is determined according to the rules of gene regulation. Of course, it is a strong limitation to consider the discrete-time in terms of changes at $t = 0, 1, 2, 3, 4, \cdots$. Still, on the other hand, this simplification makes it easy to analyze the dynamic behavior of gene regulatory networks.

Next, we consider the regulatory relationships among genes. First, we assume that each gene is regulated by randomly selected K genes (of course, this is a big simplification). The next step is to choose the type of activation or suppression

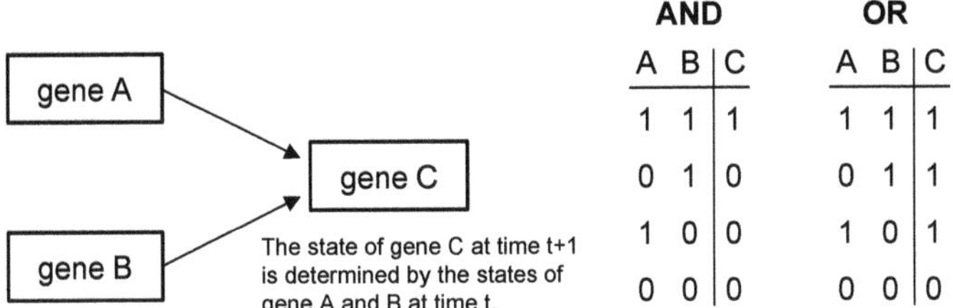

Figure 7.1 Kauffman's model of gene expression regulation. The expression state of each gene takes two discrete states: 1 (on) or 0 (off). The expression state at a certain time t determines the state at time $t + 1$ of the gene it regulates. For example, the state of gene C in the figure is determined by the states of genes A and B. Its regulation is described using Boolean functions, such as AND and OR functions.

effect of each regulation. For example, if $K = 2$, the number of possible input patterns from the regulatory genes is $2^K = 4$. The output can be determined for each input using a Boolean function. Figure 7.1 shows such Boolean functions, where genes A and B are inputs and the expression of gene C is determined by the function AND or OR (there are 2^{2^K} possible Boolean functions in total). For example, in the case of the AND function, the expression of gene C is 1 only when both genes A and B are expressed, and 0 in other cases. In other words, the expression of gene C does not depend on the current state but only on the state of genes A and B. The regulatory function is fixed over time. If K is greater than two, the number of possible input–output relations increases, but in any case, it can be expressed by a combination of Boolean functions.

What kind of dynamic behavior would this Boolean regulatory network exhibit? Figure 7.2 shows an example of the temporal evolution of a simple Boolean network for three genes A, B, and C (i.e., $N = 3$) with $K = 2$. In this example, each gene is regulated by two other genes, and each control function is represented by an AND or OR function (Fig. 7.2(a)). For example, if the expression state at time t starts from (A, B, C) = (0, 0, 0), then the state at time $t + 1$ is obtained according to the rule; if both the AND and OR functions have zero inputs, the result is the same as the first one (0, 0, 0), because both of them output 0 (Fig. 7.2(b)). In other words, this expression state did not change over time. On the other hand, if we start from (1, 0, 0), then at time $t + 1$, we have (0, 1, 1) and at $t + 2$, we have (1, 1, 1). For time $\geq t + 2$, the state no longer changes over time. Moreover, if a state starts from (0, 1, 0), it changes to state (0, 0, 1) at the next time step, and then returns to (0, 1, 0) at the next time, and the state repeats between (0, 1, 0) and (0, 0, 1) periodically after that. Figure 7.2(c) summarizes the transitions between these states. If we start from each of the eight possible initial states, there are three types of states (two fixed points and one periodically oscillating state). This simple regulatory network

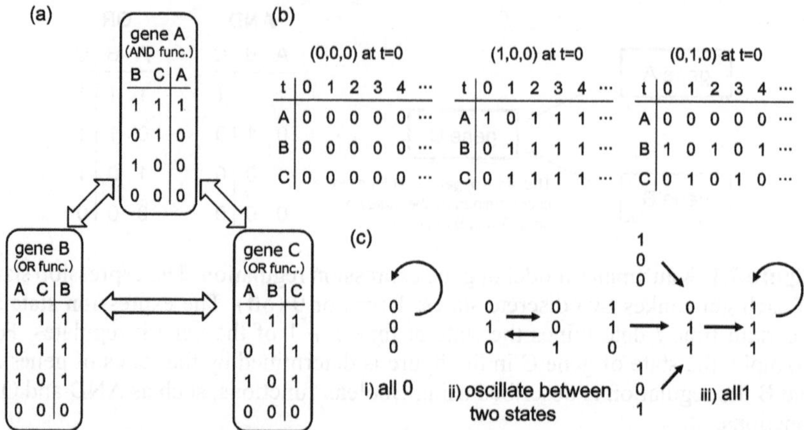

Figure 7.2 Kauffman's Boolean network (a) gene regulatory network consisting of three genes. The expression state of each gene is determined by the expression of the other two genes. (b) Example of the expression dynamics depending on the initial state. The subsequent time evolution is determined based on the expression state at $t = 0$. (c) Three attractors of expression dynamics. Depending on the initial state, one of the following three states is reached: (i) all 0, (ii) oscillation between $(0, 1, 0)$ and $(0, 0, 1)$ states with period 2, and (iii) all 1.

has three different stable expression patterns, that is, three attractors and the sets of initial states that lead to these attractors correspond to their basins.

Even if the number of genes N is large, the number of possible states is finite (i.e., 2^N), so it will eventually take the same state as before as it evolves over time. When this state is revisited, the same temporal evolution will be followed, and thus the present system will eventually fall into a periodic (or fixed-point) attractor. In the network in Fig. 7.2 there are three attractors, but as N increases, the number of attractor increases. Since different attractors have different on/off expression patterns, in this model, each attractor can be considered to correspond to a different cell type produced by the same gene regulatory network.

For $K = 1$, starting from any initial condition, the state falls on a fixed-point of a cycle with few periods after a short transient state. This is a "frozen" state. In contrast, in the case of $K \sim N$, that is, when each gene receives input from the majority of other genes, the period of cycles of the attractor is extremely large (which roughly obeys $O(2^{0.5N})$). In this regime, errors in the expression on/off of a single gene will amplify, resulting in the chaotic nature of expression dynamics; thus, it would be difficult to regard each attractor as corresponding to a different cell type. This chaotic behavior emerges at $K \geq 3$. The most interesting case is that of $K = 2$, which falls on an attractor with a relatively short period ($O(\sqrt{N})$), and the number of attractors increases with $\sqrt{N}$ (see also [19]). In this case, the number of attractors is much smaller than the number of all possible expression patterns (2^N), indicating that the possible expression states are constrained to a small number of attractors. This result might suggest that, in multicellular organisms with N genes, the number of stable cell types would be much less than 2^N and approximately $\sqrt{N}$. In the actual data, the number of cell types in each species tends to increase by approximately $\sqrt{N}$ when the number of genes is N, which might reflect the constraint of possible expression profiles, as in the case of $K = 2$. Still, it is difficult to accurately estimate the number of cell types in multicellular organisms, and the relationship between the number of genes and the number of cell types should be considered a rough estimate.

The analysis of Kauffman's Boolean network is a drastic simplification of the gene regulatory network: It considers only on/off expression levels and adopts discrete-time state updates. This simplification can significantly change the expression dynamics. For example, the number of attractors can be significantly smaller when a differential equation model is used for the same gene regulatory network (see the next section), in which an appropriate function for regulation is taken instead of the Boolean network. In particular, many attractors with cyclic expression changes, such as those shown in Fig. 7.2(c), disappear in the differential equation model.

Nevertheless, Kauffman's work on cell types in multicellular organisms as attractors of gene regulatory networks has influenced many subsequent studies. The Boolean network model has been applied to a variety of biological systems as a simple method for analyzing expression dynamics because it can construct models by knowing only the direction of expression regulation and can be easily simulated on a computer. The targets are not limited to developmental dynamics but also include analysis of network stability and evolutionary processes. For example, the Boolean model of the gene regulatory network for the cell cycle in yeast [16] showed that the temporal evolution from a variety of initial expression patterns eventually falls on the attractor corresponding to the actual cell cycle dynamics and that the cell cycle dynamics tend to be maintained even against some perturbations (e.g., random network switching), suggesting that the cell cycle dynamics are designed to be robust.

7.3 Cell Types as Attractors of Gene Regulatory Dynamics

Instead of using a discrete-state model as in the previous section, we can more directly represent cell types as different attractors by representing the temporal evolution of the cellular state as a dynamical system of continuous state and time. As a simple example, let us take the case where two genes suppress each other's expression, which is called a "toggle switch" as explained in Chapter 1. As shown in Fig. 1.2, two cell types arise: one in which gene X is expressed and Y is suppressed, and the other in which Y is expressed and X is suppressed.

The preceding example indicates that there are different states in the regulatory dynamics. The question here is how the transition between states (i.e., differentiation) occurs. One possible mechanism is the noise in intracellular reaction dynamics, which is described in Chapter 5. For example, gene expression in a cell is the result of chemical reactions, and if the number of molecules responsible for it is not large, fluctuations will occur because of the small number of molecules. In addition, gene expression is dependent on signals from the external environment, and these external influences also fluctuate in general. Here, if the noise term is introduced into the rate equation, as introduced in Chapter 6, transitions between the two cell types can occur. However, in the preceding example of the toggle switch, if the strength of noise is large, transitions occur with sufficient frequency, and each cell type will not be able to continue to exist stably. This is because the two cell types are equivalent in this model, and there is no mechanism in which either cell is less likely to transition due to noise.

On the other hand, by changing the network and reaction parameters, the degree of noise stability can be changed for each attractor. This is because the strength of the attractor stability is determined by the flow around it. For a fixed-point attractor, if it deviates from the fixed point by δx_i, the degree to which it decreases with time is determined by the eigenvalues of the Jacobi matrix at the fixed point (see Section 1.7). For example, let us introduce positive feedback on the toggle switch in which the expression of each gene activates its own expression

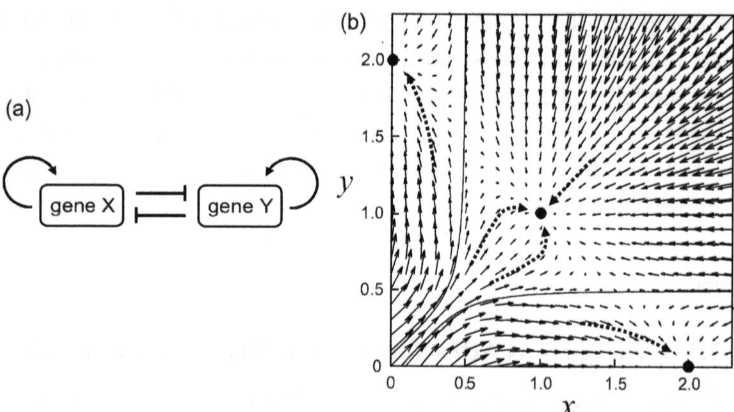

Figure 7.3 A two-gene model consisting of mutual inhibition and self-activation. (a) Schematic of the two-gene model where Genes X and Y mutually inhibit each other's expression, and each has feedback for self-activation. (b) Flow diagram of the two-gene model. There are attractors at $(x, y) = (2, 0)$, $(0, 2)$, and $(1, 1)$ (indicated by black circles). The black curve indicates the boundary of the basin of attraction. The parameters used are $a_x = b_x = a_y = b_y = 1$, $n = 4$, $\theta_{a_x} = \theta_{a_y} = \theta_{b_x} = \theta_{b_y} = 0.5$, and $k_x = k_y = 1$.

(Fig. 7.3(a)) [8]. If mRNA is again adiabatically eliminated, the dynamics of protein X and Y concentrations, denoted by x and y, are given as follows (see Example 1.4 in Section 1.6):

$$\frac{\mathrm{d}x}{\mathrm{d}t} = a_x \frac{x^n}{\theta_{a_x}^n + x^n} + b_x \frac{\theta_{b_x}^n}{\theta_{b_x}^n + y^n} - k_x x, \tag{7.1}$$

$$\frac{\mathrm{d}y}{\mathrm{d}t} = a_y \frac{y^n}{\theta_{a_y}^n + y^n} + b_y \frac{\theta_{b_y}^n}{\theta_{b_y}^n + x^n} - k_y y, \tag{7.2}$$

where θ_{a_x}, θ_{b_x}, θ_{a_y}, and θ_{b_y} represent the regulatory inputs that result in the half-maximum strength of activation or repression, respectively. Figure 7.3(b) shows the flow diagram of this system, where there are three fixed-point attractors corresponding to $x \gg y$, $x \ll y$, and $x = y$, respectively. In this case, the flow to the attractor for $x = y$ is smaller than that of the other two. Therefore, at moderate noise strength, transitions from this fixed-point state to other fixed points occur.

Exercise 7.1 For the parameters used in Fig. 7.3, find the eigenvalues of the Jacobi matrix at each fixed point and check that the attractor for $x = y$ is indeed susceptible to transition by noise.

7.4 Stochastic Differentiation Model of Stem Cells

Stem cells are defined as cells that have the ability to produce cells in the same state by self-renewal and also the ability to differentiate into different cell types. Well-known examples of stem cells are mammalian hematopoietic stem cells and epidermal stem cells, which give rise to red blood cells and white blood cells. These stem cells are necessary to supply differentiated cells that have lost their proliferative capacity and have a limited lifespan.

To provide a sustainable supply of differentiated cells without depletion, each stem cell needs to choose its fate, namely, self-renewal or differentiation. As discussed in the previous section, one simple mechanism for cell fate choice is differentiation from the attractors corresponding to stem cells, which occurs stochastically due to internal noise. Indeed, stochastic differentiation of stem cells has long been suggested in hematopoietic stem cell systems [22]. For another example, in the epidermal stem cell system, some stem cells in the basal layer and their progeny were labeled by fluorescent proteins, and the number distribution of cells produced by a single stem cell was quantified. The result showed good agreement with the properties predicted from stochastic differentiation dynamics [1, 13]. In the following section, we describe the behavior of a simple stochastic differentiation model of stem cells.

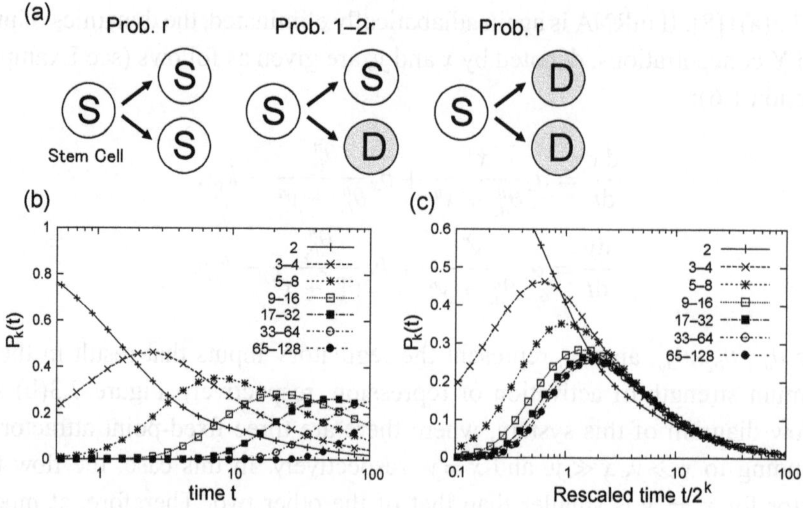

Figure 7.4 Stochastic differentiation of stem cells (a) Schematic of the model. A stem cell S self-renews during mitosis with probability r, and one daughter cell transitions to a differentiated cell D with probability $1 - 2r$ and two daughter cells with probability r. (b) Numerical simulation results of the stochastic differentiation model. Time evolution of distribution $\mathcal{P}_k^{\text{pers.}}(t)$ of one stem cell–derived clonal cell in Eq. (7.5) is plotted. (c) Scaling law for the number of cloned cells. The time evolution in (b) is plotted with time t scaled to $t/2^k$. The parameters used were those estimated from the experiment: $\lambda = 1.1/\text{week}$, $r = 0.08$, and $\rho = 0.22$.

Consider a tissue, such as the epidermis, in which the numbers of stem cells and differentiated cells produced from stem cells do not change over time and are in a steady state. We assume that a stem cell S divides λ times per unit time, and in each division S produces two stem cells with probability r without differentiation of daughter cells (Fig. 7.4(a)). Assuming that differentiated cells do not divide, for the ratio of stem cells to differentiated cells D to be stationary, their differentiation probability must be as follows:

$$
S \xrightarrow{\lambda}
\begin{cases}
S + S & \text{Prob. } r, \\
S + D & \text{Prob. } 1 - 2r, \\
D + D & \text{Prob. } r.
\end{cases}
\tag{7.3}
$$

We also assume that differentiated cells D are removed from the system at a certain rate Γ[1]. In this case, when the number of cells in the system remains unchanged after sufficient time has elapsed and ρ is the proportion of stem cells in all cells in the system, $\Gamma = \frac{\lambda \rho}{1-\rho}$ is required to maintain the steady state. When the numbers of

---

[1] For example, considering the mammalian epidermis, stem cells S and differentiated cells D correspond to stem cells and progenitor cells in the basal layer, and only the progenitor cells can leave the basal layer at a rate Γ [1, 13].

stem cells and differentiated cells are n_S and n_D, respectively, and the probability that the number of cells in the system is n_S and n_D at time t is $P_{n_S,n_D}(t)$, the master equation for the time evolution is as follows:

$$\frac{\partial P_{n_S,n_D}}{\partial t} = r\lambda[(n_S - 1)P_{n_S-1,n_D} - n_S P_{n_S,n_D}]$$
$$+r\lambda[(n_S + 1)P_{n_S+1,n_D-2} - n_S P_{n_S,n_D}]$$
$$+(1 - 2r)\lambda[n_S P_{n_S,n_D-1} - n_S P_{n_S,n_D}]$$
$$+\Gamma[(n_D + 1)P_{n_S,n_D+1} - n_D P_{n_S,n_D}]. \tag{7.4}$$

In this system, let us consider how the number of cells in a clonal population derived from a stem cell changes when there is only one stem cell at $t = 0$. Figure 7.4 shows an example of a numerical calculation. To analyze the correspondence with the experiment, instead of n_S and n_D, we consider the probability distribution $P_n(t)$ for the total number of cells $n = n_S + n_D$. Furthermore, we considered the probability distribution $P_n^{\text{pers.}}(t) = P_n(t)/\{1 - P_0(t) - P_1(t)\}$ representing the surviving clonal population, except for the states $n = 0, 1$, which are experimentally difficult to observe. In addition, the following probability distribution is considered to consolidate the data and minimize fluctuations due to counting statistics:

$$\mathcal{P}_k^{\text{pers.}}(t) = \sum_{n=2^{k-1}+1}^{2^k} P_n^{\text{pers.}}(t), \tag{7.5}$$

in which the number of cells is binned at increasing powers of 2, as plotted in Fig. 7.4(b). Interestingly, this probability distribution follows a scaling law. Figure 7.4(c) shows the results of a scale transformation of time t to $t/2^k$ for the same probability distribution, where all probability distributions roughly overlap after the initial transient state. This corresponds to the relationship between an appropriate function f and a constant representing the time scale, τ, which follows the relationship[2]

$$P_n^{\text{pers.}}(t) = \frac{\tau}{t}f(n\tau/t). \tag{7.6}$$

This scaling law is in good agreement with that found in the distribution of clonal cell populations produced by epidermal stem cells [1], suggesting that this stochastic differentiation model can adequately describe the dynamics of the epidermal stem cell system.

---

[2] We can derive the relationship (7.5) as the solution of the Fokker–Planck equation from the master equation in Eq. 7.4, under the assumption that the number of cells n_S and n_D are sufficiently large to be treated as a continuous quantity and that n_D is uniquely determined by n_S. See [1, 13] for details.

7.5 Differentiation Triggered by Cell–Cell Interaction

The viewpoint that cell types correspond to attractors of intracellular reaction dynamics has important implications for discussing the discrete nature and stability of cellular states in multicellular organisms. However, further analysis is needed to understand how transitions between cell types occur, that is, the process of differentiation and the stability of these processes. From the cell attractor's point of view, this transition is often thought to be due to noise, as described earlier. For example, in Boolean networks, noise causes transitions between multiple attractors by flipping the gene expression on and off. If we add noise to the differential equations representing the change in concentration of each component, for example, by using the Langevin equation (see Chapter 6), then the transitions between attractors will occur.

Indeed, transitions between attractors are possible as long as the noise level is adequate. However, how can the developmental process involving a large number of transitions be stable and reproducible with respect to fluctuations? The stability of a cell's state to noise is understandable: If the state is an attractor, the state, even after perturbation by noise, can return to its original state. However, for the developmental process to be stable, differentiation into different cells and the ratio of their number must be stable. For example, the stochastic differentiation model of stem cells described in the previous section assumes a constant differentiation probability of stem cells. Still, this model cannot explain stability at the population level, such as the regeneration process after tissue damage, although it can explain cell number behavior at a steady state. Therefore, it would be natural to think that the frequency of differentiation must be regulated in some way and that this would produce stability at the cell-population level.

What should be considered here are the interactions between cells and the resulting changes in the stability of the cellular state. In multicellular organisms, cells generally interact with surrounding cells, and this interaction is relevant to determine the state of the cell: The state of each individual cell depends on the states of surrounding cells. For example, transplantation experiments in frogs (*Xenopus laevis*) confirmed that the state of isolated cells differs from that of cells in a population (called the community effect [6]). Although Kauffman's Boolean network and the analysis in the previous section discussed only attractors in single cells, in multicellular organisms, there is always input to each cell by interaction with surrounding cells, and the stability of cellular states changes accordingly. In fact, it is possible that a state that is an attractor in a single cell is no longer stable due to cell–cell interactions and vice versa. From the perspective of individual cells, this can be regarded as a bifurcation phenomenon with external inputs as parameters. However, this external parameter changes due to the state of the cell population that interacts with it. Therefore, there is a circular relationship in which the state of a cell is determined by an external parameter, which is determined by the state of interacting cells; as a result, the cell and the cell population are determined in a way that satisfies the consistency of each state. In this viewpoint, the transitions between attractors are not caused by inputs from outside the system or noise but are considered as changes in cell–cell interactions.

The process of multicellular organism development begins with a small number of cells, such as a fertilized egg, and the increase in cell number through cell divisions results in the emergence of different types of cells through interactions. In this process, changes in the number of cells significantly impact the cell–cell interactions, which triggers the transitions between attractors. Here, we need a theory of how cell–state transitions (i.e., differentiation) occur under a complex interaction among intracellular dynamics (e.g., gene regulation), cell–cell interactions that affect the cellular state, and changes in cell number due to cell division (or cell death). This class of multicellular models has been analyzed over the last 20 years [10, 2, 3]. Differences in cellular states that belong to the same attractor are amplified through cell–cell interactions, resulting in transitions to distinct states. At the same time, it has been found that cells with such diverse states stabilize each other through cell–cell interactions, which in turn leads to stability at the cell–population level. In the following, we analyze the stability of the developmental process and the stability of the differentiation process by using dynamical systems modeling of interacting cells.

7.6 Dynamical Systems Model of Cell Differentiation through Cell–Cell Interactions

Consider a cell consisting of components (e.g., proteins) $m = 1, 2, \ldots, k$, and assume that the concentration of each component at time t is given by $x_i^m(t)$ for the ith cell and its time evolution is represented as $f^m(\vec{x}_i)$ with $\vec{x}_i = (x_i^1, x_i^2, \ldots, x_i^k)$. Let us assume that the cells interact with each other through the exchange of various components. For the sake of simplicity, cells interact with each other in an identical environment without considering the spatial patterns of cells.[3] We also assume that the concentration of component m in the environment is X^m. When components are exchanged between each cell and its environment, the concentration is diluted by the volume ratio V of the cell to its environment. The dynamics of the concentrations are represented as follows:

$$\frac{\mathrm{d}x_i^m}{\mathrm{d}t} = f^m(\vec{x}_i) + D^m(X^m - x_i^m), \qquad (7.7)$$

---

[3] In the early stages of development and when the cells are in a liquid medium, this approximation is valid, and the all-to-all interaction (the "mean-field approximation" in physics) is the first step in analyzing cell differentiation dynamics.

$$\frac{\mathrm{d}X^m}{\mathrm{d}t} = -D^m \sum_j (X^m - x_j^m)/V. \tag{7.8}$$

The right-hand side of Eq. (7.8) is the sum of all the cells. When the components are transported through the cell membrane, the influx is assumed to be proportional to the concentration difference between the inside and outside of the cell. Its coefficient is set to D^m, where $D^m = 0$ for components that are not transported through the cell membrane. Suppose chemical transport cannot be represented as simple diffusions, such as active transport. In that case, it is possible to use a different function $g(x_i^m, X^m)$ instead of the simple diffusion term. Assuming that concentration changes in the environment occur immediately, adiabatic elimination with $\mathrm{d}X^m/\mathrm{d}t = 0$ gives $X^m = \left(\sum_j x_j^m\right)/N$, where N is the number of cells. Therefore, we obtain

$$\frac{\mathrm{d}x_i^m}{\mathrm{d}t} = f^m(\vec{x}_i) + D^m \left(\frac{\sum_j x_j^m}{N} - x_i^m \right), \tag{7.9}$$

$$= f^m(\vec{x}_i) + I_i^m, \tag{7.10}$$

where

$$I_i^m = D^m \left(\frac{\sum_j x_j^m}{N} - x_i^m \right) = M^m - D^m x_i^m. \tag{7.11}$$

Here, I_i^m represents the effect of cell–cell interactions that the ith cell receives, and M^m represents the mean-field "interaction" that all cells commonly receive. In the following sections we discuss some basic features of the model.

I: Stability of Uniform State

If all cell states are identical, satisfying $x_i^m = x_j^m$, then the interaction term I_i^m is 0. Therefore, the uniform state in which all cells take the identical state and change commonly by $\mathrm{d}x^m/\mathrm{d}t = f^m(\vec{x}_i)$ is always the solution to this equation. Is this solution stable? For this purpose, let us see whether a small difference between the states of two cells δ^m will increase over time. Now write the equations of time evolution of these two cells and subtract the two equations; one gets the time evolution of this difference δ_m. Assuming that it is small and linearized, time evolution for δ_m is obtained by using the Jacobi matrix $J_{m\ell} = \partial f^m/\partial x_\ell$, as

$$\frac{\mathrm{d}\delta^m}{\mathrm{d}t} = \sum_\ell J_{m\ell}\delta^\ell - D^m\delta^m. \tag{7.12}$$

The stability of the uniform state can be evaluated by the eigenvalues of the matrix consisting of the Jacobi matrix minus D^m in the diagonal components. If the real parts of the eigenvalues are all negative, then the uniform state is stable. If D^m is constant for all components, this eigenvalue is simply a subtraction of D^m from the eigenvalue of the original Jacobi matrix, so the uniform state is always stable if the original fixed point is stable (all real parts of the eigenvalues of $J_{m\ell}$ are negative). However, if D^m is different for each component (e.g., if some components are zero), the interaction can destabilize the uniform state. This is discussed in the next section.

Example 7.1   Turing Instability In a two-component system, consider the case where the fixation point is stable in a single cell, where $D^2 = D > 0$ is given for the second component and $D^1 = 0$ for the first component. If the Jacobi matrix at a fixed point is $J = \begin{pmatrix} a & b \\ c & d \end{pmatrix}$, the eigenvalue λ is the solution of $\lambda^2 - (a+d)\lambda + (ad - bc) = 0$. Since this real part must be negative because it is a stable fixed point, its conditions are $a + d < 0$ and $(ad - bc) > 0$. On the other hand, the eigenvalue of the matrix $J - D\delta^2$ is the solution of $\lambda^2 - (a+d-D)\lambda + (a(d-D) - bc) = 0$. For the interaction to destabilize the uniform state, the real part of this eigenvalue must be positive. Since now $a + d < 0$, $D > 0$, and then $a + d - D < 0$. Then, destabilization requires $a(d-D) - bc < 0$, that is, $aD > (ad - bc) > 0$. Consequently $a > 0$, and since $a + d < 0$, then we would get $d < 0$. Because $bc < ad$, we obtain $bc < 0$. If $b > 0$, then it is necessary that a is positive, c and d are negative, and D is sufficiently large for the destabilization of the uniform state.

Exercise 7.2   Discuss the relationship between the preceding conditions and the Turing pattern in Section 4.3 (see reference [23]). In particular, we calculate the general case with $D^1 > 0$ and $D^2 > 0$ and discuss the conditions corresponding to the Turing pattern.

II: Differentiation as Bifurcation by Mean-Field
Next, we study what will happen in the case when a fixed-point attractor $\{x_0^{m*}\}$ is present in the one-cell state, but this uniform state is unstable, as described in (I). In this case, even though the mean-field M^m is common, the interaction I^m is different because different states arise. One possible interpretation is that the interaction I^m acts as an external parameter for a single cellular dynamical system, which leads to bifurcation. However, this external parameter is determined as a result of the cell population.

   For simplicity, we first consider the case in which two types of cells arise, whose numbers are $N_1 = \rho_1 N$ and $N_2 = \rho_2 N$, respectively. Then, we obtain

$$\frac{\mathrm{d}x_1^m}{\mathrm{d}t} = f^m(\vec{x}_1) + D^m \rho_2 (x_2^m - x_1^m), \tag{7.13}$$

$$\frac{\mathrm{d}x_2^m}{\mathrm{d}t} = f^m(\vec{x}_2) + D^m \rho_1 (x_1^m - x_2^m). \tag{7.14}$$

In this model, if $x_1^m \neq x_2^m$, then the effect of the interaction works in the opposite direction in type 1 and type 2 cells. For example, consider the case in which each cell type falls to a fixed point, $\{x_1^{m*}\}$ and $\{x_2^{m*}\}$, and the cells with different states stabilize each other. Then, in addition to the intracellular dynamics of single cells, type 1 and type 2 cells are affected by interactions $I_1^m = D^m \rho_2(x_2^{m*} - x_1^{m*})$ and $I_2^m = D^m \rho_1(x_1^{m*} - x_2^{m*})$, respectively. If we consider the interaction term I as a bifurcation parameter, the fixed point is destabilized in the uniform state with $I = 0$, and a stable state is realized by adding the interaction I. In other words, the interaction causes bifurcation. However, this term I is not given externally, but is determined by the resulting fixed point x_i^{m*} and the population ratio ρ_i of each type. That is, the determination of x by I and the determination of I by x should be consistent (in physics, this is called self-consistent). On the other hand, if we take a mean-field view, we can also write that all cells have an identical mean-field $M = \sum_j x_j^m / N$ in the identical dynamical system of $f^m(\vec{x}_i) - D^m x_i^m$. From this point of view, the same dynamical system has multiple attractors under a certain mean-field M, whereas M is self-consistent with the number distribution of each cell type.

7.7 Two-Gene Model of Cell Differentiation

As an example of the concept presented in the previous section, consider the following cell population model with two proteins X and Y, whose concentrations from the ith cell are $x_i(t)$ and $y_i(t)$, respectively. Their time developments are as follows:

$$\frac{\mathrm{d}x_i(t)}{\mathrm{d}t} = f(x_i(t), y_i(t)) - x_i(t) \tag{7.15}$$

$$\frac{\mathrm{d}y_i(t)}{\mathrm{d}t} = g(x_i(t), y_i(t)) - y_i(t) + D\{\frac{1}{N(t)} \sum_{k=1}^{N} y_k(t) - y_i(t)\}. \tag{7.16}$$

For simplification, we assume that cells interact with each other only through component y_i. Here, $f()$ and $g()$ are functions representing regulatory interactions between x_i and y_i, and the terms $-x_i$ and $-y_i$ are the effects of degradation or dilution of the proteins. $f()$ and $g()$ can be represented, for example, by the Hill function $x^\alpha / (K + x^\alpha)$ and $1/(K + x^\alpha)$ (see Section 2.1.2). For the other choice, as in Section 1.9 (Eq. 1.15), we can use the following equation:

$$f(x, y) = \frac{1}{1 + \exp\{-\beta(J_{xx}x + J_{xy}y - \theta_x)\}}, \tag{7.17}$$

$$g(x, y) = \frac{1}{1 + \exp\{-\beta(J_{yx}x + J_{yy}y - \theta_y)\}}, \tag{7.18}$$

where J_{xx} and J_{yy} are constant parameters representing the strength of self-regulation, and J_{xy} and J_{yx} correspond to mutual interactions, respectively. For

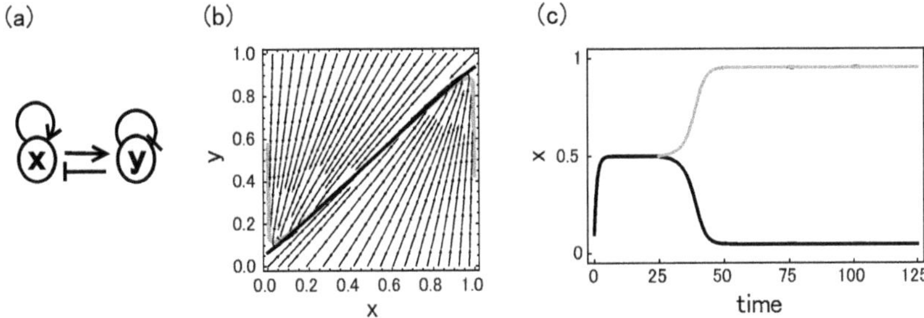

Figure 7.5 Differentiation process by Turing instability in a two-gene model (a) Example of a network exhibiting instability. (b) Nullcline of expression dynamics in one cell. (c) Time series of expression levels of gene x. As the number of cells increases through mitosis, two different stable states emerge from the cell-cell interactions. Assume that the value of J_{ij} takes the value ± 1 and that Gaussian noise with standard deviation $\sigma = 10^{-3}$ is added to x and y, respectively, as the cells divide. The parameters $(\beta, D, \theta_x, \theta_y) = (-40, 0.2, 0, 0)$ are used.

simplification, we assume that these parameters take 1, 0, or -1, representing activation, no regulatory interaction, and inhibition, respectively. In this model, instead of the Hill function, the sigmoid function $1/(1 + \exp\{-\beta(z - \theta)\})$ is used to represent the on-off switching dynamics, where z denotes the regulatory input, β is the parameter controlling the sensitivity to the input, and θ is the threshold of the on-off switching of the expression. When β is large, the on/off switch of the expression is sharp at $z = \theta$, and when β is small, the expression control is gradual.

The behavior of this two-gene model depends on the structure of the regulatory network. All possible regulatory networks, that is, all cases of 1, 0, and -1 in J, were examined, and it was demonstrated that differentiation through the cell–cell interaction is classified into the following three cases (for more information, see [4]).

Class I: Differentiation Process through Turing Instability
In this class of regulatory networks, gene X activates the expression of itself and Y, and gene Y represses the expression of itself and X (Fig. 7.5(a)). In this case, the uniform state of the fixed point is destabilized, as described in the previous section, where the diffusion interaction in the Turing pattern in Chapter 4 is replaced by a mean-field interaction (Turing discusses this example in his 1952 paper [23]). In this class, a single-cell attractor is a fixed point whose expression level does not fluctuate over time; however, when multiple cells interact, they differentiate into two different fixed-point states, as shown in Fig. 7.5(c).

The nullcline of the expression dynamics of this single cell is shown in Fig. 7.5(b). There is a stable fixed point around $(x, y) \sim (0.5, 0.5)$. When there are two cells, this fixed point is destabilized by the interaction and consequently

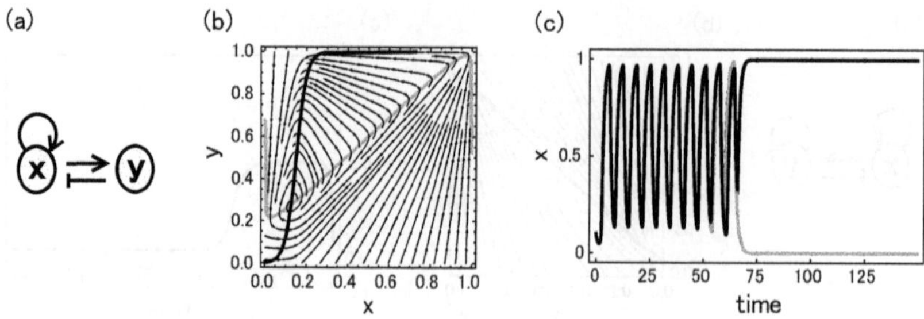

Figure 7.6 Differentiation from a limit-cycle to fixed points in the two-gene model. (a) Example of a network showing differentiation from a limit-cycle to a fixed point. (b) Nullcline of expression dynamics in one cell. (c) Time series of the gene expression levels x. The parameters $(D, \theta_x, \theta_y) = (1, -0.1, 0.15)$ were used, and the other parameters were the same as those in Fig. 7.5.

falls into two fixed points. Then, the nullclines $dy/dt = 0$ of each cell are moved upward in one cell and downward in the other by the term $I_i^y = D\rho_i(y_1 - y_2)$, resulting in the emergence of fixed. The differentiated states of the two cells correspond to the states at which these fixed points $x_i^*, y_i^* (i = 1, 2)$ become consistent through cell–cell interactions.

Class IIa: Differentiation Process from an Oscillatory State (Limit-Cycle) to a Fixed Point

As described in Section 1.9 (Eq. (1.15)), under the appropriate parameters, a single cell with this regulatory network exhibits temporally oscillating expression dynamics (limit-cycle) (Fig. 7.6(c)). Figure 7.6(b) depicts the nullcline of the expression dynamics of a single cell on the xy plane. The intersection around $(x, y) \sim (0.2, 0.3)$ is an unstable fixed point, in which the nullclines of $dx_i/dt = 0$ and $dy_i/dt = 0$ around $(x, y) \sim (1, 1)$ are not tangential, and the expression dynamics of the single cell falls to the attractor of the limit-cycle, independent of the initial conditions.

Here, bifurcation from this oscillatory state occurs by increasing the number of cells, and two different behaviors emerge: one in which the oscillation disappears and differentiates into two different fixation points (class IIa; Fig. 7.6), and the other in which some cells transition to a fixed point and the others remain in the oscillatory state (class IIb; Fig. 7.7). In the former case, due to the interaction term I, the $dy_i/dt = 0$ nullcline moves up and down in the two cell types, resulting in two fixed points. The mechanism for the emergence of two states is similar to that of class I (this mechanism is known as oscillation death [15]).

Exercise 7.3 Following the case of class I, we draw a change in the nullclines and understand how two fixed points appear self-consistently.

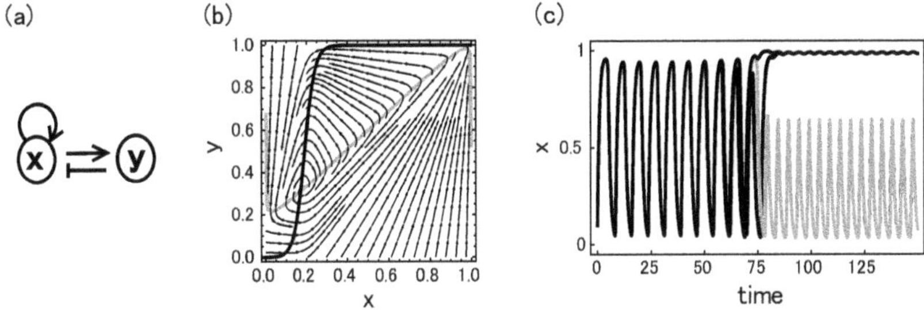

Figure 7.7 Differentiation processes that leave oscillating dynamics in the two-gene model. (a) An example network showing a differentiation process in which the oscillating dynamics remain. (b) Nullcline of expression dynamics in one cell. (c) Time series of the gene expression levels x. The parameters $(D, \theta_x, \theta_y) = (0.2, -0.1, 0.2)$ were used, and the other parameters were the same as those in Fig. 7.5.

Class IIb: Differentiation Process in Which Oscillatory Expression Dynamics Remain.

In this case, in multiple cells, the phase of the oscillation first becomes heterogeneous from cell to cell, and then some cells differentiate into a fixed-point state (Fig. 7.7). The cells with these oscillating expression dynamics are called type A, and the cells that emerge from them with a fixed-point state are called type B. In this case, the nullcline with $dy_i/dt = 0$ fluctuates in the vertical direction (y-direction) owing to the interaction term I. Figure 7.8(b) shows an enlarged view of the region around $(x, y) \sim (1, 1)$, where the x-nullcline, that is, $y = x + \frac{1}{\beta} \log(1 - 1/x) - g_x$, approaches 1. In this situation, if synchronization is lost, $I_i^y = D\left(\sum_j y_j/N - y_i\right)$ becomes negative for some cells. When the value decreases to a certain level, the x and y nullclines intersect with each other, as shown in Fig. 7.8, and the cross-point becomes the stable fixed point corresponding to type B (owing to the SNIC bifurcation described in Section 1.15). When some cells fall into the type B state by this bifurcation, the resulting increase in $\sum_j y_j/N$ leads to the inhibition of other cells from transitioning to type B cells. As a result, some cells differentiate into type B, while the rest remain in type A with a limit-cycle.

Since the state of the cells oscillates, the interaction term I changes over time. However, when we consider a time-averaged interaction, the idea of bifurcation owing to the cell-cell interaction mentioned previously holds. In addition, even though it is a fixed point, $I(t)$ oscillates because other cell expressions oscillate, and there is a small fluctuation in the type B state. However, such small fluctuations have no significant impact on the discussion. In conclusion, SNIC bifurcation by cell–cell interactions leads to differentiation from oscillating expression dynamics. Note that in the differentiation process of class IIa, similar SNIC bifurcation occurs

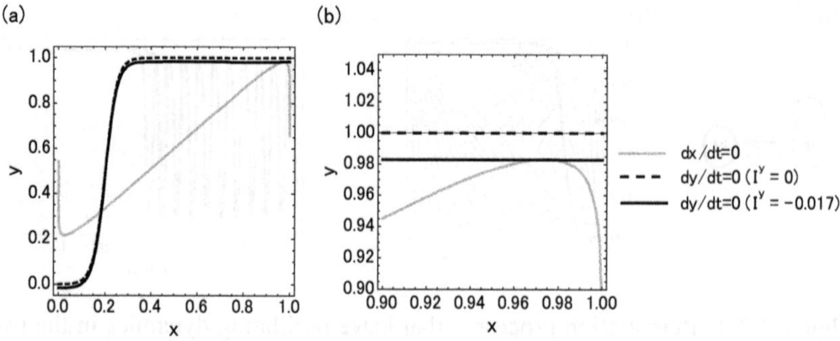

Figure 7.8 Nullcline of the differentiation process (Fig. 7.7), where the oscillating dynamics of the two-gene model remain. (a) Nullcline of $dy_i/dt = 0$ when all cell states oscillate synchronously (dashed line) and asynchronously (solid line). (b) Magnified view near $(x, y) \sim (1, 1)$. The intersection of the nullcline appears, owing to the variation in the interaction term I^y.

at $(x, y) \sim (0, 0)$ and $(1, 1)$, and the expression dynamics with limit-cycles can fall into two fixed points.

7.8 Stability at a Cell-Population Level by Regulating Differentiation Ratio

During development, the number of cells increases with cell division. When the number of cells increases from one cell to $2 \rightarrow 4 \rightarrow \cdots$, the expression levels in the two daughter cells after cell division are not exactly the same and fluctuate due to noise. Hence, if the uniform state is unstable, the difference among cells is amplified, leading to differentiation, as in the previous section. A question to be addressed here is how the ratio of each cell type is regulated through cell–cell interaction-based differentiation dynamics.

For example, suppose noise during cell division alone stochastically determines which type each cell becomes. In that case, the ratio of the number of cells of each type will be broadly distributed in each developmental process. However, the simulation results show that the ratio falls within a narrow range. Furthermore, when some cells of one type are removed from the outside, and the ratio is biased, differentiation occurs in the direction of returning the ratio to its original range. In other words, it shows stability at the cell-population level when restoring the original ratio.

Such stability at the cell-population level is a general property of interaction-based differentiation dynamics. Each cell type is stabilized by the $I_i^m(t)$ of its interaction with the surrounding cells. As an illustration, in the example of IIb in the previous section, as the cell ratio of type B increases, I_B^y of type B changes, the y-nullcline shifts upward, and the fixed point of type B becomes unstable. As a result, the number of type A cells increases, and the ratio returns to its original state. The consistency of the cell population and individual cell states results in a ratio control.

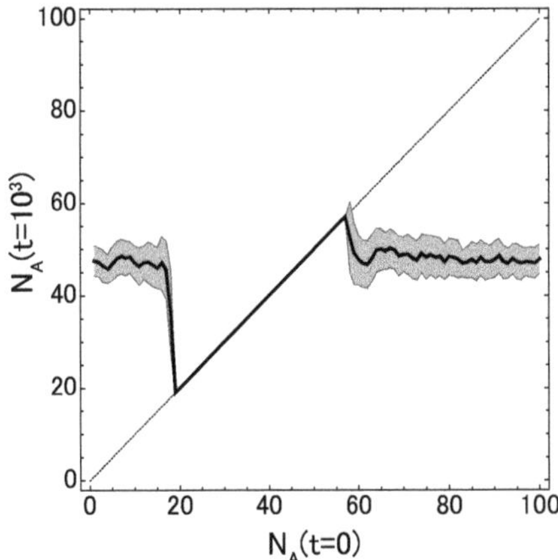

Figure 7.9 Simulation results of varying the cell-type ratio under initial conditions. The horizontal axis $N_A(t = 0)$ shows the number of cells of type A in the initial condition, and the vertical axis $N_A(t = 10^3)$ shows the number of cells of type A after 10^3 steps (assuming cells do not divide, the total number of cells is fixed at 100). In the range $20 < N_A(t = 0) < 60$, the initial cell type ratio does not change even after a certain time, but in the region where $N_A(t = 0)$ is smaller or larger, the cell type ratio becomes approximately constant. The black line represents the mean value of $N_A(t = 10^3)$, the light gray region represents its standard deviation, and the dotted line represents the line $N_A(t = 0) = N_A(t = 10^3)$ for reference.

Figure 7.9 shows the final cell-type ratio reached as a stationary state by changing the initial ratios of two cell types with a fixed number of N cells. The horizontal axis represents the number $N_A(t = 0)$ of type A at the initial condition (the number of cells of type B would be $N_B = N - N_A$), and the vertical axis represents the stationary number of type A cells stabilized after sufficient time has elapsed. When the initial N_A is large (i.e., for $N_A(t = 0) \geq 60$ in the figure), the cell states in the initial cell-number ratio become unstable because of the cell–cell interaction, and the SNIC bifurcation leads to differentiation of some cells from type A to type B, leading to the decrease in N_A. On the other hand, if the initial N_A is small ($N_A(t = 0) \leq 20$), then, conversely, some of the type B cells become unstable and "dedifferentiate" into type A cells. In the intermediate region, the transition of the cellular states does not occur. As a result, the proportion of cell types remains within a certain range.

7.9 Differentiation Dynamics of Stem Cells

Regulated proliferation and differentiation of stem cells play an essential role in the dynamics of multicellular systems, including the development, regeneration, and maintenance of tissue homeostasis. What are the properties of these stem cells? What is the difference between undifferentiated stem cells and terminally differentiated cells that have lost their differentiation potential? Following the studies by Kauffman, if we consider that each cell type corresponds to a different attractor, the differences between stem and differentiated cells can be described as properties of the corresponding attractor. In the following section, we discuss the difference between stem cells and differentiated cells in terms of their intracellular dynamics.

In the previous section, by using a simple cell model with two genes, we showed that cells with oscillatory expression dynamics (class IIb) self-replicate themselves in nearly the same state as their own. At the same time, these cells also have the ability to differentiate into another state. Thus, the cell with class IIb expression dynamics can be regarded as stem cells. These differentiation dynamics was further studied using cell models with a regulatory network of many genes [2, 20]. In these studies, regulatory networks which can generate multiple cell types were screened. Figure 7.10 shows an example of differentiation dynamics, in which cells with complex oscillatory dynamics have the potential both to reproduce themselves and to differentiate into three cell types. From thousands of simulations using different rules of regulatory interaction and cell–cell interactions, the following common features of stem cell systems are found:

1. The regulatory dynamics in a stem cell are of a complex oscillatory nature. These dynamics are responsible for generating heterogeneity in the stem cell state. The heterogeneity is necessary to achieve stem cell differentiation dynamics in which some stem cells differentiate and others remain in their original state.

2. The frequency of differentiation depends on the environment, that is, the distribution of other cell types. This regulation of the differentiation probability leads to the stability of the cell society with respect to external perturbations, as discussed in Section 7.8.

3. The irreversible loss of differentiation potential experienced in the change from a stem cell to a differentiated cell is characterized by a decrease in complexity in reaction dynamics. The reaction dynamics in stem cells are maintained by complex oscillatory dynamics involving many expressed genes. In contrast, the dynamics in differentiated cells are simpler, characterized by fixed-point dynamics or regular oscillations involving a smaller number of expressed genes.

Note that the preceding features in stem cell differentiation dynamics are independent of the details of model simulations. For example, they are essentially unchanged as long as the equations describing the intracellular expression dynamics are sufficiently complex to exhibit nonlinear oscillatory dynamics.

The importance of the oscillating dynamics in stem cells has been suggested in experiments using embryonic stem cells and neural stem cells. For example,

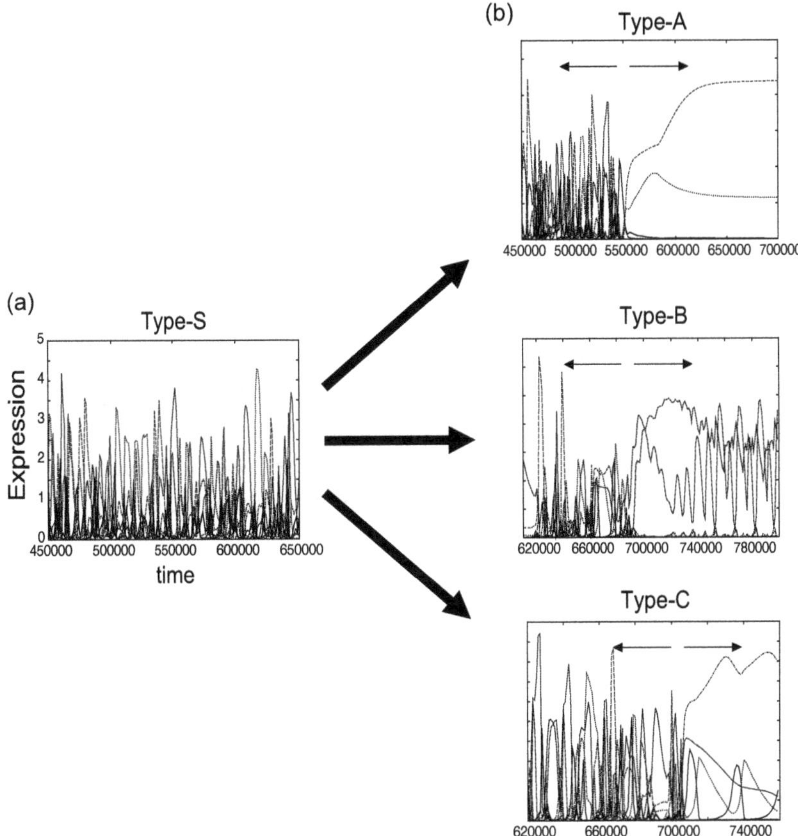

Figure 7.10 An example of stem cell differentiation dynamics. Each figure shows a time series of gene expression, while the arrows between the figures represent the potential to differentiate to the corresponding cell type. (a,b) Expression time series of a stem cell (Type-S) with complex oscillatory dynamics and differentiated cell types (Type-A, B, and C), respectively. See [2] for details.

quantitative analysis at the single-cell level has shown that the expression dynamic of the *Hes1* gene, which is involved in the differentiation of embryonic stem cells and neural stem cells, exhibits oscillations over time [14, 9]. Furthermore, it was demonstrated that oscillatory expression dynamics are essential for maintaining differentiation potential by using a technique to control the expression level externally by light stimulation (optogenetics) [9]. In this system, cells capable of differentiating into three different neurons are shown to have oscillating expression dynamics of each cell-specific transcription factor, which is in good agreement with the properties predicted from the simulations. Furthermore, a decrease in the number of expressed genes during development was demonstrated [5], which is also consistent with the preceding theoretical predictions.

7.10   Summary and Future Problems

In this chapter, theoretical approaches to the dynamics of cell differentiation in multicellular organisms have been presented. First, Kauffman's pioneering work, which introduced the concept of cell types as attractors of cellular dynamics, was discussed. In the latter half of the chapter, we explored how the transition between attractors, that is, differentiation, is caused by cell–cell interactions. The results can be summarized as follows:

1. The cellular state is not determined by individual internal dynamics alone. Instead, the cellular state is determined to satisfy the consistency between the internal dynamics and the interaction with the surrounding cells.
2. Small fluctuations can be amplified to produce large differences in cellular state.
3. The frequency of differentiation is regulated by cell–cell interactions. As a result of the regulation of differentiation frequency, stability at the cell population level emerges.

Of course, there are still many unresolved issues regarding the diversification dynamics of the cellular state. Some examples of such problems to be solved in the future are here:

- In multicellular organisms, there are undifferentiated stem cells and terminally differentiated cells that have lost their ability to differentiate, and the differentiation between them is often irreversible. How can we understand the irreversibility of differentiation when we consider it as a transition between attractors? Analysis of a multicomponent extension of the cell differentiation model suggests that differentiated stem cells have complex oscillating dynamics consisting of many components [2, 20]. Can we then extract macroscopic state variables (such as thermodynamic entropy) that describe irreversible differentiation processes? Experimental operations to reverse this irreversibility [21] need to be understood theoretically.
- Epigenetic control by various mechanisms, such as genome methylation and histone modification, is not included in the model in this chapter. Epigenetic control has been shown to result in intergenerational "memory" of cells, which may contribute to the emergence of discrete cell types. Theoretical and experimental analyses are needed to understand how these internal dynamics at different time scales are involved in the differentiation process. For some attempts toward this direction, see [18, 17], in which threshold levels for gene expression (e.g., θ_x and θ_y in Eq. (7.18)) are assumed to change slowly as a result of epigenetic modifications.
- The process of carcinogenesis is often described by the stochastic event of the accumulation of mutations in the genome. On the other hand, it has been

suggested that the cellular states reached by carcinogenesis do not necessarily differ greatly, and in some cases, they fall into similar cellular states [7]. These results suggest that the state of a cancer cell corresponds to an attractor of intracellular dynamics in the same way as a normal cell. A gene regulatory network that produces complex developmental processes may inevitably produce "surplus" attractors. How these surplus attractors exist and how their transitions can be controlled will have important implications for our understanding of carcinogenesis.

References

[1] E. Clayton *et al.* A single type of progenitor cell maintains normal epidermis. *Nature*, **446**(7132): 185–189, 2007.

[2] C. Furusawa and K. Kaneko. Theory of robustness of irreversible differentiation in a stem cell system: Chaos hypothesis. *Journal of Theoretical Biology*, **209**(4): 395–416, 2001.

[3] C. Furusawa and K. Kaneko. A dynamical-systems view of stem cell biology. *Science*, **338**: 215–217, 2012.

[4] Y. Goto and K. Kaneko. Minimal model for stem-cell differentiation. *Physical Review Journals*, E **88**(3): 032718, 2013.

[5] G. S. Gulati, et al. Single-cell transcriptional diversity is a hallmark of developmental potential. *Science*, **367**(6746): 405–411, 2020.

[6] J. B. Gurdon, P. Lemaire and K. Kato. Community effects and related phenomena in development. *Cell*, **75**(5): 831–834, 1993.

[7] S. Huang *et al.* Cell fates as high-dimensional attractor states of a complex gene regulatory network. *Physical Review Letters*, **94**: 128701, 2005.

[8] S. Huang *et al.* Bifurcation dynamics in lineage-commitment in bipotent progenitor cells, *Developmental Biology*, **305**: 695–713, 2007.

[9] I. Imayoshi *et al.* Oscillatory control of factors determining multipotency and fate in mouse neural progenitors. *Science*, **342**(6163): 1203–1208, 2013.

[10] K. Kaneko and T. Yomo. Cell division, differentiation and dynamic clustering, *Physica* D. **75**: 89–102, 1994.

[11] S. A. Kauffman. Metabolic stability and epigenesis in randomly constructed genetic nets. *Journal of Theoretical Biology*, **22**: 437–467, 1969.

[12] S. A. Kauffman. *The Origins of Order: Self-Organization and Selection in Evolution.* Oxford, Oxford University Press, 1993.

[13] A. M. Klein *et al.* Kinetics of cell division in epidermal maintenance. *Phys. Rev. E*, **76**: 021910, 2007.

[14] T. Kobayashi *et al.* The cyclic gene Hes1 contributes to diverse differentiation responses of embryonic stem cells. *Genes and Development*, **23**(16): 1870–1875, 2009.

[15] A. Koseska, E. Volkov and J. Kurths. Transition from amplitude to oscillation death via Turing bifurcation. *Physical Review Letters*, **111**: 024103, 2013.

[16] F. Li *et al.* The yeast cell-cycle network is robustly designed. *Proceedings of the National Academy of Sciences USA*, **101**(14): 4781–4786, 2004.

[17] Y. Matsushita and K. Kaneko. Homeorhesis in Waddington's landscape by epigenetic feedback regulation. *Physical Review Research*, **2**: 023083, 2020.

[18] T. Miyamoto, C. Furusawa and K. Kaneko. Pluripotency, differentiation, and reprogramming: A gene expression dynamics model with epigenetic feedback regulation. *PLOS Computational Biology*, **11**(8): 1004476, 2015.

[19] B. Samuelsson and C. Troein. Superpolynomial growth in the number of attractors in Kauffman networks. *Physical Review Letters*, **90**: 098701, 2003.

[20] N. Suzuki, C. Furusawa and K. Kaneko. Oscillatory protein expression dynamics endows stem cells with robust differentiation potential. *PLOS ONE*, **6**(11): e27232, 2011.

[21] K. Takahashi and S. Yamanaka. Induction of pluripotent stem cells from mouse embryonic and adult fibroblast cultures by defined factors, *Cell*, **126**(4): 663–676, 2006.

[22] J. E. Till, E. A. McCulloch and L. Siminovitch. A stochastic model of stem cell proliferation, based on the growth of spleen colony-forming cells. *Proceedings of the National Academy of Sciences USA*, **51**: 29–36, 1964.

[23] A. M. Turing. The chemical basis of morphogenesis. *Philosophical Transactions of the Royal Society of London Series B*, **237**(641): 37–72, 1952.

8

Spatiotemporal Patterning by the Cells

8.1 Pattern Formation by Local Interaction

In Chapter 4, we studied mathematical methods for pattern formation and propagation in space-time. In this chapter, following these methods, we further explore theoretical models to capture complex pattern formation of multicellular organisms.

The generation of patterns in animal development is attributed not solely to long-range cell–cell interactions mediated by diffusible factors but also to the short-range interplay of molecules anchored on the plasma membrane, a process termed juxtacrine signaling. The distinct cellular arrangements, dependent on cell type, characterized by mutual inhibition mechanisms involving Delta and Notch protein, lead to the formation of sharp boundaries and checkerboard patterns, serving as typical consequences of such signaling modes.

Delta and Notch proteins on the plasma membrane, constitute a pair of signaling molecules pivotal in the regulation of embryonic development. The intercellular interaction, wherein Delta from one cell interacts with Notch on an adjacent cell modulates gene expression downstream of the signaling pathway, thereby influencing cellular differentiation. This mode of interaction is classified as trans-acting signaling. Concurrently, a cis-acting repression mechanism is also recognized, characterized by mutual suppression of Delta and Notch within the same cell. Given the mutual repression between Notch and Delta expression, it is plausible to anticipate the adjacent emergence of two distinct cell types: one typified by low Notch and high Delta levels, and the other by high Notch and low Delta levels.

What types of spatially characterized patterns emerge from the Delta–Notch reaction system? To explore this question, the process of wing vein formation in *Drosophila* was examined. During the development of the wing primordium in *Drosophila* pupae, a morphogen gradient dictates the rate of Delta production, subsequently leading to the establishment of longitudinal veins at the midpoint of the gradient. In the following section, we discuss a simplified model [25] that elucidates the formation of striped patterns as a consequence of the Delta–Notch interaction dynamics.

Consider a scenario where cells are positioned on a two-dimensional plane, each represented as a vertex in a hexagonal lattice. We assume that these cells do not move on the plane. Each cell is adjacent to six others, and its state is influenced by interactions with these neighboring cells. Let X_i and Y_i denote the concentrations of Notch and Delta in the ith cell, respectively. Furthermore, let R_i represent the concentration of a reporter, whose expression is enhanced by the trans-acting interaction between its Notch and Delta of surrounding cells. Let X_i, Y_i, and R_i be uniform values within each cell.

$$\frac{dX_i}{dt} = \beta_X - \gamma X_i - \frac{X_i \langle Y \rangle_i}{k_t} - \frac{X_i Y_i}{k_c},$$

$$\frac{dY_i}{dt} = \beta_Y(x) - \gamma Y_i - \frac{Y_i \langle X \rangle_i}{k_t} - \frac{X_i Y_i}{k_c}, \qquad (8.1)$$

$$\frac{dR_i}{dt} = \beta_R \frac{(X_i \langle Y \rangle_i)^n}{k_{RS} + (X_i \langle Y \rangle_i)^n} - \gamma_R R_i,$$

where β_X, β_Y, and β_R are production rates, γ and γ_R are degradation rates, and k_c and k_t are the coefficients for cis- and trans-repression, respectively. k_{RS} is the coupling constant, and n is the Hill coefficient.

$$\langle Y \rangle_i = \frac{1}{6} \sum_{j=]i[} Y_j \qquad (8.2)$$

$$\langle X \rangle_i = \frac{1}{6} \sum_{j=]i[} X_j \qquad (8.3)$$

represent the average concentrations of the six cells in the neighborhood of the ith cell, which are denoted as $]i[$.

Assume that chemicals within the tissue establish a concentration gradient, with the resulting parameter β_Y being a function of the position x. For simplification, we refer to the linear gradient depicted in Fig. 8.1. When the production rate β_X of X exceeds that of β_Y for Y, there is a corresponding decrease in Y concomitant with an increase in X. Conversely, if $\beta_Y > \beta_X$, an excess in Y production leads to a reduction in X. It should be noted that this mutual inhibition system lacks a feedback circuit and is not bistable. Numerical simulations reveal that a stripe of R emerges at the midpoint of the concentration gradient where $\beta_Y(x) = \beta_X$. This brings us back to the issue of pattern formation in response to a gradient. The process for generating a bell-shaped pattern at the gradient's mid-point is elucidated in Section 4.2.2, focusing on the feed-forward system. In this context, it is

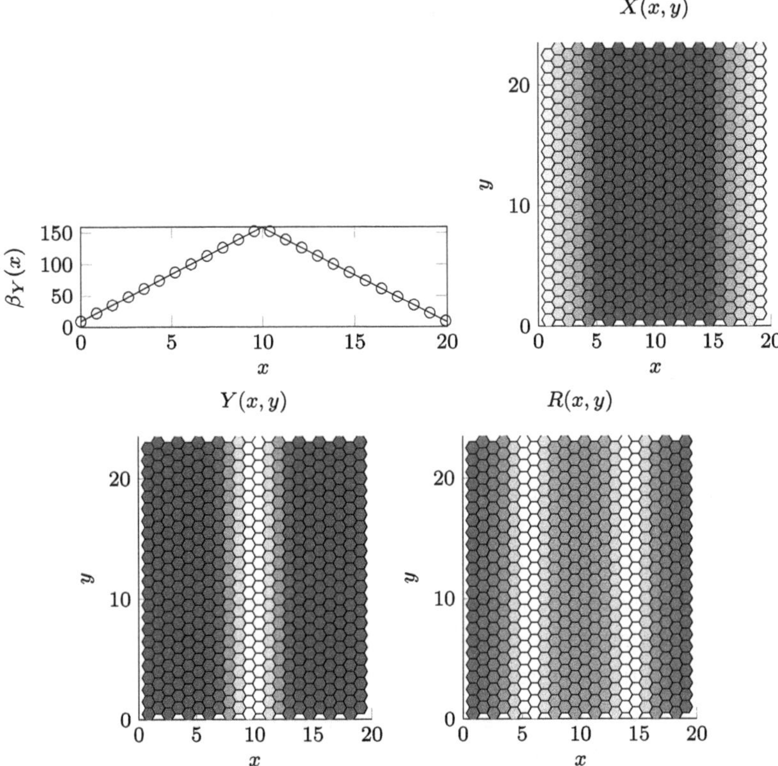

Figure 8.1 Boundary pattern formation by a local interaction of mutual inhibition. Examples of numerical simulations of Eq. (8.1). $\beta_X = 100$, $\beta_R = 75$, $\gamma = 0.1$, $\gamma_R = 1$, $k_t = 5$, $k_c = 0.1$, $k_{RS} = 1500$, $n = 1$ [25]. The slope of β_Y is 0.15.

not Delta and Notch per se, but rather the pathway downstream of Delta and Notch, which determines the variable R, that exhibits a similar structure. As discussed in Section 4.2.2, the formation of a distinct stripe by the feed-forward system necessitates a high Hill coefficient n. In contrast, this pronounced response, stemming from cis-acting mutual inhibition, allows for the emergence of a stripe even when the R response is $n = 1$.

Next, let us consider integrating a feedback mechanism into the previously described system:

$$\frac{\mathrm{d}X_i}{\mathrm{d}t} = \beta_X - \gamma X_i - \frac{X_i\langle Y\rangle_i}{k_t} - \frac{X_i Y_i}{k_c}, \tag{8.4}$$

$$\frac{\mathrm{d}Y_i}{\mathrm{d}t} = \beta_Y \frac{1}{1 + R_i^m} - \gamma Y_i - \frac{Y_i\langle X\rangle_i}{k_t} - \frac{X_i Y_i}{k_c}, \tag{8.5}$$

$$\frac{\mathrm{d}R_i}{\mathrm{d}t} = \beta_R \frac{(X_i\langle Y\rangle_i)^n}{k_{RS} + (X_i\langle Y\rangle_i)^n} - \gamma_R R_i. \tag{8.6}$$

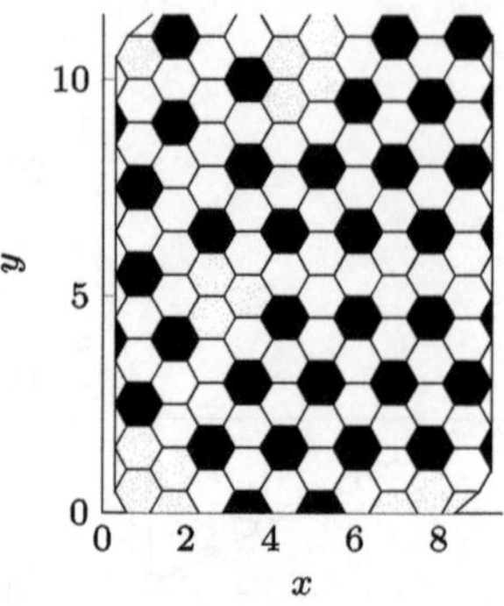

Figure 8.2 Pattern formation by adjacent interactions of mutual inhibition.

Eq. (8.5) under the presence of (8.6) represents a feedback mechanism wherein an increase in the concentration of Y in the surrounding cells leads to a decrease in Y within a given cell, mediated through an elevation in R. This phenomenon is called lateral inhibition. Moreover, the reduction in Y due to mutual inhibition prompts an increase in X, thereby facilitating the alternating appearance of cells with high X and high Y concentrations. Should cell states be determined by this mechanism, it becomes challenging to circumvent initial value defects that result in cells deviating from an equidistant arrangement, as depicted in Fig. 8.2. This issue arises because interaction from one side of a cell may augment X, whereas interaction from the opposite side might reduce it. Such "frustrated configurations" cannot be resolved without a significant reorganization of the entire system. To avoid these scenarios, one possible strategy is to establish patterning in one direction, either by spatially skewing the initial values or by directionally shifting the parameters. In real developmental systems, it is observed that patterns emerge through such progressing wavefronts [6].

The circumstances under which a spatially uniform solution becomes unstable can be elucidated via a linear stability analysis, akin to the Turing instability mechanism. While at first glance trans-acting repression, which involves interactions between different variables, may appear complex compared to diffusion processes, the Jacobi matrix of the system encompassing extended interactions can be derived. This is achieved through a tensor product of the internal dynamics and the interactions. Subsequently, the eigenvalues can be ascertained [20].

Exercise 8.1.1 Lateral Inhibition Model [2]

$$\frac{\mathrm{d}Y_i}{\mathrm{d}t} = k_Y \frac{K}{K + R_i} - \gamma_Y Y_i \tag{8.7}$$

$$\frac{\mathrm{d}R_i}{\mathrm{d}t} = k_R \frac{\langle Y \rangle_i^{\,n}}{K^n + \langle Y \rangle_i^{\,n}} - \gamma_R R_i \tag{8.8}$$

Conduct numerical simulations to investigate the destabilization and subsequent pattern formation of the uniformly stable solution in Eqs. (8.7) and (8.8). This system can be interpreted as representing a scenario where Delta Y activates the repressor R in neighboring cells through Notch interactions, and in turn, repressor R inhibits its own Delta Y. This scenario corresponds to the model described earlier, where cis-acting repression is excluded, and the changes in Notch are considered sufficiently minor.

Exercise 8.1.2 In the Delta–Notch system, long-range interactions are facilitated through elongated, thread-like projections on the plasma membrane, known as filopodia. These projections are believed to orchestrate the formation of sesamoid-like patterns [1]. In the study of skin patterns in fish, mutations in the Notch gene have been linked to the emergence of atypical patterns, with observations of filopodia-like protrusions. This signifies a pattern formation mechanism inherent to the Delta–Notch system. While this mechanism shares similarities with Turing pattern, the basis for the characteristic length of long-range correlated patterns is not exclusively attributed to diffusion. By modifying the interaction parameters within Collier's model [2], as demonstrated in Exercise 8.1.1, investigate the feasibility of generating patterns with expanded spacing.

8.2 Hybrid Model of Cellular Automaton and Continuous System

A system where discrete states evolve over time within discretized time and space, as outlined in Section 4.4.5, is commonly referred to as a cellular automaton. Cellular automata are characterized by a multitude of transition rules, leading to a wide array of emergent patterns [22, 28]. Integrating such methodologies with continuous systems has proven to be valuable in deciphering biological phenomena. An exemplary instance of this approach can be observed in the formulation of the Kessler-Levin model, which describes the evolution of cAMP waves in cellular slime molds [11].

Under starvation, cellular slime molds initiate synthesis and secretion of cAMP, which serves as a chemotactic attractant, orchestrating the aggregation of hundreds to hundreds of thousands of cells. The timing of cAMP synthesis and secretion is synchronized across the cells, resulting in the propagation of extracellular cAMP in concentric and spiral waves. In line with the approach delineated in Section 4.4.5, we consider each grid point as a cell (equating, on a realistic scale, to several cells). The state of each cell is represented by the following variables: The cell state# can assume values of 0 (stable fixed point), 1 (activation by external stimuli), or 2 (refractory period). The reaction dynamics are described by the state transition sequence $0 \rightarrow 1 \rightarrow 2 \rightarrow 0$, coupled with the diffusion equation of the attractant.

It is posited that a cell in the excited state secretes an extracellular signaling factor C at the rate of a per unit time. This secreted C diffuses and activates surrounding cells. The transition of each lattice to the excited state is contingent upon the concentration of C surpassing a threshold C_T:

$$\frac{\partial C(x;t)}{\partial t} = D\nabla^2 C(x;t) - \gamma C(x;t) + F(\text{state\#}).$$

State 0 represents the ground state of the cell, which transitions to state 1 upon receiving a stimulus exceeding the threshold C_T. From this point, we begin the temporal count with $v = 0$. The cell secretes the signal for a duration of τ. Subsequent to this period, when $v = \tau$ is reached, the cell transitions to state 2. In other words:

$$F(\text{state\#}) = \begin{cases} a \ (\text{state\#} = 1) \\ 0 \ (\text{state\#} \neq 1) \end{cases}.$$

Following this, the interval $\tau < v < \tau_{ar}$ is defined as the absolute refractory period, during which the cell remains in state 2, unaffected by the concentration of C. The subsequent time span, $\tau_{ar} < v < \tau_{rr}$, is termed the relative refractory period. In this phase, the threshold for cellular response is dynamically reduced based on the intensity of the stimulus and the elapsed time since excitation. If the reduced threshold, as determined by

$$C_T = (C_{max} - A\frac{\tau - \tau_{ar}}{\tau}), \tag{8.9}$$

is exceeded, the cell transitions again to state 1. Here, the constant A is set

$$A = \frac{\tau_{rr}}{\tau_{rr} - \tau_{ar}}(C_{max} - C_{min})$$

in such a way that $C_T = C_{min}$ at the end of the relative refractory period $v = \tau_{rr}$.

To summarize the preceding state transition rules, the state transition from a time t to a time after a small time (dt) is

$$\text{state\#}(x, t) = 0 \rightarrow \text{state\#}(x, t + dt) = 1; \ C(x;t) > C_T(= C_{min})$$
$$\text{state\#}(x, t) = 1 \rightarrow \text{state\#}(x, t + dt) = 1; \ 0 < v < \tau$$
$$\text{state\#}(x, t) = 1 \rightarrow \text{state\#}(x, t + dt) = 2; \ v = \tau$$
$$\text{state\#}(x, t) = 2 \rightarrow \text{state\#}(x, t + dt) = 2; \ \tau < v < \tau_{ar}$$
$$\text{state\#}(x, t) = 2 \rightarrow \text{state\#}(x, t + dt) = 2; \ C(x;t) < C_T; \ \tau_{ar} < v < \tau_{rr}$$
$$\text{state\#}(x, t) = 2 \rightarrow \text{state\#}(x, t + dt) = 1; \ C(x;t) > C_T; \ \tau_{ar} < v < \tau_{rr}$$
$$\text{state\#}(x, t) = 2 \rightarrow \text{state\#}(x, t + dt) = 0; \ v = \tau_{rr}.$$

In cellular slime molds, the center of a spiral wave serves as the focal point for cell aggregation, as it migrates counter to the cAMP wave's propagation direction.

Consequently, an excessive number of spiral centers results in reduced number of cells in each aggregates. In contrast, too few centers lead to overly large cell aggregates, which may adversely affect the structure of the subsequently formed fruiting body, unable to adequately support the spores. This raises the following question: What factors determine the number of spiral centers? To address this, let us discuss introducing a slow variable E as a measure of excitability. This variable influences the threshold value as described in Eq. (8.9), with adjustments made as follows:

$$C_{\mathrm{T}} = \left(C_{\max} - A\frac{\tau - \tau_{\mathrm{ar}}}{\tau} \right) (1 - E).$$

Assuming that cells increase their excitability when stimulated by cAMP,

$$\frac{\partial E(x,t)}{\partial t} = \eta + \beta C(x,t) - \alpha E(x,t) \ \ (E < 1),$$

$$\frac{\partial E(x,t)}{\partial t} = 0 \ \ (E = 1).$$

$$(8.10)$$

Even when a uniform initial value of $E = 0$ is chosen across space, each amplification of C due to noise causes a local increase in E. As E rises, there is a tendency for C to increase further. This interaction leads to spatial inhomogeneity in E, setting the stage for wave propagation in regions where it was previously hindered by low E values, thereby facilitating the formation of spiral centers. When the parameter β is large, the inhomogeneity in E increases, resulting in a greater number of spirals. Conversely, a very small β prolongs the transient state of the system, also making spiral formation more probable. Numerically, an intermediate value of β yields the minimum number of spirals, mirroring the relationship observed between developmental rates and spiral numbers in both wild strains and mutants of cellular slime molds [21]. Furthermore, in a system where E is uniformly close to $E = 1$, spiral generation is impeded. This characteristic accurately reflects the wave patterns observed in cells uniformly differentiated by agitation in suspension, as shown in Fig. 8.3. Therefore, the initial state and parameter change are critical in the actual behavior of the system.

Exercise 8.2.1 The Kessler–Levin model is represented as follows:

$$\frac{\partial u}{\partial t} = F(u,v) + D\nabla^2 u,$$

$$\frac{\partial v}{\partial t} = G(u,v).$$

Then, show the following equations hold

$$F(u,v) = -\gamma u + aH(u - u_{\mathrm{T}})\{1 - H(v - \tau)\},$$

$$G(u,v) = H(v)$$

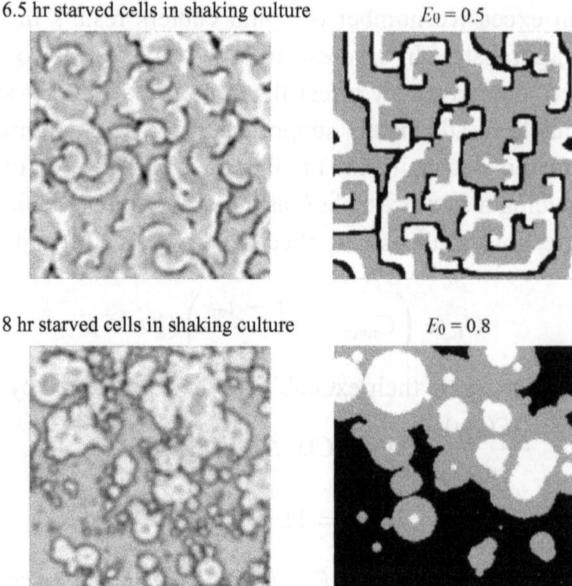

Figure 8.3 Behavior of cAMP waves in cellular slime mold and the Kessler–Levin model (snapshots). (Left column) Difference image of dark-field illuminated image z (scale: 1.2 mm × 1.2 mm). (Right column) Numerical results [21]. state#0 (gray), state#1 (black), and state#2 (white). The E_0 denotes the value of E in the initial condition.

by using the Heaviside function H [29]. Note that u_T refers to the form of C_T in Eq. (8.10), and, when $v = \tau_{rr}$ is reached, it is reset to $v = 0$.

8.3   Dynamics of Cellular Arrangement and Morphology

In the preceding sections, we discussed the cases where the cell–cell interactions contribute to the spatial patterning of organisms, encompassing aspects such as gene expression and the localization of reactants. In this section, our focus shifts to methodologies that necessitate a more explicit consideration of cellular arrangements and deformations.

8.3.1   Models Containing Cell Proliferation and Movement

Diffusion Limited Aggregation (DLA) represents a well-established model for depicting dendritic crystal patterns. The transition rule governing DLA is remarkably straightforward. Suppose that a crystal nucleus situated at the center of a two-dimensional space, functions as a trap for particles. These particles are introduced from a distant, random location and subsequently diffuse toward the nucleus. It is assumed that the particle count is limited and interactions among them are negligible. When particles adhere to a cluster of particles or a nucleus during their

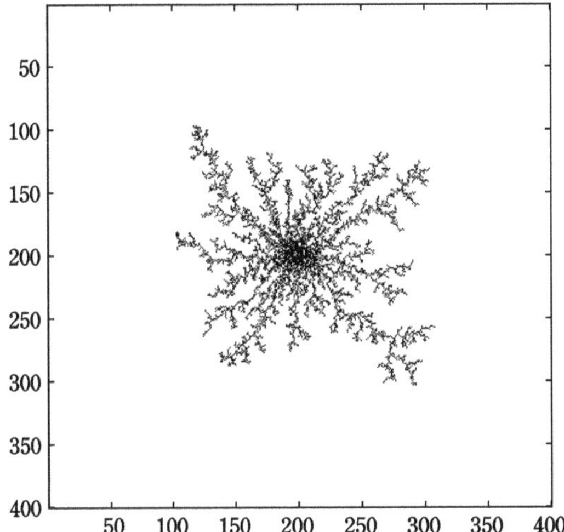

Figure 8.4 An example of DLA. Suppose that a particle engaged in a random walk on a two-dimensional plane becomes immobile upon contacting a central structure. With the repetition of this process, a greater likelihood emerges for additional particles to adhere to the extremities of the branch-like structure, consequently giving rise to such a formation.

random walk, they cease movement. Repeated iterations of this process result in the growth of outward branches, as particles are more readily adsorbed on these protrusions. Conversely, the valleys between the branches, being less accessible, exhibit minimal growth. This phenomenon creates a positive feedback loop, where the randomness induced by diffusion synergizes with the propensity for growth in areas where it has already occurred. In the pattern formation of diffusion-adsorption systems, as opposed to reaction-diffusion systems, the pattern becomes fixed through an irreversible process. Once adsorbed, the particles cannot dissociate, as depicted in Fig. 8.4. Some structures formed in bacterial colonies are known to be similar to DLA.

Bacterial growth is predominantly observed at the tips of branches in bacterial colonies, as nutrients are more efficiently diffused from distant, nutrient-rich areas to these locations. Conversely, nutrient depletion is more pronounced between the branches due to the shading effect, which impedes bacterial growth. Thus, despite the difference in mechanisms, where diffusion involves nutrient molecules and adsorption corresponds to bacterial growth, DLA is considered to appropriately represent the fundamental characteristics of bacterial colony growth [16].

The morphological characteristics of actual *Bacillus subtilis* colonies on an agar medium are known to vary depending on the concentration or firmness of the agar and the initial nutrient concentration within. The mesh apertures of the agar gel, a solid medium, are smaller than the diameter of *B. subtilis* cells, thereby preventing the bacteria from penetrating into the agar. These bacteria utilize the nutrients embedded in the agar for proliferation and division. However, due to the inability of the colonies to penetrate the surface of the agar, they predominantly exhibit two-dimensional growth along the agar surface.

In scenarios where nutrients are abundant, a circular colony pattern emerges within a few days. Conversely, under conditions of nutrient scarcity and softer agar mediums, bacterial proliferation is prolonged over weeks, resulting in a dendritic, branched morphology. To comprehend how these distinct patterns are influenced by experimental conditions, it is imperative to develop a model that incorporates both cell growth and stochastic movement processes. When the cell count is sufficiently large and the spatial scale of the pattern of interest is extensive, the discreteness of individual cells may be disregarded. Under such circumstances, the bacterial population can be approximated as a continuous entity.

Let us denote the nutrient concentration as c and the bacterial population density as N, and consider the following equations [10]:

$$
\begin{aligned}
\frac{\partial c}{\partial t} &= -kN\frac{c}{K+c} + D_c\nabla^2 c, \\
\frac{\partial N}{\partial t} &= \theta kN\frac{c}{K+c} + \nabla \cdot \{D_N\nabla N\}.
\end{aligned}
\tag{8.11}
$$

The first term on the right-hand side of the latter equation represents a Monod-type growth function and that of the former represents the corresponding decrease in the nutrient, analogous in form to the Michaelis–Menten equation. This term characterizes how the growth rate is influenced by nutrient availability. The second term in each equation describes nutrient diffusion and bacterial motility, respectively. Given the assumption that bacterial motility is contingent upon both nutrient levels and population density, we introduce the following empirical relationship:

$$
D_N = \sigma_0(1 + \eta)cN,
\tag{8.12}
$$

where η is noise. The rationale for not extending D_N outside the left operator ∇ is attributed to its dependency on both c and N, resulting in spatial variability of D_N's value. Experimentally, it has been observed that a lower concentration of agar leads to a more wetting surface, thereby facilitating bacterial motility. Consequently, under such conditions, the coefficient σ_0, representing the motility of bacteria, tends to be larger.

By linearly approximating the growth term in Eq. (8.11) and subsequently normalizing the parameters, we derive the following equations [10]:

$$
\begin{aligned}
\frac{\partial c}{\partial t} &= -Nc + D_c\nabla^2 c, \\
\frac{\partial N}{\partial t} &= Nc + \nabla \cdot \{\sigma_0 Nc\nabla N\}.
\end{aligned}
\tag{8.13}
$$

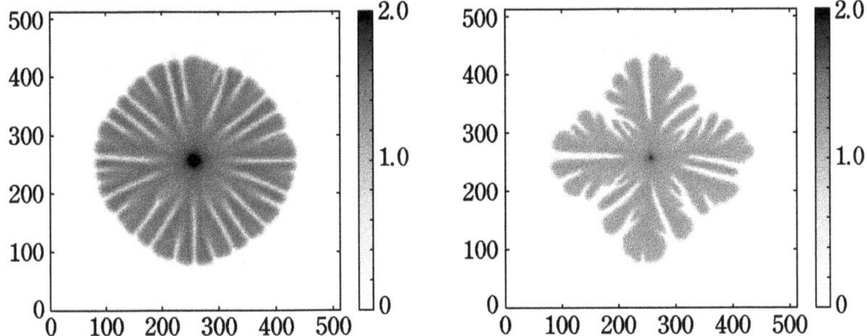

Figure 8.5 Bacteria colony pattern. A numerical simulation of Eq. (8.13) is presented. This snapshot illustrates the time evolution from an initially localized distribution at the center of the region, with $\beta = 0.8$ representing the maximum value at the center. The grayscale denotes the values of N. On the left, the initial nutritional value is $c(0) = 1.2$, and on the right, it is $c(0) = 0.7$ (courtesy of Dr. Shuji Ishihara (The University of Tokyo).

Figure 8.5 presents numerical simulations of this simplified model. A smooth pattern with few branches forms when the initial nutritional value is high. Conversely, lower initial nutritional values result in increased branching. Initially, bacteria consume c at the center of the region, spreading spatially. Areas deeper into the region are more prone to nutrient depletion, leading to a halt in bacterial growth. In contrast, the colony's periphery, having higher c and more cellular growth, is more likely to extend outwards. This positive feedback, where protruding areas grow and less protruding areas do not, intensifies density fluctuations, consequently disrupting the spatial symmetry of the growing interface. As the nutrient concentration decreases, the bacterial pattern is thinner than that in Fig. 8.5 and shows a pattern similar to that observed for DLA in Fig. 8.4. Indeed, the numerically observed pattern is in good agreement with that observed in bacterial colony experiments (as characterized, for example, by agreement in the fractal dimension) [18].

Example 8.1 Continuous Model of the Aggregation of Cellular Slime Mold. Consider the aggregation process of cellular slime molds via chemotaxis using the following system of equations [9]. Here, c represents the concentration of cAMP, v denotes the active receptor, and N shows the cell number density:

$$\frac{\partial c}{\partial t} = \phi(N)f(c,v) - (\phi(N) + \delta)\gamma c + D_c\nabla^2 c, \tag{8.14}$$

$$\frac{\partial v}{\partial t} = -k_+ cv + k_-(1 - v), \tag{8.15}$$

$$\frac{\partial N}{\partial t} = \nabla(\mu(N)\nabla N) - \nabla\chi(v)N\nabla c, \tag{8.16}$$

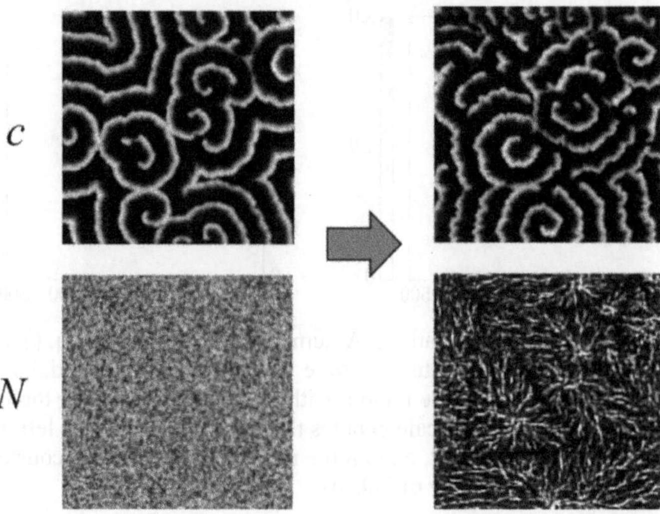

Figure 8.6 Aggregation pattern of *Dictyostelium* cells. Numerical simulation of Eq. (8.14), (8.15), (8.16) (courtesy of Dr. Daisuke Imoto (The University of Tokyo)).

where Eqs. (8.14) and (8.15) are simplified versions of the Martier–Goldbeter equation. In these equations, $f(c,v) = \lambda(bv + v^2)(a + c^2)/(1 + c^2)$ represents the production of cAMP, and $\phi(N) = N/(1 - \rho N/(K + N))$ is a function reflecting the density effect, indicating that reactions are more probable with larger values of N. The first term on the right side of Eq. (8.16) represents random motion, while the second term denotes chemotaxis. The expression $\mu(N) = \mu_1 + \mu_2 n^4/(n^4 + N^4)$ indicates that random motion diminishes as the number of cells increases. Additionally, $\chi(v) = \chi_0 v^m/(A^m + v^m)$ represents that cells are less likely to move behind cAMP waves. The simulation results of this model, as depicted in Fig. 8.6, demonstrate a good agreement with experimental observations.

8.3.2 *Cellular Potts Model*

Cellular tissues often exhibit fluid-like behavior over extended time scales. For example, long-term observation of cellular tissue or dispersed cells reveals phenomena like long-distance cell flow and the emergence of spherical cell clusters. In the context of multicellular development and morphogenesis, it is noteworthy that different cell types can organize into separate aggregates, or clusters of one cell type may be encircled by clusters of another type, a process known as cell sorting. The differential adhesion hypothesis, drawing an analogy to the phase separation of water and oil, has been proposed to explain this phenomenon [7]. This hypothesis posits that while water molecules tend to bond with each other and similarly for oil molecules, water and oil molecules are less likely to bind. The resultant force

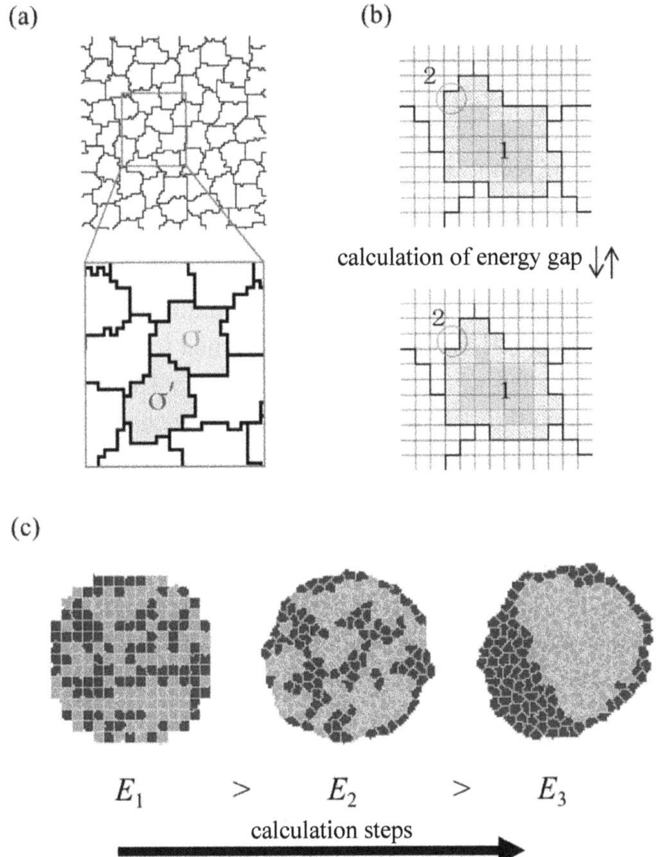

Figure 8.7 Schematics of the cellular Potts model. (a) Example of cellular boundary state. Each cell i is represented as a set of pixels with the attribute of cell type σ_i. (b) Example of pixel state transition near the boundary between cells $i = 1$ and $i = 2$. The Monte Carlo method is used to stochastically transition the attributes of a pixel so that its energy E decreases. (c) Repeated rearrangement results in cell type-dependent separation (Some of the figures are courtesy of Dr. Akihiko Nakajima (The University of Tokyo)).

(surface tension) at the water surface drives the separation of water and oil. In cellular tissues, however, surface tension is predominantly determined by the intercellular binding mediated by adhesion molecules and the tension of the membrane cortex, influenced by the cytoskeletal system [13]. In what follows, we introduce the so-called cellular Potts model[1] to explore the pattern formation of moving cells, encompassing variations in cellular adhesion force.

---

[1] The Potts model is originally an extension of the Ising model within the framework of statistical physics. It expands on the Ising model by accommodating multiple spin orientations rather than limiting itself to up/down states characteristic of the Ising model.

The space is discretized into a lattice, with each cell depicted by a collection of $N_{i\,\text{cell}}$ squares (voxels). Each voxel is assigned a cell number σ_i as an attribute. When adjacent voxels possess different cell numbers, the voxel in question is deemed to be at the cell boundary. The decision to either replace the cell number of this voxel from σ_i to σ_j or retain the original number is made based on the following probability:

$$P(\sigma_i \sigma_j') \propto \exp(-\beta \Delta E), \tag{8.17}$$

where β is the inverse temperature and ΔE is the amount of energy change when the cell occupying the voxel in question is transformed from σ_i to σ_j. Such a Monte Carlo method, as detailed in Section 5.4, that derives samples from a specified probability distribution through dynamics of a Markov chain (a Markovian probability model where a state at time $t + 1$ is determined solely by the state at time t), is known as a Markov Chain Monte Carlo method (MCMC). Markov Chain Monte Carlo methods have been instrumental in statistical physics and have recently found applications across various fields, with this model serving as one example.

The likelihood of a state change correlates to the magnitude of energy decrease; smaller reductions in energy correspond to a lower probability of state transition. The energy function

$$\begin{aligned}
E &= E_\text{J} + E_\text{V} + E_\text{L} \\
&= \sum_{\text{voxel}(i_\text{cell})} \sum_{\text{voxel}(j_\text{cell})} J(\tau(\sigma_i), \tau(\sigma_j))(1 - \delta_{\sigma(i_\text{cell})\sigma(j_\text{cell})}) \\
&\quad + \lambda \sum_{\text{voxel}(i_\text{cell})} (A - A_0)^2 + \lambda \sum_{\text{voxel}(i_\text{cell})} (L - L_0)^2
\end{aligned}$$

is often adopted. This model is known as the cellular Potts model [8]. The first term on the right-hand side, E_J, denotes intercellular adhesion, with the coefficients $J(\tau(\sigma_i), \tau(\sigma_j))$ being parameters of the system, determined by the cell type $\tau(\sigma)$ of the j_cellth cell in contact with the i_cellth cell. The symbol δ represents Kronecker's delta, which equals unity if the cells are of the same type σ. Accordingly, the interaction energy E_J is set to zero for pairs of identical cell types, whereas it is assigned a positive value for pairs of differing cell types (if $j > 0$), leading to cell-type dependent adhesion. The second term, E_V, signifies the elastic force aiming to maintain the area at A_0, while the third term, E_L, represents the elastic force serving to keep the cell-perimeter length at L_0. Here, A and L are the area and perimeter length of a cell i_cell, respectively. In three-dimensional simulations, 'area' refers to the volume, and 'perimeter' to the surface area.

Based on these assumptions, the cellular Potts model is instrumental in depicting the processes of cell-shape fluctuation, rearrangement, and the gradual decrease in energy. This model is utilized to simulate phenomena such as cell sorting [8], morphogenetic movements [14, 17], and motility in model systems like fish epidermal cells. It is also applied in simulating shape changes in single cells, such as keratocytes and amoebae of cellular slime molds [15, 19]. While thermal fluctuations drive the transition to a minimum free-energy state in phase separation, in cell sorting, nonthermal fluctuations (e.g., variations in membrane shape and cell motion) are more predominant.[2] Numerous experimental studies, including those cited in [12], have advocated for enhancements to the cellular Potts model, aiming to integrate it with more active reaction dynamics, particularly for cytoskeletal systems and related phenomena.

8.3.3 Cell-Vertex Model

The cell-vertex model [5], widely used to capture tissue morphology, represents a cell population as a collection of vertices of a polygon (in the two-dimensional case) or a polyhedron (in the three-dimensional case), and the position $x_i = (x_i, y_i)$ of each vertex (subscript i is the number of the vertex, globally assigned to the entire system) is calculated. Such simulations usually assume that the vertices are in a sufficiently viscous medium (viscosity η) and thus omit the inertia term in the evolution equations, that is,

$$\eta \frac{\mathrm{d}x_i}{\mathrm{d}t} = -\nabla E(x_i).$$

The stationary configuration of a vertex i is such that the forces on the vertex are in balance. It is convenient to describe this configuration as a minimum value of energy $E(x_i)$. For tissues such as the epithelial system, where cells are arranged like hexagonal prisms, we can approximate cells as polygons on a two-dimensional plane. Therefore, we assign numbers ($\alpha = 1, 2, \ldots, N_{\mathrm{cell}}$) to the polygons and define each polygon that consists of $j_\alpha = 1, 2, \ldots, N_\alpha$ vertices. (The subscript j_α is a local number in each cell, assigned to be a counterclockwise rotation.)

Let $\Lambda_{j\alpha, j'\alpha}$ represent the binding energy at the cell boundary, determined by the line segment (of length $l_{j\alpha, j'\alpha}$) connecting vertex j_α and its adjacent vertex j'_α. Additionally, let L_α denote the perimeter length of the cell and A_α its area. The following is a commonly employed form of energy [4]:

$$E = \sum_{\alpha=1}^{N_{\mathrm{cell}}} \left(\frac{K_\alpha}{2}(A_\alpha - A_0)^2 + \sum_{j_\alpha=1}^{n_\alpha} \Lambda_{j\alpha, j'\alpha} l_{j\alpha, j'\alpha} + \frac{\Gamma_\alpha}{2}(L_\alpha - L_0)^2 \right). \tag{8.18}$$

---

[2] The cellular Potts model is a highly expressive framework for simulating various phenomena. Initially developed for illustrating phase transitions in equilibrium systems, it has now expanded to encompass the dynamics of nonequilibrium systems. However, this application introduces theoretical complexities, such as interpreting the meaning of temperature and the ambiguity in correlating Monte Carlo simulation steps with real-time scales. Furthermore, the requirement to perform stochastic calculations at each grid point significantly increases computational time.

The first term represents the elastic force that tries to maintain the cell area at A_0. The second term encapsulates the linear tension and intercellular adhesion, attributed to the actomyosin-based cytoskeletal system along the plasma membrane. The third term represents the elastic force exerted by the boundary surface to maintain the perimeter length at L_0. The energy change about $\{N_i\}$, which is the set of indices of the cell containing the vertex i, contributes to the force applied to a vertex i, $\boldsymbol{F}_i$. Therefore, from Eq. (8.18),

$$\boldsymbol{F}_i = -\nabla E|_{x=x_i}$$

$$= -\sum_{\alpha \in \{N_i\}} \left(K_\alpha (A_\alpha - A_0)\nabla A_\alpha + \sum_{j_\alpha=1}^{n_\alpha} \Lambda_{j_\alpha,j'_\alpha} \nabla l_{j_\alpha,j'_\alpha} \right.$$

$$\left. + \Gamma_\alpha (L_\alpha - L_0)\nabla L_\alpha \right)$$

can be obtained.

Exercise 8.3.1 The perimeter length and the area of the αth polygon can be represented as follows:

$$L_\alpha = \sum_{j_\alpha=1}^{n_\alpha} l_{j_\alpha,j'_\alpha}, \tag{8.19}$$

$$A_\alpha = \frac{1}{2} \sum_{j_\alpha=1}^{n_\alpha} \left(x_{j_\alpha}^{(\alpha)} y_{j_\alpha+1}^{(\alpha)} - x_{j_\alpha+1}^{(\alpha)} y_{j_\alpha}^{(\alpha)} \right), \tag{8.20}$$

respectively. Based on them, derive the following equations:

$$\nabla A_\alpha|_{x=x_{j_\alpha}^{(\alpha)}} = \frac{1}{2} \begin{pmatrix} y_{j_\alpha+1}^{(\alpha)} - y_{j_\alpha-1}^{(\alpha)} \\ x_{j_\alpha-1}^{(\alpha)} - x_{j_\alpha+1}^{(\alpha)} \end{pmatrix},$$

$$\nabla l_{j_\alpha,j'_\alpha} = \frac{1}{l_{j_\alpha,j'_\alpha}} \begin{pmatrix} x_{j_\alpha}^{(\alpha)} - x_{j_\alpha+1}^{(\alpha)} \\ y_{j_\alpha}^{(\alpha)} - y_{j_\alpha+1}^{(\alpha)} \end{pmatrix},$$

where $x_{j_\alpha}^{(\alpha)} = \left(x_{j_\alpha}^{(\alpha)}, y_{j_\alpha}^{(\alpha)} \right)$ represents the coordinates of the j_αth vertex, locally assigned in each cell α. The vertex of interest is $x_i = x_{j_\alpha}^{(\alpha)}$ and the vertex of the clockwise neighbor is $x_{j_\alpha-1}^{(\alpha)}$, and the vertex of the counterclockwise neighbor is $x_{j_\alpha+1}^{(\alpha)}$, respectively.

Exercise 8.3.2 An alternative approach to explicitly describing the force, without the use of energy notation, is presented in [27]. By comparing the force directions, consider how the differences between these two formalisms reflect the underlying basic assumptions.

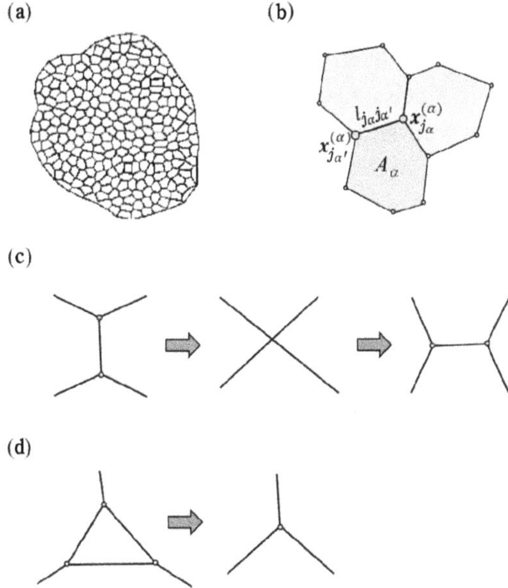

Figure 8.8 Schematics of the cell-vertex model. (a) Contour of cells constituting epithelial tissue (courtesy of Dr. Shuji Ishihara (The University of Tokyo). (b) Variables of the vertex model. Force balance among vertices is maintained, with constraints imposed on both the perimeter length and the area of the polygon composed of these vertices. (c), (d) Transition rules for vertex collisions: (c) T1 and (d) T2 transitions.

In the tissue, as depicted in Fig. 8.8(c), the shape of the cell contact surface undergoes changes. This alteration can lead to cells, which were initially not in contact, becoming adjacent to each other. A notable example is the extension of a tissue fragment (Keller's explant [24]) excised from the gastrulation site in frog embryo. This extension occurs even when the tissue is suspended in fluid, and is well known to be induced by rearrangements (termed intercalation) occurring along the axes (convergence and extension). The vertex state transition illustrated in Fig. 8.8(c) is known as the T1 transition, a fundamental process in describing cell rearrangement by vertex. Additionally, as shown in Fig. 8.8(d), there is another process where a triangular cell diminishes and vanishes, leading to the formation of a new vertex, termed the T2 transition. The vertex-model computations represent these changes by explicitly introducing rules for these recombination [5].

The vertex model offers the advantage of avoiding cell-shape distortion caused by lattice discreteness, as seen in the cellular Potts model, making it well-suited for studying the interplay between force field properties and morphology. Additionally, the ease of incorporating cell division allows for examination of the impact of cell division on tissue growth. However, this model is most effective when cell boundaries are relatively flat, as in epithelial systems, and less so when boundaries are irregular. Moreover, since it necessitates cells being in contact with each other, the vertex model is not well-suited for scenarios where cells are dispersed.

8.3.4 Phase Field Model

The phase field model is utilized to describe the spatial segregation of materials with distinct properties and the interfaces resulting from this separation. Unlike the vertex model and the cellular Potts model, this method is unique in that its spatiotemporally continuous system described by partial differential equations. The phase field model is based on the following Allen–Cahn equation:

$$\frac{\partial \phi}{\partial t} = \nabla^2 \phi - \phi(1 - \phi)(1 - 2\phi). \tag{8.21}$$

The right-hand side of this equation can also be written as $-\delta F/\delta \phi$ using the functional derivative of the Ginzburg–Landau functional

$$F = \int \left(\frac{1}{2}|\nabla \phi|^2 + G(\phi)\right) d\mathbf{r}, \tag{8.22}$$

where $G(\phi)$ is a double-well potential $G(\phi) = \phi^2(1 - \phi)^2/2$. As inferred from the double-well nature of the potential $G(\phi)$, ϕ in the Allen–Cahn equation tends to be $\phi \sim 0$ or 1, while the term $|\nabla \phi|^2/2$ in F expresses the energy penalties, where the longer the length of the interface between $\phi = 0$ and 1, the greater the energy costs. The combination of these two terms results in the formation of spatially segregated domains of $\phi = 0$ and 1 with a sharp interface between them.

A natural extension of Eq. (8.21) would be an equation where the time derivative $\partial \phi/\partial t$ is replaced by the material derivative (Lagrangian derivative) $D\phi/Dt = \partial \phi/\partial t + \mathbf{v} \cdot \nabla \phi$, where the interface velocity $\mathbf{v}$ is assumed to be proportional to the unit normal vector to the interface: $\mathbf{v} = -v\nabla \phi/|\nabla \phi|$ with $v = |\mathbf{v}|$. This leads to

$$\frac{\partial \phi}{\partial t} = \nabla^2 \phi - G'(\phi) + v|\nabla \phi|, \tag{8.23}$$

where $G'(\phi) = dG/d\phi = \phi(1 - \phi)(1 - 2\phi)$. The interfacial velocity v can be a spatial and time-dependent function.

In the modeling of the cell shape by the phase field model, the following equation is employed [23]:

$$\tau \frac{\partial \phi}{\partial t} = \eta\left(\nabla^2 \phi - \frac{G'(\phi)}{\epsilon^2}\right) - M\left(\int \phi \, d\mathbf{r} - A_0\right)|\nabla \phi| + f(r, t)|\nabla \phi|, \tag{8.24}$$

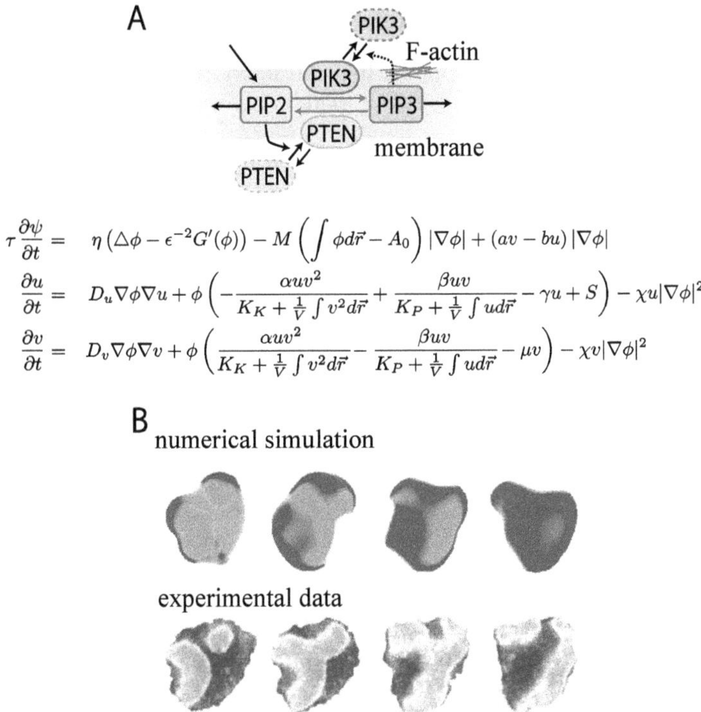

$$\tau \frac{\partial \psi}{\partial t} = \eta \left(\triangle\phi - \epsilon^{-2} G'(\phi) \right) - M \left(\int \phi d\vec{r} - A_0 \right) |\nabla\phi| + (av - bu) |\nabla\phi|$$

$$\frac{\partial u}{\partial t} = D_u \nabla\phi \nabla u + \phi \left(-\frac{\alpha u v^2}{K_K + \frac{1}{V}\int v^2 d\vec{r}} + \frac{\beta u v}{K_P + \frac{1}{V}\int u d\vec{r}} - \gamma u + S \right) - \chi u |\nabla\phi|^2$$

$$\frac{\partial v}{\partial t} = D_v \nabla\phi \nabla v + \phi \left(\frac{\alpha u v^2}{K_K + \frac{1}{V}\int v^2 d\vec{r}} - \frac{\beta u v}{K_P + \frac{1}{V}\int u d\vec{r}} - \mu v \right) - \chi v |\nabla\phi|^2$$

Figure 8.9 Simulation of cellular deformation by phase field model. A: Network and model equations of phosphorylation-modification reactions of phosphatidylinositol on cell membranes. B: Numerical simulation of the dynamics of model A and an example of deformation in a real cell [26].

where an elastic term (the second term in r.h.s.) that ensures area conservation, and a local protruding force term (the third term in r.h.s.) by the cytoskeleton defines v. Here, τ is a time scaling parameter for the dynamics in ϕ, η is the line tension, and ϵ characterizes the thickness of the boundary between cellular and noncellular regions, whereas A_0 represents a target cell area, and M denotes a parameter dictating the strength of the elastic term for the area conservation. Similar to the cellular Potts model and the vertex model, the elastic force in the second term is analogous to the force due to elastic energy, facilitating membrane relaxation to a stable state. Conversely, the third term describes the stretching and shrinking of the membrane lining due to the cytoskeletal system and is instrumental in determining the time evolution of ϕ.

The phase field model is employed to depict relatively simple morphological changes in complex structures such as Keratocytes [23] and cellular slime mold amoebae. Figure 8.9 illustrates an example of this method, where the morphology dynamics coupled with a reaction-diffusion process of the membrane-bound signals are analyzed through computer simulation, [26].

References

[1] M. Cohen *et al.* Dynamic filopodia transmit intermittent Delta–Notch signaling to drive pattern refinement during lateral inhibition. *Developmental Cell*, **19**(1):78–89, 2010.

[2] J. R. Collier *et al.* Pattern formation by lateral inhibition with feedback: A mathematical model of delta-notch intercellular signalling. *Journal of Theoretical Biology*, **183**(4):429–446, 1996.

[3] S. Danø, P. G. Sørensen and F. Hynne. Sustained oscillations in living cells. *Nature*, **402**(6759):320–322, 1999.

[4] R. Farhadifar *et al.* The influence of cell mechanics, cell–cell interactions, and proliferation on epithelial packing. *Current Biology: CB*, **17**(24):2095–2104, 2007.

[5] A. G. Fletcher *et al.* Vertex models of epithelial morphogenesis. *Biophysical Journal*, **106**(11):2291–2304, 2014.

[6] P. Formosa-Jordan *et al.* Regulation of neuronal differentiation at the neurogenic wavefront. *Development*, **139**(13):2321–2329, 2012.

[7] R. A. Foty and P. L. Townes. The differential adhesion hypothesis: A direct evaluation. *Developmental Biology*, **278**(1):255–263, 2005.

[8] F. Graner and J. A. Glazier. Simulation of biological cell sorting using a two-dimensional extended Potts model. *Physical Review Letters*, **69**(13):2013–2016, 1992.

[9] T. Höfer, J. A. Sherratt and P. K. Maini. *Dictyostelium discoideum*: Cellular self-organization in an excitable biological medium. *Proceedings Biological Sciences/The Royal Society*, **259**(1356):249–257, 1995.

[10] K. Kawasaki *et al.* Modeling spatio-temporal patterns generated by Bacillus subtilis. *Journal of Theoretical Biology*, **188**(2):177–185, 1997.

[11] D. A. Kessler and H. Levine. Pattern formation in *Dictyostelium* via the dynamics of cooperative biological entities. *Physical Review E, Statistical Physics, Plasmas, Fluids, and Related Interdisciplinary Topics*, **48**(6):4801–4804, 1993.

[12] M. Krieg *et al.* Tensile forces govern germ-layer organization in zebrafish. *Nature Cell Biology*, **10**(4):429–436, 2008.

[13] M. L. Manning *et al.* Coaction of intercellular adhesion and cortical tension specifies tissue surface tension. *Proceedings of the National Academy of Sciences*, **107**(28):12517–12522, 2010.

[14] A. F. Marée and P. Hogeweg. How amoeboids self-organize into a fruiting body: Multicellular coordination in *Dictyostelium discoideum*. *Proceedings of the National Academy of Sciences of the United States of America*, **98**(7):3879–3883, 2001.

[15] A. F. M. Marée *et al.* Polarization and movement of keratocytes: A multiscale modelling approach. *Bulletin of Mathematical Biology*, **68**(5):1169–1211, 2006.

[16] T. Matsuyama and M. Matsushita. Fractal morphogenesis by a bacterial cell population. *Critical Reviews in Microbiology*, **19**(2):117–135, 1993.

[17] R. M. H. Merks *et al.* Contact-inhibited chemotaxis in de novo and sprouting blood-vessel growth. *PLoS Computational Biology*, **4**(9):e1000163, 2008.

[18] M. Miura, H. Sakaguchi and M. Matsushita. Reaction-diffusion modelling of bacterial colony patterns. *Physica A*, **282**(1-2):283–303, 2000.

[19] S. I. Nishimura and M. Sasai. Modulation of the reaction rate of regulating protein induces large morphological and motional change of amoebic cell. *Journal of Theoretical Biology*, **245**:230–237, 2007.

[20] H. G. Othmer and L. E. Scriven. Instability and dynamic pattern in cellular networks. *Journal of Theoretical Biology*, **32**(3):507–537, 1971.

[21] S. Sawai, P. A. Thomason and E. C. Cox. An autoregulatory circuit for long-range self-organization in *Dictyostelium* cell populations. *Nature*, **433**(7023):323–326, 2005.

[22] J. L. Schiff. *Cellular Automata: A Discrete View of the World*. Hoboken, John Wiley and Sons, 2008.

[23] D. Shao, W.-J. Rappel and H. Levine. Computational model for cell morphodynamics. *Physical Review Letters*, **105**(10):108104, 2010.

[24] J. Shih and R. Keller. Cell motility driving mediolateral intercalation in explants of *Xenopus laevis*. *Development*, **116**(4):901–914, 1992.

[25] D. Sprinzak *et al.* Mutual inactivation of Notch receptors and ligands facilitates developmental patterning. *PLoS Computational Biology*, **7**(6), 2011.

[26] D. Taniguchi *et al.* Phase geometries of two-dimensional excitable waves govern self-organized morphodynamics of amoeboid cells. *Proceedings of the National Academy of Sciences of USA*, **110**(13):5016–5021, 2013.

[27] M. Weliky and G. Oster. The mechanical basis of cell rearrangement. I. Epithelial morphogenesis during Fundulus epiboly. *Development*, **109**(2):373–386, 1990.

[28] S. Wolfram. *Cellular Automata and Complexity: Collected Papers*. Boca Raton, CRC Press, 2018.

[29] V. S. Zykov *et al.* Selection of spiral waves in excitable media with a phase wave at the wave back. *Physical Review Letters*, **107**, 254101, 2011.

9

Theoretical Approaches Concerning the Origin of Life

The origin of life remains one of the most fascinating and difficult questions in biology. Unfortunately, we cannot directly observe the events that led to the origin of life on Earth. Therefore, what we can do is make inferences about how living systems arose and try to recreate them in experiments.

In order to discuss the origin of living systems, it is necessary to define what life is. This definition is not simple and is still under discussion. However, it is widely accepted that living systems need to have the following two features:

i). A living system can replicate itself.
ii). The replication is not perfect, and a slight state change occurs. The change is inherited in succeeding generations.

It should be stressed that systems satisfying (i) and (ii) have evolvability. When replicating units that have a higher "fitness" emerge, they can increase their population. The repetition of reproduction with slightly changed, inheritable states and selecting those units with higher fitness will lead to the evolution of replicating units. Although such evolvability is regarded as a necessary condition for a living system, the mechanism of how systems satisfying (i) and (ii) emerge is unclear. In this chapter, we will discuss the origin of life problem, focusing on the theoretical aspects.

9.1 Chicken-and-Egg Problem in the Origin of Life

The living systems on Earth consist of two types of components. One is the component involved in genome replication, which transmits genetic information to the next generation. The other is various chemical components that support the cells' growth and division, including proteins that catalyze the chemical reactions, membrane molecules that compartmentalize the cell, and small molecules concerning the metabolic reaction which produces those components. For simplicity, the former is referred to as the "genetic information system" and the latter as the "metabolic system." There is an elaborate replication mechanism in which genetic information is precisely transmitted to the next generation. In contrast, the numbers of many molecules involved in the metabolic system are not rigidly fixed but are distributed, which means they are not precisely controlled in the division. Of course, both systems are interdependent. The genetic information codes protein sequences that control the metabolism, and the dynamics of the metabolism are indispensable for replicating the genetic information. That is, metabolism cannot be maintained without genetic information, and genetic information cannot be maintained without metabolism. Then in discussing the origin of life, there exists the so-called "chicken-and-egg problem" between genetic information and the metabolic system. The question is, in what order did both of them emerge in the early stage of living systems?

A catalytically active RNA polymer discovered in the 1980s, called ribozyme, greatly impacted the chicken-and-egg problem of the origin of life. It has been discussed that this RNA polymer has a flexible three-dimensional structure like proteins and therefore may have catalytic activity. The group of Cech and colleagues found that the self-splicing of the ribosomal RNA of Tetrahymena proceeds without enzymes and further determined in their analysis that the partial sequence of the RNA catalyzes the transfer reaction of nucleotides [3]. The discovery that this RNA polymer can have catalytic activity as a protein had brought about the possibility that RNA polymer coding genetic information has metabolic activity maintaining its replication, which will solve the previously mentioned chicken-and-egg problem. Here, we can assume that an RNA polymer with catalytic activity that can replicate itself emerges by chance in the prebiotic world. Then, when mutational changes to the RNA sequence are produced by replication error and are selected with higher fitness, for example, replication activity, evolution to such RNA polymers can be achieved. After that, genetic information encoded by catalytic RNA polymers is replaced by more stable molecules without catalytic activity, that is, the DNA polymer and the system sustaining genetic information and metabolic activity are separated, as in organisms we know.

This scenario, in which the origin of life was based on RNA polymers with replication activity, is often referred to as the "RNA world" scenario, which has been intensively studied as a central hypothesis in the origin of life researches. Several observations support this RNA world scenario. For example, several RNA polymers, such as messenger RNA, transfer RNA, and ribosomal RNA, are involved in transcription and translation reactions. In fact, ribosomal RNA is located at the active center of the ribosome. Since the transcription and translation systems are fundamental reaction networks maintaining intracellular catalytic activities, RNA polymers play essential roles in these reactions. Hence the importance of RNA catalytic activity in the ancient metabolic system is suggested.

In spite of the preceding arguments supporting the RNA world hypothesis, this scenario has several unsolved problems. First, it is still being determined how likely it is that RNA polymers that can catalyze self-replication would emerge in the prebiotic condition. Several experimental studies have suggested that the catalytic activity of RNA polymers is limited. A ribozyme that can replicate itself has not been constructed yet.[1] Moreover, there is currently little data to support the stable formation of RNA polymers in prebiotic environments. For example, various molecules, including isomers, are thought to be produced simultaneously in the prebiotic production of RNA molecules. Still, incorporating such molecules inhibits RNA polymer's replication reaction. At present, how a RNA polymer that catalyzes the self-replication assumed in the RNA world scenario can appear has not been explained.

In addition to the physiochemical difficulty of the RNA world scenario, as discussed earlier, another difficulty was caused by an error during the self-replication of RNA polymers. In the next section, after explaining the RNA evolution experiment of Spiegelman, we will discuss a mathematical model of the RNA replication system introduced by Eigen. Then, using the model, we will analyze the conditions that the self-replication dynamics of RNA polymer need to satisfy in the RNA world scenario.

9.2   Spiegelman's Experimental Evolution of RNA Polymer

In the 1960s, Spiegelman studied the replication mechanism of RNA phages (viruses) that infect *Escherichia coli*. DNA replication was clarified at that time, but how the RNA phage is replicated was not clear, and reverse transcriptase from RNA to DNA had not yet been discovered. However, Spiegelman found that one of the proteins encoded by RNA phages is an RNA replicase, replicating only those RNAs with specific recognition sequences among the various RNA molecules in the cell. In addition, he purified the enzyme named $Q\beta$ replicase and proved that it has the ability to amplify RNA phages in vitro.

Next, Spiegelman demonstrated that an *in vitro* evolution experiment of RNA polymer is possible by using this $Q\beta$ replicase [13]. Figure 9.1 shows a schematic representation of the experimental method. First, the purified $Q\beta$ replicase, the ribonucleotides required for replication, and the RNA phage genome (about 4,000 base pairs), which serves as a template, are put into a test tube. Then, the replication reaction is carried out. An important point here is that the accuracy of the

---

[1] However, there is a possibility that advances in experimental research will solve this problem. For example, in the research of Holliger and colleagues, a ribozyme that can catalyze the elongation reaction of an RNA chain of about 200 bases was constructed by using screening by evolution experiment in vitro [1]. Furthermore, Lincoln and Joyce constructed RNA polymers capable of self-replication using four types of RNA molecules with several tens of base pairs as substrates [12].

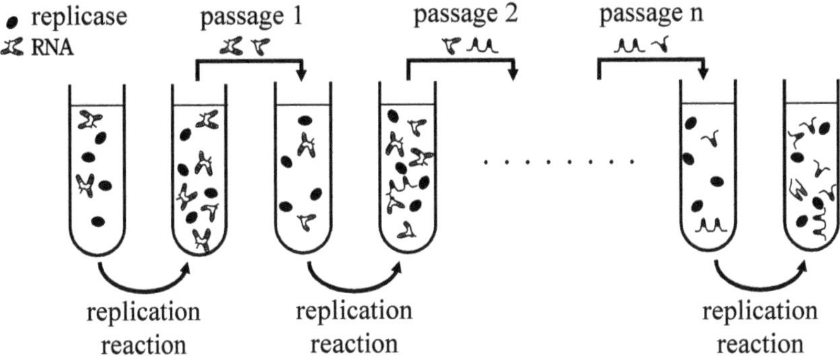

Figure 9.1 Spiegelman's experimental evolution of RNA polymer.

replication of RNA molecules is not perfect, and base substitution and deletion of partial sequence occur stochastically, resulting in varying replication efficiency (i.e., phenotype). By this variation in replication efficiency, after the replication reaction, the RNA molecules with higher efficiency than others increase their relative population. Then, a part of the RNA population was transferred to the next test tube. This cycle of replication reaction and selection was iterated.

As a result of iterating the transfer experiment several times, the RNA polymer, whose length was shortened to several hundred bases, occupied the majority in the test tube. This result can be explained by the fact that the shorter the RNA polymer, the faster it replicates. By this correlation, the shorter RNA was chosen during the iteration of replication and selection. Then why isn't a shorter RNA molecule gained further? This occurs because $Q\beta$ replicase can replicate only RNA species with appropriate recognition sequences; therefore, RNA species whose recognition sequences have been deleted are not replicated.

In this experiment, Darwinian evolution by the cycle of mutation fixation and selection of RNA species with higher fitness is constituted only by the components identified. It enables us to quantitatively catch what is happening in the process of evolution. Spiegelman performed this experimental evolution under different environmental conditions (i.e., different selection pressures) and observed its evolutionary dynamics. For instance, he found that by adding a replication inhibitor to the environment, the RNA molecule with resistance to the inhibitor emerges. Sometimes, RNA molecule species that depend on this inhibitor, that is, whose replication is slower without it, had also been observed. This experiment opened the field of molecular evolution experiment afterward. It also played a significant role as a basis of the theoretical analysis of replicating dynamics, as will be discussed in the next section.

9.3 Theoretical Model of Self-Replicating Systems

After Spiegelman's experimental evolution of RNA polymers, Eigen and colleagues started theoretical analysis of self-replicating systems. This theory is applicable not only to RNA polymers but also to self-replicating systems in general. Thus, this section discusses the theoretical analysis independent of RNA molecules. First, the behavior of the theoretical model without mutation during replicating dynamics is explained by focusing on the case in which multiple kinds of replication units compete. In the next section, we discuss how mutations in the polymer's sequence affect the population dynamics of replicating units.

Let us assume that there are k species of replicating chemical species. The terms x_i and f_i represent the ratio of ith species to the total population and the replication rate of ith species per unit time, respectively. The f_i term generally depends on the concentration of replicating species and types of replicating mechanisms. For example, when a species replicates itself using its structure as a template, f_i will be proportional to x_i. When two identical species need to associate with replicating themselves, f_i will be proportional to x_i^2. For another example, if ith replicating species is produced from a substrate with a constant concentration, the term f_i is independent of x_i and will be a constant value. Of course, the production of ith species can depend on other species; for instance, if jth species catalyze the replication reaction of ith species. We will later discuss such cases involving catalytic reaction networks, but here let us assume that the replication rate depends only on its population ratio, that is, the case $f_i(x_i)$. In this case, the dynamics of self-replicating chemical species of the preceding examples can be described by the following equation:

$$f_i(x_i) = a_i x_i^\alpha, \tag{9.1}$$

where $\alpha = 1$ and $\alpha = 2$ correspond to the cases where one or two chemical species are involved in the replicating reactions, and $\alpha = 0$ corresponds to the case with constant production of the species, respectively. The replication rate should depend on the characteristics of each chemical species, which is represented by the constant a_i for ith species.

By this replication reaction, the total amount of chemical species increases and eventually will diverge with time. The preceding discussion only considers the ratio of ith chemical species to the total amount of chemicals to avoid such divergence. This is natural because the experiments are carried out in a test tube with a finite volume. When the total amount of chemicals increases, it will be diluted by transferring it to the next test tube, as in Spiegelman's experiment. In this representation, the sum of ratio x_i overall species is kept at unity, that is, $\sum_i x_i = 1$. To maintain this condition, we should consider that when the total amount of chemicals increases, the ratio of each chemical species decreases by dilution. This dilution effect should be proportional to the ratio x_i. By considering the dilution effect, the population dynamics of each replicating chemical species is represented by the following differential equation:

$$\frac{dx_i}{dt} = a_i x_i^\alpha - x_i \sum_j a_j x_j^\alpha. \tag{9.2}$$

By setting an initial condition satisfying $\sum_i x_i = 1$, we can confirm that the sum of x_i is kept at unity by summing the preceding differential equation for all species, that is, $d(\sum_i x_i)/dt = 0$.

Let us consider the cases $\alpha = 0$, 1, and 2 by using the preceding population dynamics of replicating chemicals. Suppose there are k replicating chemical species, each of which has a different replication rate a_i. We can name the species arbitrarily, so we choose the index i in descending order of the replication rate, that is, $a_1 > a_2 > a_3 > \cdots > a_k$. We assume that there is no pair of chemical species with the same replication rate. From this equation, in the case of $\alpha > 0$, the fixed-point solution x_i^* satisfies either of the following:

(1) $x_i^* = 0$
(2) $a_i(x_i^*)^{\alpha-1} = \sum_j a_j(x_j^*)^\alpha$

(We will discuss the case of $\alpha = 0$ later.) Because there is the constraint $\sum_i x_i = 1$, it is impossible to satisfy the condition (1) for all i. Then, for the cases $\alpha = 1$ and 2, let us analyze the combination of conditions (1) and (2) that satisfies the fixed-point solution for each replicating species.

(A) $\alpha = 1$: Survival of the Fittest

In this case, condition (2) is $a_i = \sum_j a_j x_j^*$. Here, the right-hand side is independent of i, whereas the left-hand side depends on i. Therefore, condition (2) cannot be satisfied for more than one chemical species. In other words, only one species can have a nonzero population ratio. Thus, the possible fixed-point solution is that, for one chemical species m, the population is unity, that is, $x_m = 1$, and for other species $\ell \neq m$, $x_\ell = 0$. This means that there are k possible fixed-point solutions with different surviving species m. We can then check the stability of these fixed-point solutions by using the linear stability analysis. That is, after adding a small perturbation δ_ℓ to x_ℓ of the fixed-point solution, the time development of the population dynamics will be analyzed. Note that the total population of the chemical species is constant, that is, $\sum_i x_i = 1$. Thus, substituting $x_\ell = \delta_\ell$ $(\ell \neq m)$ and $x_m = 1 - \sum_j \delta_j$ into the dynamics of x_ℓ and considering only the first order of δ_ℓ, we obtain

$$\frac{d\delta_\ell}{dt} = a_\ell \delta_\ell - \left(\sum_j a_j \delta_j + a_m\right)\delta_\ell = (a_\ell - a_m)\delta_\ell. \tag{9.3}$$

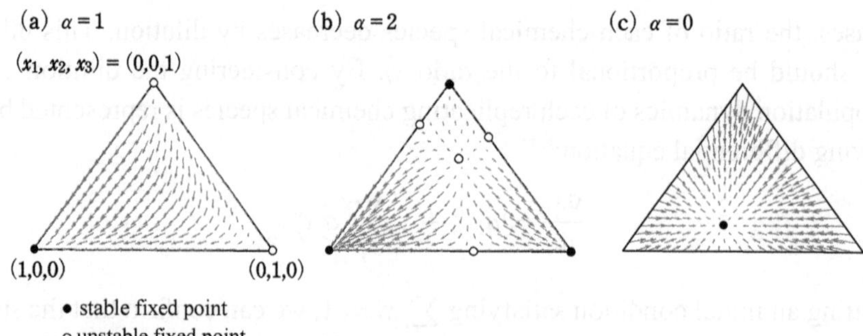

Figure 9.2 Dynamics of self-replicating systems with three species. The parameters $k = 3$, $a_1 = 3$, $a_2 = 2$, and $a_3 = 1$ are used in these diagrams. The triangle vertices correspond to states $(x_1, x_2, x_3) = (1, 0, 0)$, $(0, 1, 0)$, and $(0, 0, 1)$, respectively. Filled circles and open circles represent stable and unstable fixed points, respectively. The arrows represent the flow in the state space.

Here, if $a_\ell < a_m$ is true for all $\ell \neq m$, the solution is stable because the coefficient before δ_ℓ is negative; otherwise, it is unstable. Only the fixed-point solution where the mth species having the largest replication rate ($m = 1$ in this case) has $x_m = 1$, and the others having $x_j = 0$ ($j \neq m$) are stable. This result agrees with the concept of "survival of the fittest" through natural selection.

The dynamics of x_i can be described by the flow in the state space. For example, in the case of $k = 3$, we can describe the dynamics by the flow on a two-dimensional plane: Note that even though there are three chemical species, the dimension can be reduced due to the constraint $\sum_{j=1}^{3} x_j = 1$. With this constraint, we can describe the population dynamics by the flow within the state space of equilateral triangles, in which the three edges correspond to x_1, x_2, and x_3, respectively. Figure 9.2(a) shows the population dynamics in case (A). In this case, we can see the dynamics that all initial states fall into the fixed-point attractor $x_1 = 1$ (except for the initial condition $x_1 = 0$).

(B) $\alpha = 2$: Survival of the First Comer

In this case, condition (2) is $x_i^* = (1/a_i) \sum_j a_j (x_j^*)^2$. As in case (A), the solution $x_m^* = 1$ and $x_\ell^* = 0$ for ($\ell \neq m$) satisfies this condition. However, by taking the population ratio of species i as (const.)$/a_i$, the coexistence of multiple species can also satisfy this condition. This means that there can be more fixed-point states in the population dynamics. Now we need to analyze the stability of these. For simplicity, let us consider the case $k = 3$, in which case there can be the following three cases.

(i) All species satisfy condition (2): The population ratio of these species is

$$x_i^* = \frac{1/a_i}{\sum_j (1/a_j)}. \tag{9.4}$$

(ii) Two species satisfy condition (2): The population ratio of two species m and n is

$$x_i^* = \frac{1/a_i}{\sum_j (1/a_j)} \tag{9.5}$$

for $i = m, n$. The other species have zero population.

(iii) Only one species m satisfies condition (2), and other species have zero population.

Now we check the stability of the fixed-point state of case (iii), that is, $x_m = 1$, by adding a small perturbation δ_ℓ to ($\ell \neq m$) species. By considering only the first order of δ_ℓ, the time development of the perturbation is given by

$$\frac{d\delta_\ell}{dt} = -a_m \delta_\ell. \tag{9.6}$$

As $-a_m < 0$, the fixed-point solution where the one dominant species m is always stable. Figure 9.2(b) shows the flow of population dynamics in a two-dimensional state space. As shown in the figure, the solutions with coexistence of two or three species are unstable. (This can be checked straightforwardly by taking the fixed points in case (ii) or (iii) and applying linear-stability analysis, even though the calculation is a bit tedious.) In contrast, all solutions with single dominant species are stable. The flow of the state in Fig. 9.2(b) indicates that a species with a higher population ratio in the initial condition tends to dominate the whole population.

The stability of the fixed points for the existence of any single species $x_m^* = 1$ and the instability of all the other fixed points with coexisting multiple species are generally true for any values of k. As long as x_m is close to unity, the state is attracted to the fixed point $x_m = 1$. Hence the first comer wins. When the exponent α is larger than one, that is, the dependence of replication per unit time (i.e., the first term of r.h.s of Eq. (9.2)) is larger than the first order, the first comer always survives. For instance, suppose a product or software that is not necessarily the best is released in the beginning, but it has a higher-order effect of increasing the number of products (i.e., other, different products use this first software). In that case, even if there could be better software, the latecomer would not be able to break its stronghold. The case with the present $\alpha > 1$ case represents this kind of first-comer advantage.

(C) $\alpha = 0$: Coexistence of All Species
In this case, the condition for the fixed-point solution is

$$a_i - x_i^* \left(\sum_j a_j \right) = 0, \tag{9.7}$$

which has a unique solution

$$x_i^* = \frac{a_i}{\left(\sum_j a_j \right)}. \tag{9.8}$$

To analyze the stability of this solution x_i^*, here we consider a small perturbation, $x_i = x_i^* + \delta_i$. Then, we obtain

$$\frac{\mathrm{d}\delta_i}{\mathrm{d}t} = -\delta_i \left(\sum_j a_j \right), \tag{9.9}$$

indicating that the fixed-point solution is always stable. For example, Fig. 9.2(c) shows the population flow of three species. In this case, they coexist with a population ratio corresponding to the ratio of replication rate a_i.

Exercise 9.1 Let us consider the case of two species $k = 2$, which is simpler than the preceding example of $k = 3$. In this case, the dynamics can be described by a single variable x_1, because $x_2 = 1 - x_1$. Discuss the difference in the dynamics among the conditions $\alpha = 0$, 1, and 2.

Exercise 9.2 In the population dynamics described by Eq. (9.2), consider what will happen when $\alpha = 1/2$. Although the case $\alpha = 1/2$ seems unlikely at first glance, it can occur, for example, when there are single-strand and double-strand RNA molecules, and a replication reaction occurs from only single-stranded RNA [17]. Consider where the difference between the preceding cases (A) and (C) appears in the condition $0 \le \alpha \le 1$, and where the difference between (A) and (B) appears in the range $1 \le \alpha \le 2$.

Exercise 9.3 In the population dynamics described by Eq. (9.2), consider what will happen in the case $f_i(x_i) = a_i x_i / (1 + a_i x_i / K)$. First, confirm that the dynamics behave in the same way as the case $\alpha = 0$ when $K \ll a_i$ for all i, and behave in the same way as the case $\alpha = 1$ when $K \gg a_i$ for all i. Then, analyze the dynamics in the intermediate condition, for example, $K \sim a_i$.

This model describes the following case. The substrate chemical S_i is necessary for the replication of ith species, which is modeled by the reaction $S_i + X_i \rightarrow 2X_i$. The substrate S_i is supplied to the system from the outer environment with the constant substrate conecentration $\bar{S}$. The flow of the substrate from the environment is assumed to be proportional to the concentration difference, that is, $D(\bar{S} - S_i)$, where D is the diffusion coefficient. Then, the dynamics of substrate concentration are given by $\mathrm{d}S_i/\mathrm{d}t = -a_i S_i x_i + D(\bar{S} - S_i)$. By assuming the steady state $\mathrm{d}S_i/\mathrm{d}t = 0$, we obtain $S_i = \bar{S}/(1 + a_i x_i/D)$. This means that the production rate of x_i is proportional to $a_i x_i \bar{S}/(1 + a_i x_i/D)$. By replacing coefficients according to the preceding model, we obtain $f_i(x_i) = a_i x_i/(1 + a_i x_i/K)$. Here, when D is sufficiently large, by replacing $a_i \bar{S}$ by a_i, the equation can be regarded as the case $\alpha = 1$ in Eq. (9.2).

9.4 Error Catastrophe

In the previous section, we considered the case without replication errors to simplify the analysis. Here, let us consider the case where replication errors, namely mutations as substitution and deletion of monomer, occur under a certain probability. We assume that these errors change the replication efficiency of the polymer. If a replicating polymer with high replication efficiency appears by mutations, the ratio of this polymer will increase by iterating replication. Evolutionary dynamics emerge, in which polymers with high replication efficiency are selected. On the other hand, replication errors may cause deviation from this evolutionary dynamics toward the dominance of the species with the high replication rate. Such polymer species with high replication efficiency will be rare within a variety of possible monomer sequences. Then, the dominance of the original (functional) polymer is lost as the monomer sequence in the polymer is altered by mutational changes. In other words, the replication errors erase the information that makes the polymer functional. Therefore, when the error rate is too high, the information cannot be maintained because of the errors, even when a polymer with high replication efficiency is dominant in the population initially. To solve this problem, Eigen introduced the concept of "error catastrophe" [5]. In what follows, we will discuss the relationship between the error rate and evolutionary dynamics using a simple model of replicating polymers.

Consider a population of replicating polymers with length N. We assume that the polymers replicate themselves, that is, the population dynamics are described by the case $\alpha = 1$ in Eq. (9.2) in the previous section. As discussed previously, when there is no replication error, the population is occupied by the polymer with the highest replication efficiency, which we call the "dominant species." In what follows, to simplify the analysis, let x_m be the proportion ratio of the dominant species and x_ℓ the proportion ratio of other (nondominant) species in the population, respectively ($x_m + x_\ell = 1$). The term a_m represents the replication rate of the dominant species, and we assume that the replication of nondominant species can be characterized by the average replication rate over nondominant species, denoted by $a_\ell(< a_m)$. When μ is the probability that an error occurs in the replication of one monomer per replication reaction, the probability of a polymer consisting of N monomers making the same sequence is $Q = (1 - \mu)^N$. Taking these facts into account, the population dynamics of dominant and nondominant species, respectively, are as follows:

$$\frac{dx_m}{dt} = a_m Q x_m - x_m(a_m x_m + a_\ell x_\ell),$$ (9.10)

$$\frac{dx_\ell}{dt} = a_\ell x_\ell + a_m(1 - Q)x_m - x_\ell(a_m x_m + a_\ell x_\ell).$$ (9.11)

The first term in r.h.s. of Eq. (9.10) represents the replication of the dominant species, whereas the first and second terms in r.h.s. of Eq. (9.11) represent the replication of nondominant species and the increase of nondominant species by the dominant species replication error, respectively. Because the probability that an error causes a transition from nondominant to dominant species is much smaller than the transition between two nondominant species, we neglect the effect of errors that generate the dominant species (recall that there is only one dominant (functional) species. In contrast, there are many more nondominant species).

In this population dynamics, when no error occurs (i.e., $Q = 1$), the dominant species fully occupies the population, as discussed in the previous section. In contrast, as the error rate increases, the transition from the dominant species due to the error increases, leading to a decrease in the dominant species' population ratio. Here, let us analyze the conditions of a fixed-point solution in which the dominant species is maintained, that is, $x_m > 0$. First, from the condition that Eq. (9.10) is zero, we obtain that $x_m = 0$ or $x_m = Q - (a_\ell/a_m)x_\ell$. Substituting the second equation with $x_m > 0$ into Eq. (9.11), the following equation can be derived:

$$x_m = \frac{a_m Q - a_\ell}{a_m - a_\ell}, \quad x_\ell = \frac{a_m(1 - Q)}{a_m - a_\ell}.$$ (9.12)

Then, considering the conditions $a_m > a_\ell$ and $x_m > 0$, we obtain

$$Q\frac{a_m}{a_\ell} > 1.$$ (9.13)

In this equation, the ratio a_m/a_ℓ represents how much the replication efficiency of the dominant species is higher than the average efficiency of the nondominant species. To maintain the population of the dominant species, this ratio needs to be larger than the population loss by the errors denoted by Q. By assuming that μ is small enough, Q is approximated as $Q = (1 - \mu)^N \sim \exp(-N\mu)$. Substituting it into Eq. (9.13), we obtain

$$N\mu < \log\left(\frac{a_m}{a_\ell}\right).$$ (9.14)

In this relationship, the r.h.s represents the logarithm of the ratio of replicating efficiencies between dominant and nondominant species, which is not large unless the efficiencies differ by several orders of magnitude. That is, $\log(a_m/a_\ell)$ is typically

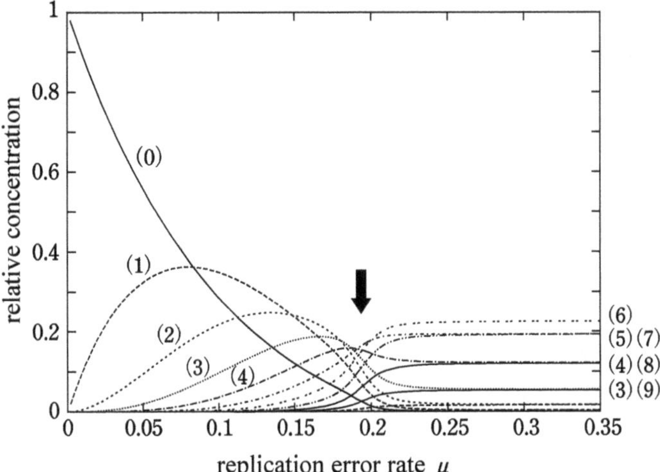

Figure 9.3 Information loss by replication error: error catastrophe. For the case with $N = 12$, the change in population ratio is plotted as a function of a error rate. The number in the parentheses indicates the Hamming distance from the sequence of the dominant species. The replication rate of the dominant species is set to 1, while the replication rate of other $2^{12} - 1$ species is set to 0.1. In this simulation, we consider an error from nondominant species to dominant species neglected in Eq. (9.10). As expected from the analysis, when the error rate exceeds $\log(a_m/a_\ell)/N \sim 0.19$, all possible sequences become almost uniformly distributed regardless of the sequence of the dominant species.

of order $O(1)$. Then, the preceding equation shows that to maintain the population of the dominant species, the polymer length N is constrained by the error rate μ, that is, $N\mu \lesssim O(1)$.

Figure 9.3 shows how the population ratio of the dominant and nondominant species in a fixed point depends on the error rate μ in the case of $N = 12$. Here we assumed that there are two monomer species resulting in total of 2^{12} polymer species. (For instance, by denoting the monomer by 0 or 1, the dominant polymer with a high replication rate is chosen as $00\ldots0$.) In this analysis, we assume that nondominant species have an identical replication efficiency (smaller than the dominant species). The numbers in the parentheses, as (1), (2), $\ldots$, indicate the population ratio of the species, which have 1, 2, $\ldots$, errors from the dominant species. In other words, they represent the population ratio of species that has Hamming distance 1, 2, $\ldots$, from the sequence of the dominant species. With the increase of μ, first, the species with a single error increase their population. Then, with a further increase in μ, species with two errors start to increase their population, and then species with three and four errors increase, and so on. The species whose sequences are close to the dominant species are often called "quasi-species" [5]. By increasing μ further, at the point $\mu \sim \frac{1}{N} \log \left(\frac{a_m}{a_\ell} \right)$ (denoted by an arrow in

Fig. 9.3), the population ratios of all species turn out to be almost identical, which means that the sequence information of the dominant species is lost. In this parameter region of μ, even when a species with a higher replication efficiency than other species appears by random mutations, it cannot increase its population due to the excessively high error rate. This phenomenon, which is a kind of phase transition, was called "error catastrophe" by Eigen.

The upper limit of the mutation rate below which the system is free from the error catastrophe raises an important question about the origin of life, particularly with regard to the scenario of how self-replicating molecules, such as in the RNA world scenario, first appeared, as the error rate should be rather high by considering just chemistry. Various enzymes have been developed in modern organisms to repair replication errors, thereby maintaining a low mutation rate. For example, the mutation rate of *E. coli* cells is about 10^{-9} per base per generation, which is about two orders of magnitude lower than the mutation rate at which error catastrophe occurs. (The *E. coli* genome consists of about 4×10^6 base pairs.) Although the mechanism of how errors are repaired to maintain the low mutation rate is an interesting problem (see also Section 10.5), it is unlikely that such an elaborate mechanism can emerge by chance at the early stage assumed in the RNA world scenario. For example, the error rate of the known ribozyme, which has replication activity of RNA sequence, is about 0.08 [8], while the replicase used in Spiegelman's experimental evolution is about 10^{-3}. Considering these facts, the error rate of ribozyme, which appeared by chance, is anticipated to be about 10^{-2}, even by adopting a lower estimate. With this error rate, only replicating RNA molecules up to 100 bases in length can maintain their sequence information. On the other hand, a longer RNA molecule will be required to produce a replication system with a lower error rate. A paradox arises here. Although low error rates are needed to maintain the long-sequence information for a self-replicating system, self-replicating systems achieving such low error rates (e.g., by some proofreading mechanism as in Section 10.5) cannot be constructed without long-sequence information. This dilemma, known as Eigen's paradox, is one of the fundamental problems in considering the origin of life.

To sum up, the emergence of self-replicating RNA molecules does not necessarily mean they can be sustained and evolve. It is necessary to avoid Eigen's paradox in some way to maintain self-replicating molecules so as to achieve a higher replication rate and function through evolutionary dynamics. Then, how can this paradox be avoided? In the following sections, we discuss two possible mechanisms for avoiding this. One is the reaction network among self-replicating molecules, and the other is the compartmentalization within which replicating molecules are confined.

9.5 Hypercycle

Because of the error catastrophe, it is hard to maintain replicated molecules with long sequences in the early stages of life, which makes it difficult to explain the emergence of replication mechanisms and other elaborate functions in vivo with high replication accuracy. As a method for solving the problem of this error catastrophe, Eigen and Schuster proposed the reaction network consisting of the replication molecules, which mutually catalyze each other's replication, which is termed "hypercycle" [6]. Figure 9.4 shows an example of a hypercycle consisting of four species of replicating molecules. Here, replicating molecule 1 catalyzes the replication of molecule 2, molecule 2 catalyzes the replication of molecule 3, 3 catalyzes 4, and 4 catalyzes 1, forming a loop of the chained catalytic reactions of replicating molecules. In this case, all rates of the replicating reactions in the loop will be proportional to the second-order term, such as the increase of molecule 2 is proportional to $x_1 x_2$. During the replication, the mutation is generated by the replication error, as in the previous section. However, the phenomenon due to the replication error becomes a term proportional to the first order of x_i. Once this hypercycle is established, the replicating reactions are maintained because the amplification with the second-order terms overcomes the decrease due to the error, which is linear in the concentration.

As long as the structure of this hypercycle is maintained, Eigen's paradox can be avoided. However, this replication system is not always stable if replication errors change the structure of the hypercycle. For example, computer simulation of the hypercycle suggested that it generally has vulnerability with respect to the following changes in the loop structure [16].

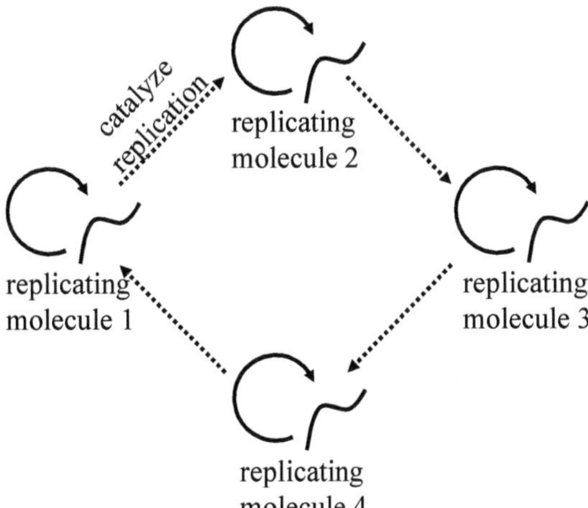

Figure 9.4 Schematic figure of the hypercycle. Self-replicating molecules catalyze the replication of other molecules, forming a loop of catalytic reactions.

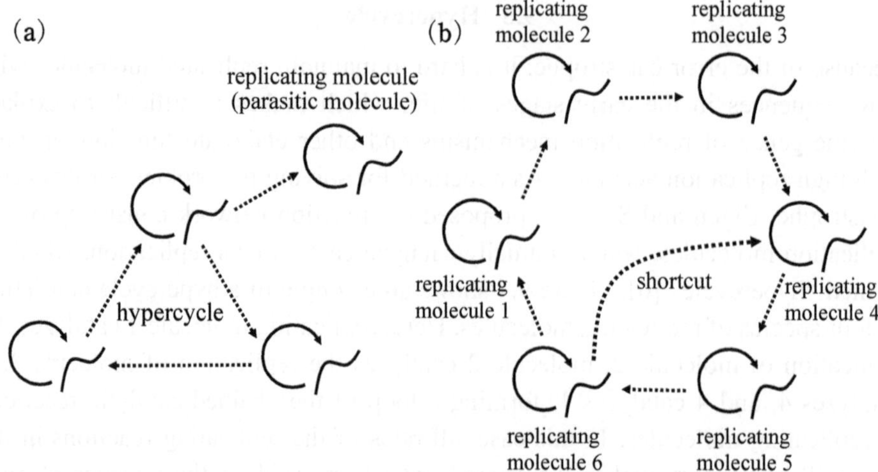

Figure 9.5 Vulnerability of hypercycle. (a) The emergence of a parasite molecule. (b) The emergence of a shortcut reaction.

(1) Vulnerability to parasites

In the hypercycle shown in Fig. 9.4, we considered the case in which four replicating molecules catalyze the replication of another molecule. Here, molecules with different sequences may arise by replication errors, whose replication is catalyzed by the molecules in the cycle. Then, some molecules will emerge whose replication is catalyzed by other molecules in the hypercycle but which do not catalyze other replication reactions (Fig. 9.5(a)). These molecules correspond to "parasites" to the hypercycle. These parasitic molecules are also amplified in proportion to the second-order term as other replication molecules composing the hypercycle. On the other hand, there can be a large variety of such parasitic molecules than that of composing the hypercycle. Thus, as replication errors accumulate, parasitic molecules can be the majority of the population, and the number of replication molecules which maintain the hypercycle decreases. Accumulation of these parasitic molecules eventually results in the loss of the ability to replicate the hypercycle.

(2) Vulnerability to shortcut reactions

When replication errors occur in the molecules composing the hypercycle and begin to catalyze other replication reactions in the loop, the loop length of the hypercycle can be reduced (Fig. 9.5(b)). The short loop may have a faster amplification rate by the emergence of such a shortcut reaction. In this case, the shortened hypercycle takes over the entire population. With the advent of such shortcut molecules, the hypercycle is reduced to a simpler hypercycle and eventually to a single replication molecule.

(3) Vulnerability to fluctuations in molecular number
In the hypercycle, the solution in which the number of molecular species varies (oscillates) with time often emerges by increasing the loop length, for instance, given by a limit-cycle. In fact, such oscillatory (or time-varying) dynamics generally appear due to the time delay in the positive feedback of the replication process as the loop length increases. This often results in a large variation in the number of molecules with time. Then, there is a possibility that the number of molecules constituting the hypercycle declines to zero. Once a replicated molecule with a molecular number of zero (i.e., extinct species) appears, the hypercycle consisting of that molecule collapses because such extinct species cannot be recovered by a catalytic reaction with another molecule due to the absence of itself (i.e., the template).

Computer simulation of the hypercycle shows that the collapse of the hypercycle by cases (1) and (2) can easily occur when the number of replication molecule species is large. This is because one replicating molecule, which appears due to replication error, causes collapse by cases (1) and (2), and the higher the total number of molecules, the higher the probability of such molecules. Alternatively, it is natural that collapse due to case (3), that is, the number of molecules in the hypercycle being zero, is more likely to occur as the number of total molecules decreases. The computer simulation of the hypercycle suggests that the region of the total number of molecules in which the hypercycle can be maintained is narrow, and the lifetime is finite even in such a region.

9.6 Compartmentalization

Revisiting the shortcomings of the hypercycle described in the previous section, the problem is that, eventually, the hypercycle will not be replicated. It is important to note that the hypercycles in the solution discussed so far are not units of natural selection. The Darwinian process is applied only to replicating molecules, that is, the molecules that replicate faster increase their population so that whether such molecules are advantageous for the replication of hypercycle itself is not subject to natural selection. Even if a better hypercycle (e.g., more efficient replication, more resistant to the vulnerabilities mentioned earlier, and so on) emerges due to replication errors, it would not necessarily be favored by natural selection. For example, in the hypercycle consisting of four replicating molecules (Fig. 9.4), let us consider that molecule 1' emerges by mutations, which catalyzes the replication of molecule 2 with higher efficiency than molecule 1. It is assumed that molecules 1 and 1' have the same replication efficiency (catalytic activity by molecule 4). Here, although molecule 1' is an "altruistic" replicating molecule that can contribute to increasing whole activity in the hypercycle, it will not increase its ratio in the population. (Recall that when a molecule is catalyzing some others, it cannot be replicated.) Instead, selfish replicating molecules that are more capable of being catalyzed than others are more likely to appear in the evolutionary dynamics of the hypercycle. In this situation, hypercycles with higher efficiency or stability cannot be expected to emerge through replication errors of molecules. This is because a hypercycle is not a unit of natural selection.

A simple way to give evolutionary capability to a hypercycle is to make it a unit of natural selection. For this purpose, we should consider the situation in which a group of molecules competes with other groups of molecules, i.e., hypercycles competing, with each other. For example, consider the proliferation of a "proto-cell" within which molecules consisting of a hypercycle are confined by a membrane surrounding them. The proliferation and division of the membrane are catalyzed by self-replicating molecules in the hypercycle (Fig. 9.6(a)). Here, the growth of the proto-cell is stopped when the hypercycle within the proto-cell is collapsed, as in Section 9.5, for instance, due to the emergence of parasitic molecules. Then, such proto-cells cannot grow and are eliminated by the selection process. On the other hand, a proto-cell containing a hypercycle with a higher overall replication rate, such as the appearance of altruistic molecules by replication errors, will be able to have a greater number of offspring (Fig. 9.6(b)). Here, by the compartmentalization of hypercycles, a higher-level unit of natural selection emerges, in which the state of the hypercycle is transmitted to the next generation. The compartmentalization enables the evolution of hypercycles with higher replication efficiency.

This compartmentalization does not necessarily have to be a cell-like structure surrounded by a membrane, like present organisms. For example, consider a case where a network of self-replication reactions such as a hypercycle is formed on a mineral's surface, as much attention has been gathered in the studies of prebiotic evolution. If diffusion of molecules on the surface is slow enough compared with replication reactions, properties similar to membrane compartmentalization can appear. Boerlijst and Hogeweg have carried out computer simulations of the reaction-diffusion system to analyze such dynamics, in which a hypercycle replicates and diffuses. They found the emergence of spatial patterns such as spiral waves formed by replicated molecules, which suppresses the spread of parasitic molecules [2]. Here, the suppression of the emergence of parasite molecules and shortcut reactions by compartmentalization is achieved when the number of replicating molecules per compartment is small. If the number of molecules in the compartment increases, selfish replicating molecules in the compartment will be amplified sufficiently.

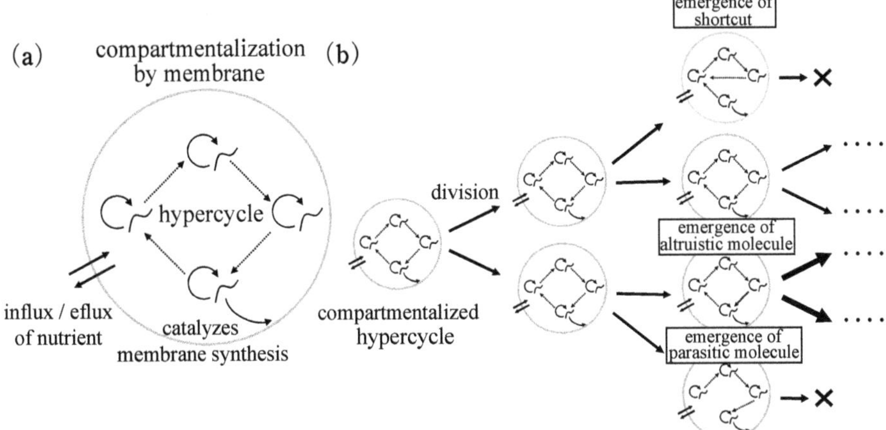

Figure 9.6 Compartmentalization of the hypercycle. (a) Schematic representation of the compartmentalized hypercycle within a membrane. One of the replicating molecules composing the hypercycle catalyzes the synthesis of the membrane, while nutrient components used for replication reaction can permeate the membrane. (b) Evolution of compartmentalized hypercycle. Consider a system in which a proto-cell, containing a hypercycle, divides through membrane synthesis. The proto-cell stops dividing when the emergence of parasitic molecules or shortcut reactions disrupts the hypercycle. In contrast, proto-cells, in which the emergence of altruistic molecules increases the replicating efficiency of the hypercycle, will be able to divide faster and leave more offspring. As a result, the evolutionary dynamics of hypercycles with high replication efficiency emerge.

Ichihashi and colleagues have experimentally confirmed the small-number effect of replicating molecules in the compartments. This study used a self-replication system consisting of the RNA encoding $Q\beta$ replicase and the cell-free protein translation system. The RNA is replicated by $Q\beta$ replicase, which is encoded by the RNA itself. Then, mutations are introduced into the RNA, and units consisting of RNA and replicase molecules that achieve higher replication efficiency are selected, where the molecules in the unit are encapsulated into a test tube or water-in-oil emulsion. The cycle of iterating these processes leads to Darwinian evolution. Here, when the cycle is carried out in a system including many RNA molecules within a test tube, short parasite RNA molecules that do not encode the replicase appear, and the replication reaction eventually stops. Then, when this replication system was enclosed in a small compartment constructed by water-in-oil emulsion, and the replication reaction was carried out there, the replication reaction was stably maintained in which the parasite molecules did not appear. As a result, the evolution of RNA and replication enzyme with high replication efficiency was observed in this compartmentalized replication system [9]. This result indicated the importance of compartmentalization and keeping a small number of molecules in a self-replicating system.

9.7 Dyson's Catalytic Reaction System

As discussed so far, the hypothesis that "replication molecules appeared first" is widely accepted as a clear and easy-to-understand story, as also supported by the discovery of ribozymes, that is, RNA molecules with catalytic activity. However, considering the stability against replication error discussed earlier, catalytic networks among replication molecules and compartmentalization of the catalytic network seem to be required to sustain the evolutionary dynamics of the system. This means that when only the self-replicating molecules appeared alone on the primitive Earth, the evolution process toward higher function may not have been realized only by such replicating molecules. Instead, the appropriate network of catalytic reactions should have emerged simultaneously to achieve the evolutionary dynamics of replicating systems.

From the preceding discussions, the question arises that if a network of catalytic reactions exists in the prebiotic environment, even if there is no replication molecule, information can be inherited to the next generation in some form, leading to the evolvability of the catalytic reaction network. In this case, a complex catalytic reaction (metabolism) network is assumed to appear before the replication molecule. For example, polypeptides produced by heating amino acid solutions have been shown to have various catalytic activities. Although these polypeptides generally do not replicate themselves, and thus they do not become units of natural selection, catalytic networks in which polypeptides catalyze each other's production may be able to reproduce themselves. Such autocatalytic reaction networks can ultimately be a unit of natural selection. Theoretical physicist Freeman Dyson discussed that it would have been difficult for the replicating molecules to have appeared first and evolved in the prebiotic environment due to the problem of error catastrophe. Then, he discussed an alternative hypothesis: that the catalytic reaction networks which can maintain themselves and are reproduced inexactly appeared first. Using an abstract model consisting of catalytically active and inactive polymers, Dyson examined how the catalytic network catalyzes its reproduction in the prebiotic condition. Here, we will briefly sketch Dyson's catalytic reaction network model.

Let us consider a cell that contains polymers consisting of N monomers (e.g., amino acids) of a different monomer species. For simplification, cellular growth and division are ignored. The polymers in the cell catalyze each other's synthesis. However, the reproduction of each molecule does not always occur exactly. The wrong monomer is incorporated into the polymer with a certain probability. When the correct monomer is incorporated, it is called an "active" monomer because it contributes to its catalytic activity. When the incorrect monomer is incorporated, it is called an "inactive" monomer. To simplify the model, we disregard the details of the individual reactions and assume that the proportion of monomers correctly incorporated into the polymer (in the active state) varies depending on the proportion of active monomers in the polymer. Specifically, let $\phi(x)$ be the probability of a new active monomer being incorporated into a polymer as a function of x, where x is the ratio of active monomers in the polymer. The function $\phi(x)$ represents how the state of catalytic activity is inherited in the intracellular dynamics. For example, $\phi(x) = x$ corresponds to the case where x in an initial state is inherited without changing the ratio of active monomer x.

Dyson first assumed that there are two stable states of the system. One corresponds to a state in which the activity is maintained, and another corresponds to a state in which the activity is lost. To achieve such bistability, he also assumed that $\phi(x)$ obeys an S-shaped curve intersecting at three points α, β, and γ between 0 and 1, as shown in Fig. 9.7, to represent the transition between the states. He then considered the conditions under which such a postulated S-shaped curve would emerge. First, let us consider the case in which the activation energy required for incorporating an active monomer is different from that for the incorporation of an inactive monomer and that the difference is Ux with U as a constant. In this case, the probability of incorporation of an active monomer is greater by a factor b^x than the probability of incorporation of an inactive monomer, where $b = \exp(U/k_{\mathrm{B}}T)$.

Since there is only one type of active monomer and $(a - 1)$ types of inactive monomers, the probability of incorporating an active monomer $\phi(x)$ is as follows:

$$\phi(x) = (1 + (a - 1)b^{-x})^{-1}. \tag{9.15}$$

This function is an S-shaped curve, as shown in Fig. 9.7, with the appropriate parameters a and b. Since the state of catalytic activity takes an equilibrium state when $\phi(x) = x$ holds, there are three equilibrium states corresponding to the intersections α, β, and γ in Fig. 9.7. The equilibrium state is stable when the slope of $\phi(x)$ at the intersection point is less than 1 and unstable when it is greater than 1. For example, if the ratio x is slightly changed from the state of the intersection point α, the system evolves to its original state. On the other hand, the state of the intersection point β is unstable; given a small perturbation in x, it is amplified further, and the state deviates from the intersection. The intersection point γ is stable, as in the case of α. Hence, in this system, two stable states α and γ are separated by an unstable saddle point β. In state α, the polymer is composed mostly of inactive monomers, and thus the self-reproduction reaction occurs only rarely. Dyson considered that state α corresponds to the "death" state. On the other hand, in state γ, the active monomer is incorporated into the polymer, corresponding to the "living" state where the self-reproducing reaction is maintained actively.

Here we address the question under what conditions the transition from the death state to the living state can occur. Since these two states are stable, no transition between states occurs when the polymer length N is sufficiently large, and there

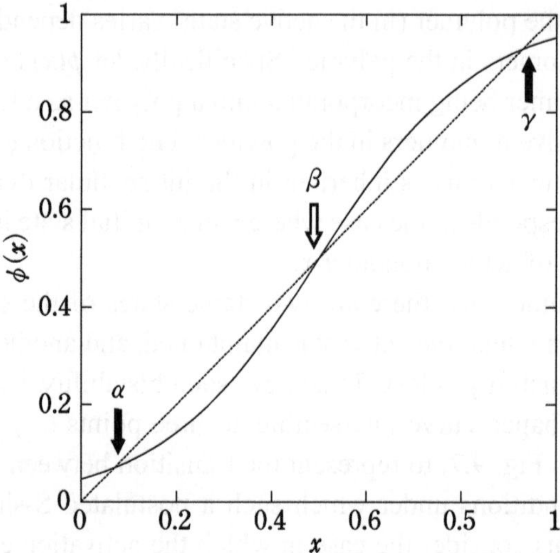

Figure 9.7 Function $\phi(x)$ used in Dyson's model of catalytic reaction. The term x represents a ratio of active monomer in the polymer, while $\phi(x)$ is the probability of a new active monomer being incorporated into the polymer with the ratio x. Among the three states satisfying $\phi(x) = x$, α and γ are dynamically stable, while β is unstable.

exist only tiny perturbations by stochasticity due to small numbers. On the other hand, when N is small, stochastic fluctuations may cause transitions between these states (see Chapter 6). Using this model, Dyson estimated the range of parameters within which the transition could occur as follows [4]:

$$8 < a < 10, \qquad 60 < b < 100, \qquad 2,000 < N < 20,000 \qquad (9.16)$$

Of course, this is an estimation from a very simplified model, and thus it is necessary to consider how it corresponds to a real system. However, the ranges of these parameters are consistent with a system in which amino acid polymers with primitive catalytic activity are synthesized in a prebiotic condition. The range of the number of monomer species $8 < a < 10$ is not much different from the 20 amino acids used in modern organisms. Current enzymes, such as polymerases, are estimated to have b ranging from 5,000 to 10,000, representing the difference in discriminating between correct and incorrect monomers. Since this range of b results from fine-tuning by evolution, it is not surprising that primitive catalytic polymers have a much lower range of b. Although it is difficult to estimate the polymer length N of the primitive catalytic polymer, when the polymer length N is in the order of tens of thousands, it is possible to maintain a sufficiently complex catalytic reaction network and transition between states by fluctuations in a realistic timescale.

An important feature of this model is that the state of γ where active self-reproduction occurs has a high error tolerance. This error tolerance is not sustained by the exact replication of polymers. For example, in the preceding conditions of a, b, and N, the probability of incorporating the wrong monomer is about 20 to 30% in the state of γ. The replicating molecules discussed in Section 9.4 cause error catastrophes and lose sequence information at such high error rates. In contrast, even though each polymer does not replicate perfectly in Dyson's model, the reproducing activity is maintained as a group of polymers. Thus, the system allows for a high error rate, which facilitates the transition from death state to living state. This high error tolerance can be an essential property in primitive biological systems where elaborately designed enzymes do not exist.

It should be noted that Dyson's model is highly abstract, and it is difficult to verify experimentally. At present, the experimental verification of this model has yet to be advanced as compared with the RNA world hypothesis, and further studies, including theoretical and experimental approaches, are required.

9.8 Catalytic Reaction Networks

In Dyson's model and other "metabolism appears first" models, it is assumed that the network in which the catalytic molecules, such as the polymer of amino acid, mutually catalyze their synthesis, and the molecules are reproduced autocatalytically. For example, in Fig. 9.8, P_i represents a molecule with such catalytic

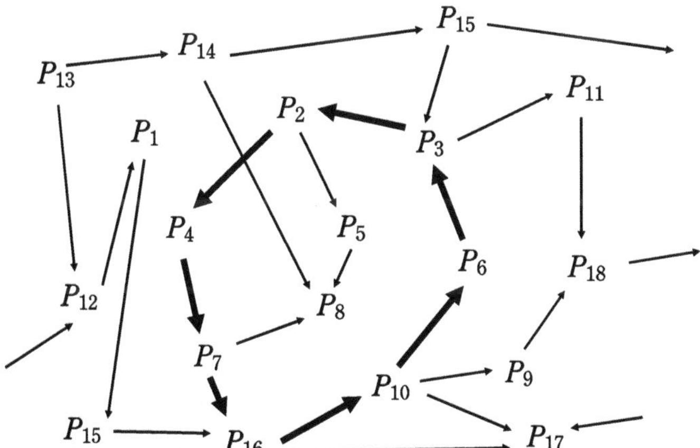

Figure 9.8 Schematic representation of a catalytic reaction network. The term P_i represents molecules with catalytic activities, while the arrows show the relationship between catalysts and molecules they synthesize. Although each molecule has no self-replicating activity in this network, a loop consisting of seven catalytic reactions (denoted by thick arrows) exhibits self-replicating activity.

activities, and the arrows represent the catalytic relationships. In the network loop drawn by the thick arrows, the reactions that synthesize the molecules in the network are catalyzed by the molecules in the same network. That is, these molecules are synthesized autocatalytically. Once such an autocatalytic network emerges, and if it is linked to the growth and division of membranes, such a system can be regarded as a primitive organism whose internal state can be inherited to some extent. Dyson's model in the previous section assumed such a network of molecules sustaining autocatalytic reactions as a whole. However, it remains unclear whether such autocatalytic reaction networks can emerge in a prebiotic environment. In the present organism, catalysts (enzymes) are synthesized through transcription and translation machinery based on genomic sequences, and there is no reaction in which the catalyst is directly synthesized. So then, how can such an autocatalytic network emerge in prebiotic environmental conditions?

For this problem, Kauffman used a simple model of polymer reaction and argued that such autocatalytic reaction networks can emerge stochastically [11]. In this model, two monomers, A and B, form polymers through the concatenation reactions as AB+BAB $\rightarrow$ ABBAB and cleavage reactions as AAABB $\rightarrow$ AAA+BB. Each reaction is catalyzed by other polymers, forming a catalytic reaction network, where P represents the probability that a polymer will catalyze an arbitrarily chosen reaction. Figure 9.9 shows an example of an autocatalytic reaction network that emerges in this polymer model. Here, in addition to the monomers A and B, AA and BB are provided to the system as nutrients. An autocatalytic network emerges from these nutrients, in which the same polymer set catalyzes all reactions to synthesize all polymer species in the network. The probability for the emergence of such an autocatalytic network is sufficiently high; for example, Kauffman demonstrated that when the maximum polymer length $N = 13$ and $P = 10^{-9}$, the probability for the existence of the autocatalytic reaction network is more than 0.999. Although the analysis is based on a simplified model, Kauffman argued that by linking such autocatalytic networks with membrane synthesis and other mechanisms of proliferation and division, primitive cells might emerge.

Furusawa and colleagues studied a cell model with an intracellular chemical reaction network that divides depending on its state to explore the universal nature of the reaction dynamics of replicating cells [7]. In this model, the time development of chemical abundance is governed by a network of catalytic reactions, which is determined randomly (Fig. 9.10(a)). Some chemicals are supplied as nutrients from the environment by diffusion through the membrane (with a diffusion coefficient D), increasing total chemical abundance in the cell. When the total chemical abundance exceeds a threshold, the cell divides into two daughter cells, where the mother cell's chemicals are evenly split among the daughter cells.

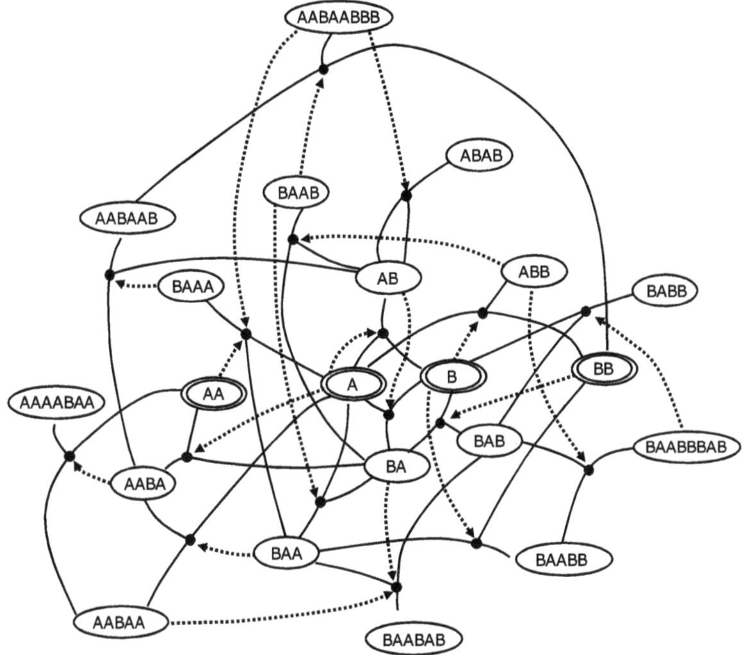

Figure 9.9 Kauffman's polymer model of a catalytic reaction network. A network of reactions among polymers composed of two types of monomers, A and B is shown. The solid lines show the reactions among polymers (e.g., AABA+AB → ABABAAB). Each reaction is catalyzed by another polymer, which is represented by the dotted arrows. The polymers and monomers enclosed in the double circles are supplied from the outside as nutrients.

In Fig. 9.10(b), the growth rate (inverse of doubling time) and replication accuracy (similarity of chemical composition between mother and daughter cells) are plotted as functions of the diffusion coefficient D. As D increases, both the growth rate and replication accuracy increase until continuous division becomes impossible at a critical diffusion coefficient D_c (indicated by a dotted line in Fig. 9.10(b)). Beyond this critical point, the uptake of nutrients is so fast that the intracellular chemical reactions transforming them into chemicals sustaining "metabolism" cannot keep up. Notably, at the critical point $D = D_c$, the rank-ordered distribution of chemical abundances obeys a power-law relationship, as shown in Fig. 9.10(c). The slope of the rank-ordered distribution increases with an increase in diffusion coefficient D, approaching to a power law with an exponent of -1 at $D = D_c$. This inverse relationship between rank and quantity, known as Zipf's law, has been observed in a variety of quantities, such as the frequency of words in natural language and the population of cities. The power-law distribution of chemical abundances as described by Zipf's law is a property that arises independently of model specifics and is anticipated to be a universal characteristic of self-replicating

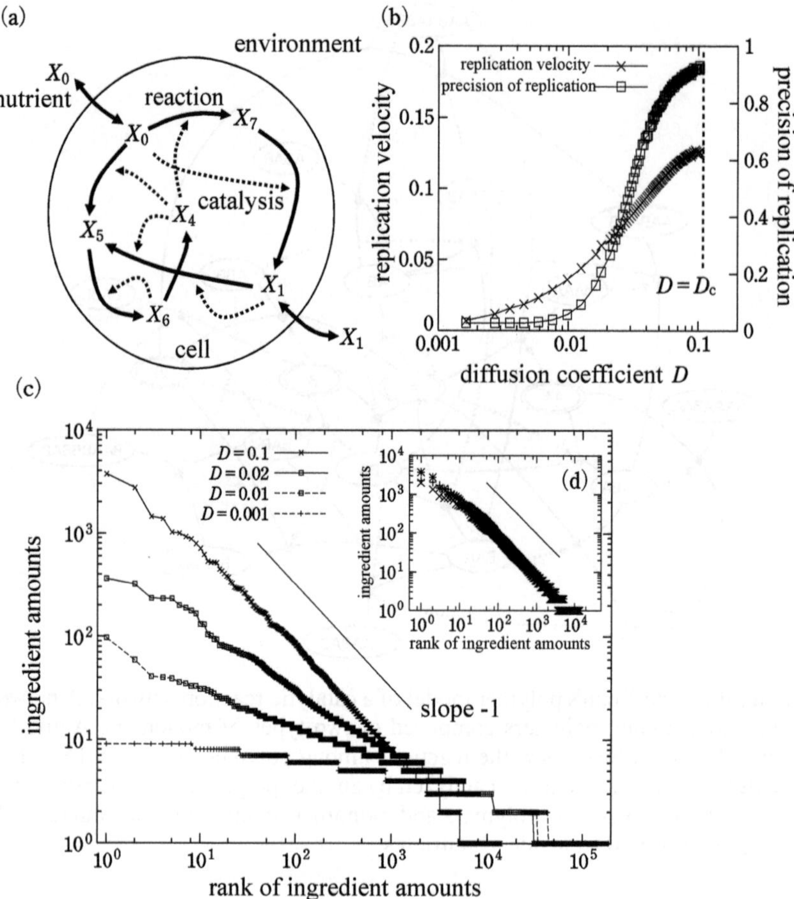

Figure 9.10 A cell model with an intracellular chemical reaction network. (a) Schematic figure of the model. Each reaction is assumed to be a catalytic reaction of the form $X_i + X_j \rightarrow X_\ell + X_j$, where X_j acts as a catalyst. Nutrient chemicals are taken in from outside the cell by diffusion, with a diffusion coefficient D. The cell divides when the total number of chemicals in a cell exceeds a threshold value due to nutrient uptake. (b) The growth rate of a cell and replication accuracy, plotted as a function of the diffusion coefficient D. The replication accuracy between mother and daughter cells is calculated as the scalar product of k-dimensional vectors, where each element corresponds to the abundance of a particular chemical species and k denotes the number of chemical species. (c) Rank-ordered abundance distributions of chemical species. The vertical axis indicates the chemical abundance, and the horizontal axis displays the rank of chemical species determined by the abundance. Distributions for different diffusion coefficients D are overlaid. In this case, the critical diffusion coefficient D_c is approximately 0.1.

cells capable of rapid growth and high replication accuracy. Indeed, the rank distribution of gene expression levels follows a power-law distribution with an exponent close to -1 in various transcriptome samples, including human and mouse tissues, yeast, and bacterial cells [7].

9.9 Minority Control in Catalytic Reaction Networks

One of the key criticisms of the "metabolism network appears first" hypothesis is that the mechanism by which the catalytic network has evolvability is unclear. For a self-replicating system to have evolvability, it needs to have both plasticity and robustness of state, that is, it can change its state with each replication and this change can be inherited stably by the next generation. As discussed in the previous section, let us consider a proto-cell, a self-replicating system with an autocatalytic chemical reaction network. Here, the reaction network is separated by a membrane-like compartment, and the catalytic reactions contribute to membrane synthesis leading to cell divisions. Due to environmental fluctuations, inaccurate chemical replication, and the loss of a small number of components during division, accurate self-replication is difficult in this proto-cell. It is natural to assume that the state of the proto-cell will change in various ways with each division. In such a replicating system, even if a state with relatively high fitness (e.g., a high replication rate) emerges due to fluctuations, the state with the high fitness is not selected because the information is lost during replication. That is, this system cannot have evolvability. Then, what kind of mechanism can give evolvability to such systems with inaccurate replication?

One possible scenario is that a catalytic reaction network with inaccurate replication is merged with another system that has evolvability. For example, it has been proposed that the self-replicating information molecules, as discussed in the RNA world hypothesis, invaded as "parasites" in the primitive metabolic system with catalytic reactions and gradually came to control the entire system. A. G. Cairns-Smith has also proposed a theory of the origin of life, in which self-replicating clay crystals became catalytically active, which control the primitive metabolic system. Then, replicating organic polymers, such as RNA and DNA, took over the system and started to control the whole reaction network, as they do in the present organisms. These scenarios are based on the process of a self-replicating system that has evolved independently and interacts with a catalytic network with inexact reproduction, bringing accurate replication and evolvability to the entire interacting system. However, it remains unclear how an exogenous replication system can control and change the catalytic reaction networks to bring about accurate replication.

As an alternative scenario, it has been argued that, unlike the interaction with these exogenous replication systems, dynamics in which some components control the whole system can emerge from autocatalytic reaction networks [10]. The key here is the small-number effect of the molecules. In an autocatalytic network system with stochastic reactions, the number of some components (e.g., those with low synthesis rates) would be zero at the event of the division of the proto-cell. When such minority components exist in an autocatalytic network, the system's behavior can change drastically depending on whether the component is present or absent, that is, whether the number of molecules is zero or one (see also Section 5.4 for the fluctuation of abundances of chemical components). As a result, the minority components will control the behavior of the entire network. The changes caused by the fluctuation of the number of minority components will create the plasticity of the state, and the inheritance of the components to the next generation will create the recursive production of the state. Then it is proposed that such minority components will turn out to carry primitive genetic information that can bring evolvability.

Here, let us consider a simple catalytic network within a proto-cell to discuss the emergence of minority components that control the whole system (see Fig. 9.11). For simplification, we assume that in this proto-cell, there are only two autocatalytic molecular species X and Y, where X catalyzes the synthesis of Y and Y catalyzes the synthesis of X, respectively. In other words, the catalytic reactions $X + Y \rightarrow 2X + Y$ and $Y + X \rightarrow 2Y + X$ occur, resulting in an autocatalytic network. Here, as discussed in Section 9.4, due to error in replication, there is a large probability that the synthesized molecule may lose the catalytic activity so that the fraction of active X or Y molecules can be smaller than that of inactive ones. The growth and division of this proto-cell are assumed to be controlled by these two molecular species (e.g., consider the case that the cell membrane is synthesized from molecules X and Y). We also assume that the synthesis rates per unit time are different between these molecules, with X for the larger synthesis rate and Y for the smaller one. Considering the situation in which the proto-cell containing an autocatalytic network divides, the relative proportion of the slowly synthesizing molecule Y will decrease with each division. Eventually, a cell will emerge where the number of active Y molecules is zero. On the other hand, for this proto-cell to continue to divide, at least one molecule each of active X and Y molecules must be present. A cell that loses active Y molecules through division loses its proliferative capacity. That is, the presence or absence of the active molecule Y determines the proliferation of the proto-cell. (In contrast, many X molecules exist, so a certain fraction of them are active molecules.) The important point here is that the minority molecules, Y, rather than the majority molecules, X, play a role in controlling the state of the proto-cell. When the number of molecules in X and Y fluctuates stochastically, fluctuations in the number of molecules in X will not significantly affect the cell state. In contrast, changes due to fluctuations in Y can significantly change the state.

Such control of the state by minority molecules also appears in more complex autocatalytic networks, including a larger number of molecular species. If the synthesis rates are different among molecular species, there will always be

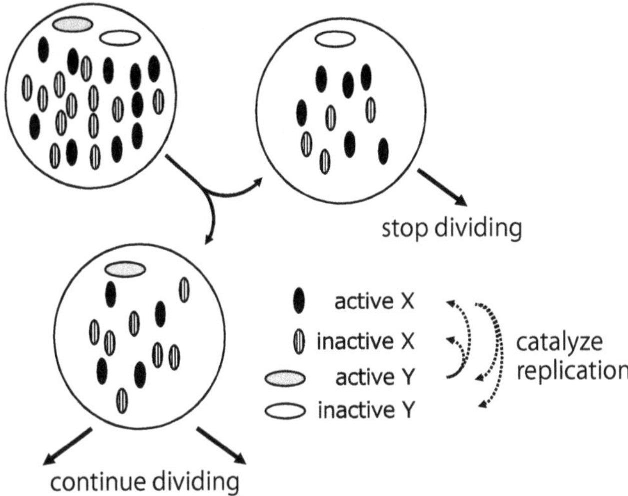

Figure 9.11 Reproduction system with mutual catalytic reactions of replicating molecules. Once an active molecule is lost at either X or Y, growth cannot continue.

minority molecules, and if these molecules are necessary for division, they will eventually control the whole dynamics. Such minority molecules can significantly change the state of the cell, as their number changes with each division, resulting in the plasticity of the cellular state. On the other hand, when the same set of minority molecules is inherited by replication, it results in a similar cellular state, and consequently, recursive production is achieved. As a result, evolutionary dynamics may emerge in the population of such proto-cells with autocatalytic reactions involving essential minority molecules. Here, competition at the cellular level leads to a population dynamics in which cells that cannot proliferate much will go extinct, and cells that can maintain their proliferation activity will remain. In fact, analysis using multicomponent autocatalytic network models has demonstrated that, by introducing appropriate mutations (e.g., random changes in catalytic activity), a network with suppressed fluctuations in the number of minority molecules controlling the state can achieve high fidelity in reproduction and then they are selected during the evolutionary dynamics. These results suggest that control by minority molecules, which can maintain evolvability, can emerge from autocatalytic networks with inaccurate reproduction without self-replicating molecules.

Note that the preceding discussion started from the assumption that there is a sufficient difference in the synthesis rates of X and Y. As another scenario, Takeuchi considered the reproduction of a proto-cell consisting of a population of molecules X and Y, which initially have both information-carrying template functions and catalytic activity. In this case, it was demonstrated that spontaneous symmetry breaking between the characteristics of two molecular species occurs, and the roles of the two molecular species differentiate into information-carrying molecules, while majority molecules with catalytic activity [18]. This study suggested that separating functional and informational molecules, as in proteins and DNA, is an inevitable course for proto-cells to sustain reproduction and attain evolvability.

9.10 Perspective of Research on the Origin of Life

In this chapter, we have discussed theoretical approaches concerning the origin of life. Two hypotheses on the origin of life, that is, (1) replicating molecules appeared first, and (2) the metabolic networks appeared first, have been discussed, whereas each of them has a weak point. For hypothesis (1), low error tolerance, which is known as error catastrophe, is an essential problem. In contrast, for hypothesis (2), it is difficult to explain how the system has both recursive production and evolvability. Of course, theoretical scenarios are not limited to either (1) or (2). It is also possible that self-replicating molecules, primitive metabolic systems that supply the raw materials, and dividing membrane molecules appeared simultaneously and established a relationship of control/controlled through interaction with each other. In such a case, how the plasticity and recursiveness of the state, which are necessary for evolvability, emerge should be analyzed theoretically.

One of the difficulties in understanding the origin of life is that the origin of biological systems, as we know them, occurred only once in the past and cannot be accessed directly. However, the theoretical aspects discussed in this chapter will provide a basis for discussing what properties need to be satisfied in the origin of life and what mechanisms are possible to sustain the process.

Experimental studies have also challenged the understanding of the origin of life. One such system is the self-replicating system using $Q\beta$-replicase described in the previous sections. In this system, evolvable artificial cells are constructed by encapsulating the RNA molecules that encode the replicase to replicate them into a compartment such as emulsions. Such experimental studies will reveal what is possible and what is difficult to achieve for continuous self-replication and evolution.

Another approach involves extraterrestrial exploration to find life-forms that may be different from what we know on Earth. For example, Enceladus, a satellite of Saturn, has an ocean of liquid water beneath its icy surface. It is planned to send a probe there to collect samples. If appropriate samples are obtained, one must ask which properties are regarded as traces of biological systems. Further advancements of the theoretical approaches will be necessary to address such a question, which will open a new door to the problem of the origin of life.

Column VIII: von Neumann's System of Self-Replicating Automata

In the 1940s, John von Neumann, whose work spanned mathematics, physics, economics, and computer science, discussed the properties that self-replicating and evolvable systems should possess. Here, we briefly introduce von Neumann's self-replicating cellular automaton.

A cellular automaton is a model in which cells, each with an internal states, update based on the states of their neighbors. For instance, consider cells arranged on a two-dimensional lattice. The evolution of a cell's state is determined by the states of the cell and its neighbors to the right, left, above, and below, referred to as the Neumann neighborhood, as shown in Fig. 9.12(a). That is, $x_{i,j}(t+1)$, the state x at the position (i, j) at time $t + 1$, can be written as follows:

$$x_{i,j}(t+1) = f(x_{i,j}(t), x_{i+1,j}(t), x_{i-1,j}(t), x_{i,j+1}(t), x_{i,j-1}(t)), \tag{9.17}$$

where $f(\cdots)$ is a function common to each cell and is called the state transition function. Although it is a simplified model of discrete space and discrete time, it

(a)   (b)

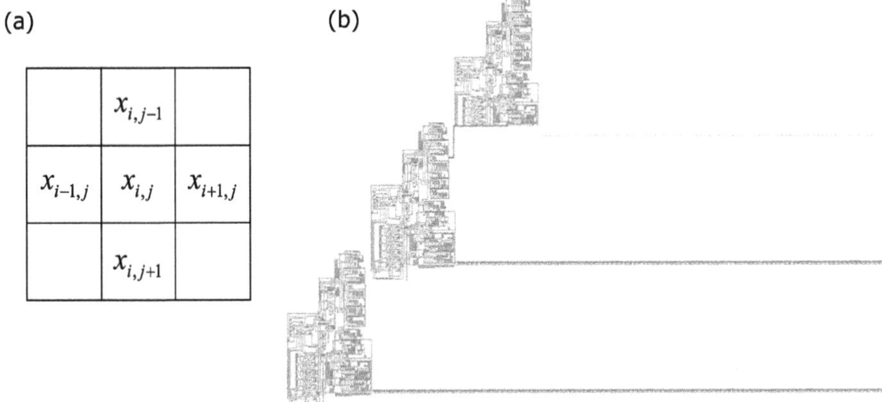

Figure 9.12 von Neumann's self-replicating automaton. (a) The von Neumann neighborhood. The state of a cell at time $t + 1$ is determined by the states of that cell and its neighbors above, below, left, and right at time t. (b) The structure of von Neumann's self-replicating automaton. The structure that extends horizontally and looks like a bar is called a tape, which encodes the information about the automaton's structure. The complex-looking structure is a Universal Constructor (UC), which creates the same UC and tape in a neighboring region based on the information on the tape. Adapted from Wikipedia: von Neumann universal constructor.

is used to analyze not only biological processes but also various physical phenomena such as crystal growth and turbulence.

von Neumann was interested in the dynamics of biological systems that could replicate their own structures. After experimenting with various models, such as dynamical systems, he arrived at discrete-state and discrete-time cellular automata to achieve this. He wasn't focused on dynamics such as crystal growth, where identical structures replicate (which can be easily implemented even with cellular automata). Instead, von Neumann aimed to create dynamics where the state of certain cells could change randomly, like a "mutation," but still be inherited and replicated. In other words, he wanted to see if cellular automata could construct patterns capable of evolution.

After trial and error, von Neumann found a self-replicating automaton with 29 internal states and approximately 200,000 cells (Fig. 9.12(b)) [14]. Although the state transition function of this system is complex and not discussed in detail here, it implements various functions, such as signal transfer and identification by multiple cells. The remarkable feature of this self-replicating automaton is that it functions as a Universal Constructor (UC) that can produce any automaton. This self-replicating cellular automaton consists of two main components: tapes, which encode the automaton's structure, and UC, which creates automata in adjacent regions based on the information encoded in the tapes. This separation between the tapes and the UC allows for open-ended evolutionary dynamics that can continuously produce diversity. For example, if a tape contains a few mutations, it may replicate itself while being inherited. This separation of tape and UC reminds us of the genetic code on DNA and its replication system in biological systems, but von Neumann's self-replicating automaton was proposed before the genetic code system was elucidated by Watson, Crick, and others. This case can be regarded as a prediction of the properties that biological systems should have based on logical considerations
alone.

In an era when computers were not yet developed, von Neumann is said to have analyzed the behavior of this complex cellular automaton with pieces of graph paper and pencils. Since the 1990s, this self-replicating cellular automaton has been implemented on computers. However, it is difficult to observe open-ended evolutionary dynamics in practice. This is because a small mutation to the tape causes the generated automaton to lose its function as a UC. In other words, this self-replicating cellular automaton is fragile to noise, which makes stable evolutionary dynamics difficult (see the earlier discussion of error catastrophes). This could pose a fundamental challenge to the idea that information-replicating molecules based on logical elements emerged first. Further research is needed to clarify how robust self-replicating automata can be achieved or whether they can emerge from loosely replicating systems such as the one studied by Dyson.

Column IX: The Game of Life

The Game of Life, developed in the 1970s by John H. Conway, popularized cellular automata beyond academic research [15]. The Game of Life is a cellular automaton designed to represent the dynamics of populations, such as growth and selection, in ecosystems. The rules of the Game of Life are relatively simple. At any given time, each cell can be in one of two states: "live" or "dead," and its state at the next time step is determined by the state of the eight neighboring cells (top, bottom, left, right, and diagonal, collectively referred to as the Moore neighborhood). There are four basic rules. If a cell is "dead" at time t: (i) it becomes "live" at time $t + 1$ if there are three "live" cells surrounding it, and (ii) it remains "dead" otherwise. If a cell is "live" at time t: (iii) it remains "live" if there are two or three "live" cells surrounding it, and (iv) it becomes "dead" if there are fewer than two or more than three "live" cells surrounding it. When starting from an initial condition (e.g., randomly placing "live" and "dead" cells), a dynamic pattern of "live" and "dead" cells emerges when observing the time evolution with the preceding rules (Fig. 9.13 shows a snapshot.) However, just looking at the snapshot patterns does not give you any idea of the fun of the Game of Life. It is easy to implement this simple rule, and it would be good to implement it as a computer programming exercise and see how it behaves (or try programs available on the Internet, etc.).

Simulations of the Game of Life often settle on stable patterns consisting of elements that do not vary in time or oscillate periodically. For example, a pattern consisting of a 2×2 square arrangement of "live" cells surrounded by "dead" cells is stable and does not vary in time (several such patterns can also be seen in Fig. 9.13).

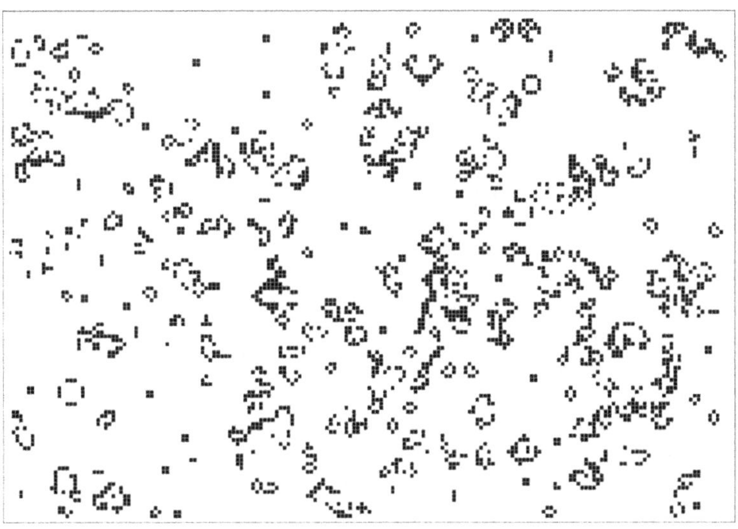

Figure 9.13 A snapshot of the Game of Life.

There are also many patterns that change periodically; for example, a pattern of three "live" cells arranged vertically or horizontally repeatedly changes from vertical to horizontal, vertical to horizontal, . . . , at period 2. A famous and interesting one is a structure of five "live" cells called "glider," which can move diagonally while maintaining the pattern in period 4. Conway was interested in the question of whether an infinitely proliferating pattern exists in the Game of Life without settling into a stable pattern and offered a $50 prize to anyone who proved or disproved it. As a result, various infinitely proliferating patterns were discovered. For example, a pattern consisting of dozens of cells called a "glider gun" has the dynamics of continuously firing one glider at period 30. Another pattern has been found in which "live" cells are left at regular intervals as they move, which is called "puffer train" because it seems to leave behind smoke. The discovery of the glider gun made it possible to perform complex operations in the Game of Life. Gliders shot out at regular intervals from a glider gun can be used as unit pulses for logic operations to construct logic circuits such as AND and NOT gates. In fact, the Game of Life has proved that the patterns that act as universal Turing machines can be constructed. This means that, given the initial conditions of the appropriate pattern, the Game of Life can perform all computationally feasible calculations. For example, emulating the Game of Life by the Game of Life or constructing a self-replicating cellular automaton in von Neumann's style has already been implemented. Despite its simple rules, the Game of Life continues to fascinate researchers due to its ability to generate complex behaviors.

References

[1] J. Attwater, A. Wochner and P. Holliger. In-ice evolution of RNA polymerase ribozyme activity. *Nature Chemistry*, **5**: 1011, 2013.

[2] M. C. Boerlijst and P. Hogeweg. Spiral wave structure in prebiotic evolution: Hypercycles stable against parasites. *Journal of Theoretical Biology*, **48**(1): 17–28, 1991.

[3] J. A. Doudna and T. R. Cech. The chemical repertoire of natural ribozymes. *Nature*, **418**: 222, 2002.

[4] F. Dyson. *Origins of Life* (2nd Edition). Cambridge, Cambridge University Press, 1999.

[5] M. Eigen, J. McCaskill, and P. Schuster. Molecular quasi-species. *The Journal of Physical Chemistry*, **92**(24), 6881, 1988.

[6] M. Eigen and P. Schuster. *The Hypercycle: A Principle of Natural Self-Organization.* Berlin, Springer, 1979.

[7] C. Furusawa and K. Kaneko. Zipf's law in gene expression. *Physical Review Letters*, **90**: 088102, 2003.

[8] D. P. Horning and G. F. Joyce. Amplification of RNA by an RNA polymerase ribozyme. *Proceedings of the National Academy of Sciences of USA*, **113**(35): 9786–9791, 2016.

[9] N. Ichihashi *et al.* Darwinian evolution in a translation-coupled RNA replication system within a cell-like compartment. *Nature Communications*, **4**: 2494, 2013.

[10] K. Kaneko and T. Yomo. On a kinetic origin of heredity: Minority control in a replicating system with mutually catalytic molecules. *Journal of Theoretical Biology*, **214**(4): 563, 2002.

[11] S. Kauffman. *The Origins of Order: Self-Organization and Selection in Evolution*. Oxford, Oxford University Press, 1993

[12] T. A. Lincoln and G. F. Joyce. Self-Sustained Replication of an RNA Enzyme. *Science*, **323**(5918): 1229–1232, 2009.

[13] D. R. Mills, R. L. Peterson and S. Spiegelman. An extracellular Darwinian experiment with a self-duplicating nucleic acid molecule. *Proceedings of the National Academy of Sciences of USA*, **58**(1): 217, 1967.

[14] J. von Neumann and A. W. Burks. *Theory of Self-Reproducing Automata*. Champaign, University of Illinois Press, 1966.

[15] W. Poundstone. *The Recursive Universe: Cosmic Complexity and the Limits of Scientific Knowledge*. Mineola, Dover Publications, 2013.

[16] J. M. Smith and E. Szathmary. *The Major Transitions in Evolution*. Oxford, Oxford University Press, 1998.

[17] E. Szathmary. Sub-exponential growth and coexistence of non-enzymatically replicating templates. *Journal of Theoretical Biology*, **138**: 55, 1989.

[18] N. Takeuchi and K. Kaneko. The origin of the central dogma through conflicting multilevel selection. *Proceedings of the Royal Society B*, **286**(1912): 20191359, 2019.

10

Information and Biology

10.1 Introduction of Information Quantity

What is information in living organisms [19]? Schrödinger, in his monumental book *What is Life*, asserted that genetic information is given by a sequence of atoms (aperiodic crystals) in a very long chain of macromolecules. Subsequently, Watson and Crick proposed the double helix structure of DNA, and the molecular entity and mechanism of genetic phenomena in evolution. As a result, it has become clear that genetic information is coded as digital information in the DNA base sequence. Currently, the term "bioinformatics'" generally refers to a research field in which a huge amount of life science data, such as genomes and protein structures, is analyzed using information science methods. For example, the development of various databases and data analysis methods, identification of genes, determination of protein structures, homology searches, and creation of phylogenetic trees have become widespread as research tools. The importance of such developments is not disputed, but when "bioinformatics" was first proposed by Hogeweg et al. in the early 1970s, it broadly meant "the study of information processing in biological systems" [4]. Here, we would like to consider information in living organisms in the broadest sense of the term. Biological organisms sense the external world, change their own state, and take action, that is, there is a flow: (observation) $\rightarrow$ (state change) $\rightarrow$ (selection among possibilities). This is often thought to be associated with autonomy and spontaneous activity [14]. Although these properties are intrinsic to the organism, they are not quantitatively expressed as they are. In this chapter, we will first discuss the "information quantity" that quantifies the selection of such possibilities.

We use the word "information" in our daily life. Then, how can we quantify this everyday sensation? Shannon has succeeded in quantitatively characterizing such information. We get information when we can decide on one from different possibilities. If we open our palm to know whether the coin hidden in our hand is the front or the back, we know one of the two possibilities. The "degree of surprise" of knowing that one is chosen from such a bundle of possibilities is the information.

Let us define the "information quantity" in such a way that the function $I(M)$ expresses the information quantity when one event is chosen from among M equal probability alternatives. If the choice is made twice independently, the total information content is expressed as $I(M \times M)$, which should be equal to the sum of the two times of the information content $I(M)$ that can be obtained for each choice,

$$I(M \times M) = I(M) + I(M). \tag{10.1}$$

This shows the additivity of information. In addition, in the case of selecting one of the M alternatives and one of the N alternatives, we consider the following:

$$\textit{If } N > M, \textit{ then } I(N) > I(M). \tag{10.2}$$

This shows the monotonic increase of information. The function that satisfies the preceding two conditions is a logarithmic function. (More strictly speaking, we can show that a logarithmic function is uniquely determined by imposing continuity and differentiability on the function.) Let us then assume that

$$I(M) = a \log_e(M) + b. \tag{10.3}$$

Here we can see that $I(1) = 0$, and accordingly $b = 0$ because no information can be obtained when choosing only a unique one of the alternatives. As the minimum unit of information, information to select one of the two states (as either 1 or 0) is assumed to be 1 [bit], then $I(2) = 1$ and $a = 1/\log_e(2)$, so that

$$I(M) = \frac{\log_e(M)}{\log_e(2)} = \log_2(M) \text{ [bit]}, \tag{10.4}$$

where $P = 1/M$ since all M events occur with the same probability P. Therefore, when one of the events occurring with probability P is selected, the amount of information obtained is formulated as

$$I\left(\frac{1}{P}\right) = \log_2\left(\frac{1}{P}\right) = -\log_2(P) \text{ [bit]}. \tag{10.5}$$

For example, in the case of obtaining information on the face or the reverse with two coins, the amount of information per coin tossing is $-\log_2(1/2) = 1$ [bit], so that the total amount of information is 2 [bit], and in the case of obtaining information on two coins at once, the amount of information is $-\log_2(1/4) = 2$ [bit], and it can be confirmed that both are consistent.

In the preceding discussion, we have discussed equal-probability events, but more generally, let us consider the information quantity I one obtains by observing the event A that occurs with probability $P = k/M$. Then, to obtain the information

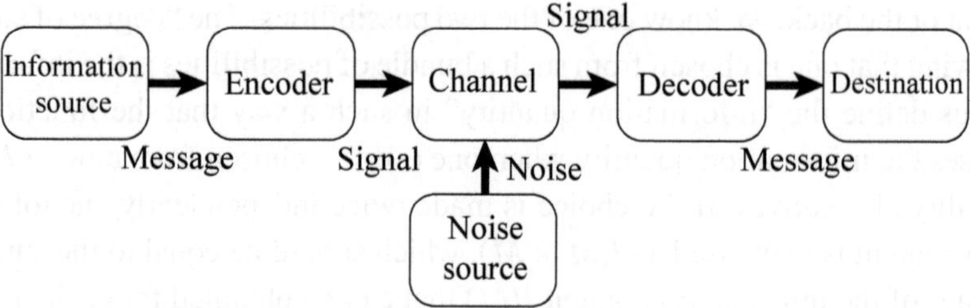

Figure 10.1 Shannon's communication channel model.

quantity I one notes the following relationship between the information quantity $\log_2(M)$, which is obtained by knowing which of the M events occurred, and the information quantity $\log_2(k)$, which is obtained by knowing which of the k events of event A occurred,

$$\log_2(M) = I + \log_2(k). \tag{10.6}$$

Therefore, the amount of information is

$$I = \log_2(M) - \log_2(k) = -\log_2\left(\frac{k}{M}\right) = -\log_2(P)\,[\text{bit}] \tag{10.7}$$

and the same formulation as (10.5) is obtained.[1] For example, if one card is taken from 52 cards and the card is an ace of diamonds, the information quantity I when knowing only that the card is a diamond is expressed as $I = \log_2(52) - \log_2(13) = -\log_2(13/52) = 2\,[\text{bit}]$ using the information quantity $\log_2(52)$ when knowing only that the card is an ace of diamonds and $\log_2(13)$ when knowing only that the card is an ace. Since the probability P_j is $0 \le P_j \le 1$ and $-\log_2(P_j) \ge 0$, the amount of information is always nonnegative, and the information that can be obtained by knowing that an event has occurred is greater as the probability of the event is lower. This is natural, since we are more likely to be surprised when we know that an infrequent event has occurred than when we know that it happens frequently.

Next, let us consider cases in which such stochastic events continue to occur. For example, Shannon considered the amount of information that can be obtained when a message consisting of several symbols is sent. In general, let us set up the following communication channel model. That is, the message (information source) of the sender is first encoded (encoding), transferred through a communication channel with noise, decoded into the original message (decoding), and then passed to the receiver (Fig. 10.1). Now, suppose that there are K kinds of events (or words) and the probability of their occurrence is given by P_j $(j = 1, 2, \ldots, K)$.

---

[1] Even if P is an irrational number, this expression can be obtained by taking a sufficiently large M to approximate it.

Considering the situation in which a sequence of symbols is encoded by assigning one of the symbols to each event and is transmitted through a communication channel, the amount of information obtained when the receiver obtains the symbol j is $-\log_2(P_j)$ if the noise of the communication channel can be ignored. Since the occurrence probability of the symbol itself is P_j, the average information obtained for each character received can be expressed as $E(-\log_2(P_j)) = -\sum_{j=1}^{K} P_j \log_2(P_j)$, considering the expected value. Therefore, if a sequence of symbols has a total length N, the average information is

$$H = -N \sum_{j=1}^{K} P_j \log_2(P_j). \tag{10.8}$$

This is the "Shannon entropy." This expression can also be derived as follows: if N is sufficiently larger than K, then the number C in the case of all possible different sequences of symbols will be

$$C = \frac{N!}{(NP_1)!(N - NP_1)!} \cdot \frac{(N - NP_1)!}{(NP_2)!(N - (NP_1 + NP_2))!}$$
$$\cdots \frac{(NP_{K-1} + NP_K)!}{(NP_{K-1})!(NP_K)!}$$
$$= \frac{N!}{(NP_1)!(NP_2)! \cdots (NP_K)!} \tag{10.9}$$

and substituting $P = 1/C$, then the amount of information is

$$H = \log_2(C) - -\log_2\left(\frac{1}{C}\right) = -\log_2\left(\frac{(NP_1)!(NP_2)! \cdots (NP_K)!}{N!}\right). \tag{10.10}$$

By applying Stirling's formula $\log N! \sim N \log N - N$, we can derive Eq. (10.8). Note that since $-P_j \log_2(P_j) \geq 0$, Shannon entropy is nonnegative ($H \geq 0$). Using this Shannon entropy, we can go back to the definition of information quantity, and say that information is quantified by "the degree of decrease in the uncertainty of the situation due to the occurrence of the event." In other words, when the Shannon entropy of a situation changes from prior H to posterior H' ($H \geq H'$)) by obtaining the information, the "information quantity" in this case is

$$I = H - H'. \tag{10.11}$$

As mentioned previously, if the situation is fixed uniquely by obtaining information and the degree of uncertainty is $H' = 0$, then $I = H$. In this way, Shannon quantified the "information" as a function of probability, using only statistical properties such as the frequency of occurrence of symbols, while entirely leaving out the semantic content of "information." For example, the amount of information in a Chinese character is greater than that in an English character, but this is not because of the difference between phonetic and ideographic characters or the complexity of

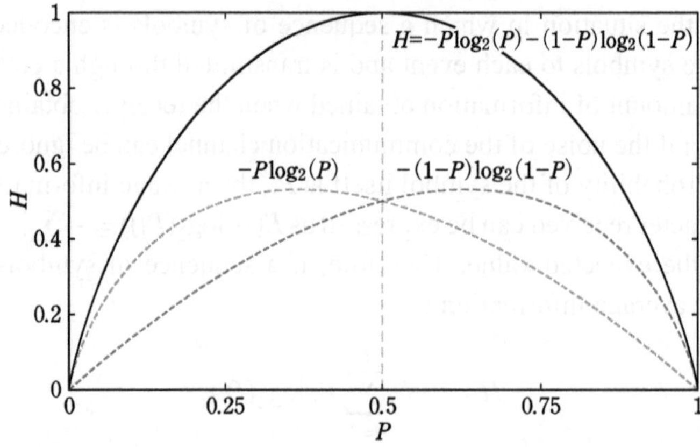

Figure 10.2 Shannon entropy for the case where two types of symbols are used and one letter is received.

the characters, but because there are thousands of Chinese characters while there are only 26 in the alphabet. This allows us to study the information as a digitally encodable, communicatable, and computer-processable object.

10.2 Characteristics of Shannon's Information Theory

How can we maximize the Shannon entropy per character if we can change the occurrence probability of the symbol here? We can do this by maximizing $-\sum_{j=1}^{K} P_j \log_2(P_j)$ under the condition that $\sum_{j=1}^{K} P_j = 1$. This can be solved by the Lagrange multiplier method,

$$\frac{\partial}{\partial P_i}\left(-\sum_{j=1}^{K} P_j \log_2(P_j) + \lambda \sum_{j=1}^{K} P_j\right) = 0, \tag{10.12}$$

where i is any natural number between 1 and K. As a result, for all i, $-1 - \log_2(P_i) + \lambda = 0$ holds, and hence P_i is constant. Noting that $\sum_{j=1}^{K} P_j = 1$, it means that the maximum Shannon entropy is obtained when $P_i = 1/K$, that is, any symbol appears with equal probability. This is natural, since the greater the prior uncertainty between symbols, the greater the amount of information. For example, if there are two types of symbols and one letter is received, and the occurrence probability of one symbol is P and that of the other symbol is $1 - P$, then the Shannon entropy is

$$H = -P \log_2(P) - (1 - P) \log_2(1 - P). \tag{10.13}$$

It can be confirmed to be the maximum when $P = 1/2$ (Fig. 10.2).

Next, let us consider the minimum length of the signal required for communicating an arbitrary sequence of symbols. In general, there is a difference in the appearance probability of symbols. Then, assigning short (long) code for a symbol that appears with a high (low) probability, respectively will shorten the average signal length of the entire symbol sequence. In order to confirm this quantitatively, let us define Kullback–Leibler divergence (also called relative entropy; see also Section 10.3), a measure of the difference between two probability distributions, by

$$D[P||Q] = \sum_{j=1}^{K} P_j \log_2 \left(\frac{P_j}{Q_j}\right) = \sum_{j=1}^{K} P_j \log_2 (P_j) - \sum_{j=1}^{K} P_j \log_2 (Q_j). \qquad (10.14)$$

Then, using the inequality $-\log_e(x) \geq 1 - x$, we get

$$D[P||Q] = -\sum_{j=1}^{K} P_j \log_2 \left(\frac{Q_j}{P_j}\right)$$

$$\geq \sum_{j=1}^{K} P_j \left(1 - \frac{Q_j}{P_j}\right) = \sum_{j=1}^{K} (P_j - Q_j) = 0, \qquad (10.15)$$

so that it is always nonnegative (Shannon's lemma). This shows that an event with a probability of P_j that is transferred by assigning a signal of arbitrary length $-\log_2(Q_j)$ is always more redundant than that when a signal of length $-\log_2(P_j)$ is assigned. In other words, the Kullback–Leibler divergene represents the number of bits that are expected to be additionally needed when a code corresponding to a probability distribution Q is used instead of a true distribution P. It shows that when an event j occurs with probability P_j, a code of length $-\log_2(P_j)$ can be assigned to minimize the total code length to be communicated. In fact, in the case of noise-free communication, it can be proven that the average code length of the information source is as close as possible to the Shannon entropy (shortest code, compact code). For more details, please refer to information theory textbooks (e.g., [22, 25]) to see "Shannon's first theorem" (Noiseless coding theorem) and "data compression."

In the preceding, we have dealt with the coding of information sources (memory-less sources) where there is no correlation between a symbol and the next symbol and everything occurs with a single character occurrence probability P_j. On the other hand, there may be a correlation between one symbol and the next symbol in many cases. Let us consider our natural language (as an example of a Markov source) in which the occurrence probability of the next symbol depends on the first symbol. Then, the probability of occurrence of a character is not equal, and we can immediately see that there is also a correlation between neighboring characters in which the symbol next to a given symbol is unlikely to appear (or easy to appear). For example, if we calculate the Shannon entropy of a 27-character English sentence (26 letters of the alphabet and a space symbol) from the occurrence probability P_j, it is reported as $H = 4.03$ [bits] [17], which is small compared to the maximum Shannon entropy $H_0 = 4.75$ [bits] in the case of $P_j = 1/27$ and equal probability. In addition, the Shannon entropy of each of two characters as a unit of two characters is reported as $H = 3.32$ [bits] [17], which is even smaller.[2] If we introduce the redundancy of the information source, r, which is defined as

$$r = \frac{H_0 - H}{H_0}, \tag{10.16}$$

$r = (4.75 - 4.03)/4.75 = 0.152$ for each letter of the alphabet. Hence, a sentence is about 15% redundant in terms of information content. This is because there is a correlation between two letters, such as the tendency for u to appear after q. There is also the effect of the bias in the frequency of word occurrence (see Zipf's law, Section 9.8), due to the presence of frequent words such as "the" and "of," as well as the effect of repetition of grammatical structures and expressions in sentences. There could be several reasons why Shannon entropy is not maximum, but one reason may be that redundancy allows us to recover (error-correct) a sequence of symbols even if we make a mistake or miss a signal. If the shortest coding, where the average code length of the information source is infinitely close to Shannon entropy, can be achieved, then the code has no redundancy and is vulnerable to noise contamination. Since noise is always present in actual communication, it is necessary to add an appropriate amount of redundancy to the code in order to ensure robustness to noise so that the amount of information received will not be smaller than the amount of information sent by it. In fact, it can be proved that a suitable coding scheme can be used to reduce the error probability as much as possible with redundancy within a certain range. For more details, please refer to "Shannon's second theorem" (Channel coding theorem) and "error-correcting code" in information theory textbooks (e.g., [22, 25]).

10.3 Mutual Information

Let us consider not only the sender of the signal but also the receiver's prior knowledge in communication. This is because the amount of information that can be obtained as a result of a communication also depends on what the receiver knows beforehand. For example, if we know that the result of the roll of a dice is even numbers, the amount of information we are informed that it is 6 is not one of the six choices but one of the three choices, so that the amount of information we can get is small. This is a simple example, but the amount of information obtained in the presence of prior knowledge is always smaller than the amount of information obtained without prior knowledge. In general, if the receiver has a prior probability

---

[2] The frequency analysis of co-occurrence relations where N symbols appear next to each other in a symbol sequence is called "N-gram analysis." This was also introduced by Shannon.

distribution $P(x)$, we can define the amount of information obtained by considering a conditional probability distribution (posterior distribution) $P(y|x)$ from it. Then, "conditional entropy" is defined as

$$H(y|x) = -\sum_{x,y}^{K} P(x)P(y|x)\log_2(P(y|x)) = -\sum_{x,y}^{K} P(x,y)\log_2(P(y|x)), \quad (10.17)$$

where $P(x,y)$ is a simultaneous probability distribution, and Bayes' theorem,

$$P(y|x) = \frac{P(x,y)}{P(x)}, \quad (10.18)$$

is used to transform the equation. In the preceding example of the dice, the probability that the number is even is $P(x)=\frac{1}{2}$, the probability that the number is 6 is $P(y)=\frac{1}{6}$, and the probability that the number is 6 when the number is even is $P(y|x)=\frac{1}{3}$, so the Shannon entropy and the conditional entropy are $H(y)=\log_2(6)$ [bit] and $H(y|x)=\log_2(3)$ [bit] respectively, and we can confirm the decrease in the amount of information. Therefore, let us introduce "mutual information" to quantify the "degree of decrease in the amount of information based on prior knowledge." The mutual information is defined as

$$I(x,y) = \sum_{x,y}^{K} P(x,y)\log_2\left(\frac{P(x,y)}{P(x)P(y)}\right) = D[P(x,y)||P(x)P(y)] \quad (10.19)$$

by using the Kullback–Leibler divergence that we introduced earlier. This shows the difference between the probability distributions $P(x,y)$ and $P(x)P(y)$. If x and y are completely independent, $I(x,y)=0$ from $P(x,y)=P(x)P(y)$, which is the same with or without prior knowledge. If y is completely dependent on x, $I(x,y)=H(x)$ with $H(x)$ as the Shannon entropy from $P(x)=P(y)$ and $P(x,y)=P(x)$, which is all determined by prior knowledge. In general, $I(x,y)$ can be transformed as

$$I(x,y) = \sum_{x,y}^{K} P(x,y)\log_2\left(\frac{P(x,y)}{P(x)P(y)}\right)$$

$$= \sum_{x,y}^{K} P(x,y)\log_2\left(\frac{P(y|x)}{P(y)}\right)$$

$$= \sum_{x,y}^{K} P(x,y)(\log_2(P(y|x)) - \log_2(P(y)))$$

$$= -\sum_{x,y}^{K} P(x,y)\log_2(P(y)) + \sum_{x,y}^{K} P(x,y)\log_2(P(y|x))$$

$$= H(y) - H(y|x). \quad (10.20)$$

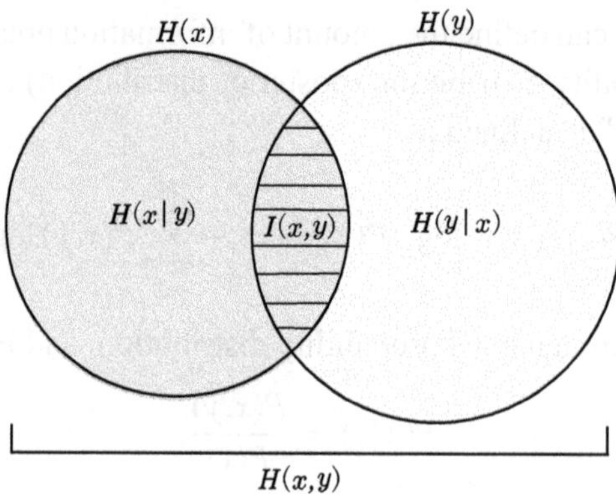

Figure 10.3 Venn diagram representation of mutual information.

Similarly it is represented as

$$I(x,y) = H(x) - H(x|y) = H(x) + H(y) - H(x,y), \qquad (10.21)$$

where $H(y|x)$ is the conditional entropy and

$$H(x,y) = -\sum_{x,y}^{K} P(x,y) \log_2(P(x,y)) \qquad (10.22)$$

is the "joint entropy."

From this formula, it can be seen that the mutual information formulates the "degree of reduction of information based on prior knowledge" (see Fig. 10.3). In addition, since $I(x,y) = D[P(x,y)||P(x)P(y)] \geq 0$ from the nonnegativity of the Kullback–Leibler divergence, the mutual information is always nonnegative. That is, the information gain is reduced or is zero with prior knowledge. In addition, the mutual information can be considered as an indicator to capture the correlation between symbols. Now, $I(x,y) \geq 0$ results in $H(x) + H(y) \geq H(x,y)$. Hence, if there is some kind of correlation between symbol x and symbol y, it is possible to shorten the coding length by encoding x and y together rather than encoding them individually. This is also useful for the quantification of intersymbolic correlations and redundancy in natural languages and gene sequences as described earlier, and is also used for the quantification of correlations between inputs and outputs in communication.

In general, because noise is inevitable in communication channels, it is questioned how accurately the input signal is transmitted to the output signal, and it is possible to quantify this using mutual information. The Shannon entropy of passing a signal through the communication channel is measured in bits per second, and the mutual information between the input signal and the output signal corresponds

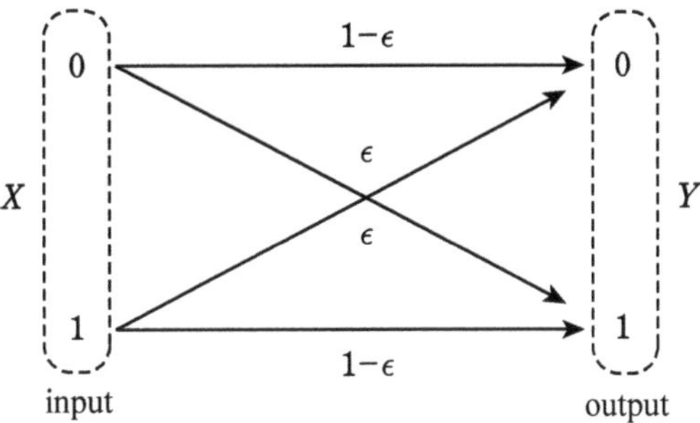

Figure 10.4 The binary symmetric communication channel.

to the average amount of information passing through the communication channel in a unit time. Therefore, as an example, we formulate the mutual information in a binary symmetric communication channel, taking into account the influence of noise in information transmission (Fig. 10.4). $P_X(i)$ and $P_Y(i)$ ($i = 0, 1$) are the probabilities that the input and output take an i value, respectively, and $P_{Y|X}(i|j)$ ($i, j = 0, 1$) is the probability that the output takes an i value when the input is a j value and is given as

$$P_Y(0) = \sum_{i=0,1} P_{Y|X}(0|i)P_X(i)$$

$$= P_{Y|X}(0|0)P_X(0) + P_{Y|X}(0|1)P_X(1)$$

$$= (1 - \epsilon)p + \epsilon(1 - p) \equiv q, \tag{10.23}$$

$$P_Y(1) = 1 - P_Y(0) \equiv 1 - q \tag{10.24}$$

and the Shannon entropy of output and the conditional entropy between the input and output are

$$H(Y) = -\sum_{i=0,1} P_Y(i) \log_2(P_Y(i))$$

$$= -q \log_2(q) - (1 - q) \log_2(1 - q), \tag{10.25}$$

$$H(Y|X) = -\sum_{i=0,1} \sum_{j=0,1} P_Y(i)P(Y(i)|X(j)) \log_2(P(Y(i)|X(j)))$$

$$= -\sum_{j=0,1} (P_Y(0)P(0|X(j)) \log_2(P(0|X(j)))$$

$$\qquad + P_Y(1)P(1|X(j)) \log_2(P(1|X(j))))$$

$$= -q((1 - \epsilon) \log_2(1 - \epsilon) + \epsilon \log_2(\epsilon))$$

$$\qquad - (1 - q)(\epsilon \log_2(\epsilon) + (1 - \epsilon) \log_2(1 - \epsilon))$$

$$= -\epsilon \log_2(\epsilon) - (1 - \epsilon) \log_2(1 - \epsilon), \tag{10.26}$$

respectively. Then, the mutual information between the input and output is

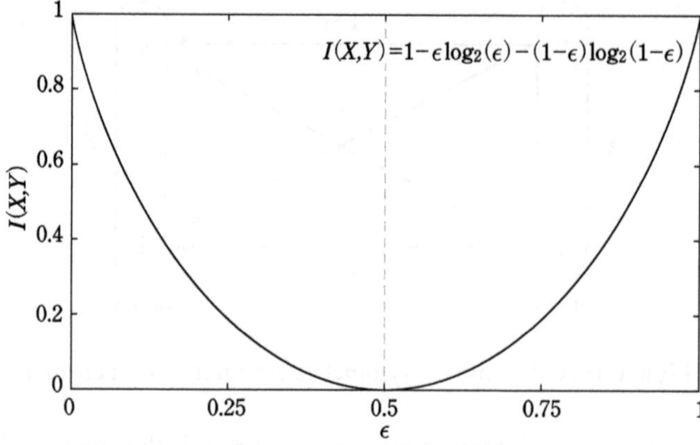

Figure 10.5 Mutual information in the binary symmetric communication channel.

$$I(X, Y) = -q \log_2(q) - (1-q) \log_2(1-q) + \epsilon \log_2(\epsilon) + (1-\epsilon) \log_2(1-\epsilon). \quad (10.27)$$

Considering the case of $p = 1/2$ and $q = 1 - q = 1/2$,

$$I(X, Y) = 1 + \epsilon \log_2(\epsilon) + (1 - \epsilon) \log_2(1 - \epsilon), \quad (10.28)$$

with $I(X, Y) = 1$ for both cases of $\epsilon = 0$ (all correct answers) and 1 (all errors) (Fig. 10.5). When $\epsilon = 1/2$, the lowest value $I(X, Y) = 0$ is taken, and this is because no input information can be obtained from the random sequences. The maximum amount of mutual information in this communication is the maximum amount of information that the communication channel can transmit, and it is called the "channel capacity" under the noise strength ϵ.

Now, we discuss possible connection between the (stochastic) dynamical systems in the previous chapters and the information quantity in this chapter, as information flow in dynamical systems. In noisy dynamical systems, the predictability of their behavior can be quantified by using information theory. When Shannon's communication channel model is applied to a stochastic dynamical system, probability distributions that represent the past and future states of the dynamical system are assigned for transmission and receival of information, respectively. The communication channel can be mapped to the dynamics (map) of the dynamical system. Then, the temporal development of the system forms an iterative map, and the dynamical system can be regarded as a "device that transmits information in time" [18]. If there is a stationary distribution in the system, one can quantify the relationship between the current state of the system and the stationary distribution, as well as the amount of information conveyed by the system, by using mutual information.

In particular, as seen in Section 1.16, a chaotic dynamical system can be expressed as a symbolic dynamical system by expressing the sequence of symbols as time series. The diversity of the symbol sequence is quantified by Shannon information, and chaos is quantitatively characterized by Shannon information. When elements exhibiting chaotic dynamics interact with each other, it is possible to obtain the information flow from one element to another by calculating the mutual information of time series $x(t)$ and $y(t + \tau)$ with time differences between $x(t)$ and $y(t)$. In particular, when the time decay of the mutual information is slow, the information flow is transmitted while retaining the information. From this point of view, the significance of chaos in cell-to-cell communication has also been discussed [7].

10.4 Information in DNA Sequence

In the Sections 10.1 and 10.2, we have given a brief introduction to Shannon's information theory framework. Let us consider the amount of information contained in the A, T, G, and C base sequences of DNA. For example, the human genome has a total length of 3.2 billion base pairs (3.2 Gbp, 1 G (*Giga*) $= 10^9$). If the occurrence probability of the four types of bases is equal, and the base pair complementarity is perfect so that one sequence of the two strands of DNA is defined, then the other sequence is uniquely defined, from a possible $4^{3.2G} = 2^{6.4G}$ sequences. This indicates that the human genome can encode up to 6.4 G [bits] $= 0.8$ [GB] (1 [B] $= 1$ [byte] $= 8$ [bits]) of information. How does a reader feel to know that there is less than 1 [GB] of information in the genome to make a human being viable? Moreover, the information coded could be even smaller because each base does not occur with the same probability, but is redundant. Let us now examine the factors that create the redundancy in the sequence.

Difference in the Frequency of AT and GC Occurrence: In terms of the complementary strands of DNA, the frequencies of A and T, and those of G and C are the same, respectively (Chargaff's rule), but the frequencies of AT and GC are not the same. The GC content of the genome varies from organism to organism; for example, the ratio of AT content to GC content in the human genome is about 6 : 4, and it is known that the difference between species is particularly large in prokaryotes. One of the major reasons for this bias in frequency is the existence of mutational pressures (GC and AT pressures) in genome evolution. In addition, spatial heterogeneity in the GC content in each genome has also been found, with the tendency for more genes to be present in regions with high GC content. Also, since GC is thermodynamically more stable than AT, the higher the GC content, the higher the melting point of DNA tends to be.

Genetic Code: The genetic code that maps from the bases in the DNA sequences to amino acids is a three-digit (triplet). Then, there is a correlation between successive bases in the DNA sequence (in the coding region that is transcribed into the mRNA). The genetic code is degenerate code as $4^3 = 64$ codons specify 20 amino acids and translation initiation and termination (i.e., $64 \rightarrow 22$). There are multiple amino acids that correspond to as many as 6 synonymous codons, such as leucine. Furthermore, the frequency tendency of which synonymous codon is used differs greatly between species and genes within a homologous organism (codon bias). This is widely known experimentally because it has a significant effect on genetic engineering when a recombinant protein is expressed in a species different from the original one. For example, the base of the third letter of the codon tends to fluctuate (wobble rule). Thus, when the pressure to enhance GC frequency is present in evolution, the third letter tends to be G or C, which may contribute to the codon bias. In addition, the higher the gene expressions are, the more pronounced the codon bias becomes, and the one with a large number of corresponding tRNAs tends to be used among the synonymous codons. The presence of codon bias can lead to increased translation efficiency, robustness to errors such as point mutations, and changes in the mechanical properties of DNA molecules such as chromatin structure. There is also a bias in the frequency of amino acids in the cell, with a tendency to increase the frequency of amino acids such as glycine and alanine, which are small in size and allow for rotation of the side chains and soften proteins.

Long-Range Correlation: There is a nonnegligible correlation in the long-distance bases of DNA sequences. Let the site of each base be i, and denote A and T as 1, and G and C as 0; then the base sequence can be represented by the binary sequence $s_i = 0, 1$. Then, the correlation with sites m away from each site can be expressed as $C(m) = \langle s_i s_{i+m} \rangle - \langle s \rangle^2$, where $\langle \cdots \rangle_i$ is the average over all i. Let us examine how this $C(m)$ decays with distance m. If there is no correlation with any of the following characters, then $C(m) = 0$ for $m \geq 1$. If the influence from the distant sites decays successively by random perturbations over sites, the standard form of the correlation is expressed as $C(m) \sim \exp(-m/\ell)$, where ℓ is the correlation length. Analysis of the data shows that this correlation decays more slowly than the exponential function for m, and is given approximately as $C(m) \sim m^{-\gamma}$ (with $0 < \gamma < 1$) [9, 15]. In other words, $C(m)$ does not have a finite correlation length and there exists a long-range correlation. In this sense, when we consider DNA sequences as the transfer of genetic information, they are far from being optimal for the amount of communication.

However, when explored in more detail, there are some features that suggest the approach to optimal information transmission. Comparative analysis of the DNA sequences of eukaryotes and prokaryotes and the DNA sequences of introns and exons (see Column X) in eukaryotes shows that the degree of long-range correlation is higher in eukaryotes, and is higher in introns than in exons [1, 2]. This means that protein coding regions have lower correlation than noncoding regions, so it is possible that the former is approaching optimal coding to allow for large amounts of information through evolution, while the latter is not. Given the central dogma, that is, the flow of genetic information from DNA $\to$ (transcription) $\to$ RNA $\to$ (translation) $\to$ protein, it is inevitable that mutations due to replication errors are introduced into the transcription and translation process, albeit at a low probability. Therefore, a certain amount of redundancy is required to retain information, and long-range correlation of DNA sequences may be useful for this purpose. On the other hand, it has been pointed out that the redundancy may have occurred as a byproduct of the evolution of gene duplication[3] and mutation [8].

The preceding are some of the factors that create redundancies in nucleotide sequences. In general, the information encoded in a DNA base sequence includes not only the amino acid composition of the genes but also various physical information such as the regulatory structures of the genes, and even the evolutionary history of the genes. This information cannot be assembled by analyzing the amount of information as a simple symbolic sequence, and it is necessary to have a complex and dynamic view of the genome in order to read and interpret it.

Column X: Intergenic Sequence

The more "complex" an organism is, the larger its genome size tends to be (although there are many exceptions): eukaryotes rather than prokaryotes, multicellular rather than unicellular organisms, and vertebrates rather than invertebrates. However, the number of genes does not increase in proportion to the genome size, and the density of genes on the genome decreases as the genome size increases. Thus, the proportion of intergenic sequences that do not encode genes increases.

In addition, the so-called introns are also present within gene sequences in eukaryotes, especially in multicellular organisms: After the gene sequence is transcribed into mRNA, it is reorganized into mRNA in which only a part related to protein information is left and the part in between is removed (splicing). The former is called exon and the latter is called intron. After this splicing, the mRNA is translated into protein. The introduction of introns is associated with the existence of a mechanism in which there is diversity in the way introns are taken up within a single gene, and their "editing" results in the production of multiple proteins from a single gene (selective splicing). This allows us to create a diversity of proteins without

---

[3] Gene duplication is a process where a segment of DNA is replicated, resulting in an additional copy of a particular gene or set of genes within an organism's genome. This duplication can occur through various mechanisms and is a significant driver of evolutionary change, as it can lead to the emergence of new genes with modified or novel functions.

increasing the number of genes. For example, the total number of genes in humans is about 20,000, but the number of protein types is estimated to be 50,000 to 100,000.

Furthermore, intergenic sequences have traditionally been called "junk DNA," which is thought to have no function [13]; but recent progress in the genome project has revealed that there are a large number of sequence regions that are transcribed into RNA and perform various functions even though they are not translated into proteins (noncoding RNA (ncRNA)). For example, in the human genome, gene coding regions occupy about 25% of the genome, most of which are intron, and the exon region encoding proteins occupies only about 2% of the genome,[4] while about 70% of the genome regions are transcribed as RNAs, the majority of which are ncRNAs [3]. ncRNAs are also involved in the regulation of gene expression, so the differences in species and phenotypes may be due to differences in the regulation of gene expression involving ncRNAs rather than to the presence or absence of gene sequences encoding proteins.

There are two types of intergenic sequences: single sequences and repetitive sequences. The former type includes pseudogenes, while the latter type includes tandem sequences in a single row and interspersed sequences scattered throughout the genome. In the mammalian genome, intergenic sequences occupy a very large area, and scattered forms are particularly common [6]. Thus, the genome is more intrinsically complex and dynamic than just a static collection of genes, and the genome project calls for a greater understanding of the genome as a whole, which also concerns chromatin dynamics.

10.5 Kinetic Proofreading

The function of a cell consists of biochemical reaction networks with molecular recognition based on a small free energy difference, which is not much different from thermal fluctuations, as described in Chapter 5 and the previous section. However, as is typical for DNA replication and mRNA translation, these reactions are required to function with high accuracy. As an example, let us recall the DNA replication reaction discussed in Section 9.4. Assuming the mutation rate in the reaction to be μ and the genome size to be L, and leaving out the repair mechanism for the sake of simplicity, the probability of accurate replication is $(1 - \mu)^L \approx \exp(-L\mu)$ when μ is small. Therefore, when the mutation rate μ is greater than $1/L$, the functional gene sequences are lost, a phenomenon discussed as "error catastrophe" in

---

[4] This is confirmed by the following estimates. Since proteins consist of an average of about 400 amino acids, the average length of one gene is about 1200 bp. The total number of genes is about 21,000, so its average length is estimated to be about 3×10^7 bp. It indicates that the exon region is 1/100th of the total length of the genome.

Section 9.4.[5] For example, a low mutation rate of 10^{-9} has been achieved in yeast DNA replication reactions with base pairs of genome size L in the order of 10^7.

Then, how can such a high accuracy of molecular recognition be achieved from a combination of wobbly and unreliable elements? Hopfield and Ninio provided an answer to this question [5, 12], by proposing a mechanism called "kinetic proofreading." As a preliminary step, let us first estimate the error rate by thermodynamic considerations from the model of a simple molecular recognition mechanism of the Michaelis–Menten type.

$$N + C \leftrightarrow NC \rightarrow \text{correct product,}$$
$$N + E \leftrightarrow NE \rightarrow \text{wrong product.} \tag{10.29}$$

Let k_1^C and k_1^E be the binding rate constants of substrates C and E to N, respectively; k_{-1}^C and k_{-1}^E be the dissociation rate constants of substrates from NC and NE, respectively; and k_2^C and k_2^E be the final product formation rate constants of NC and NE, respectively. In the DNA replication reaction, the base of the replication site of the template DNA is N, the base complementary to it is C, and the base not complementary to it is E. In the translation reaction, the codon of the ribosomal binding site of mRNA is N, the tRNA with the corresponding anticodon is C, and the tRNA without the corresponding anticodon is E. These can be expressed as differential equations, which are denoted by

$$\frac{d[NC]}{dt} = k_1^C[N][C] - (k_{-1}^C + k_2^C)[NC],$$
$$\frac{d[NE]}{dt} = k_1^E[N][E] - (k_{-1}^E + k_2^E)[NE]. \tag{10.30}$$

In the steady state, $\frac{d[NC]}{dt} = 0$, $\frac{d[NE]}{dt} = 0$, and

$$[NC] = \frac{k_1^C[N][C]}{k_{-1}^C + k_2^C}, \quad [NE] = \frac{k_1^E[N][E]}{k_{-1}^E + k_2^E}. \tag{10.31}$$

Here, since both of the coupling rate constants are considered to be diffusion-limited (see Section 5.10), we can assume that $k_1^C = k_1^E = k_1$. Furthermore, assuming that $k_2^C = k_2^E = k_2$ and $[C] = [E]$, the error rate is expressed as

$$P = \frac{k_2^E[NE]}{k_2^C[NC]} = \frac{\frac{k_2 k_1[N][E]}{k_{-1}^E + k_2}}{\frac{k_2 k_1[N][C]}{k_{-1}^C + k_2}} = \frac{k_{-1}^C + k_2}{k_{-1}^E + k_2}. \tag{10.32}$$

From this, when $k_{-1}^C, k_{-1}^E \ll k_2$, then $P \sim 1$ is obtained, and neither C nor E can be discriminated. On the other hand, when $k_{-1}^E > k_{-1}^C \gg k_2$, the discriminability is improved and the error rate is reduced to

---

[5] The irreversible accumulation of these chance-based deleterious mutations in asexual reproductive populations is referred to as "Muller's ratchet." It is thought that one of the advantages of sexual reproduction is to stop this ratchet.

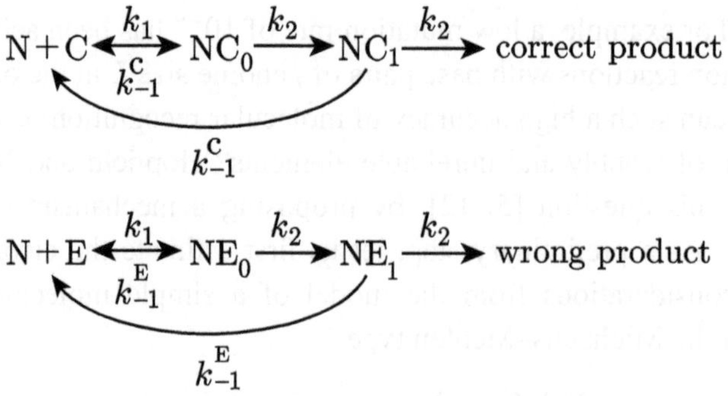

$$N + C \xrightleftharpoons[k_{-1}^{C}]{k_1} NC_0 \xrightarrow{k_2} NC_1 \xrightarrow{k_2} \text{correct product}$$
$$k_{-1}^{C}$$

$$N + E \xrightleftharpoons[k_{-1}^{E}]{k_1} NE_0 \xrightarrow{k_2} NE_1 \xrightarrow{k_2} \text{wrong product}$$
$$k_{-1}^{E}$$

Figure 10.6 Kinetic proofreading model.

$$P_0 \simeq \frac{k_{-1}^{C}}{k_{-1}^{E}} = \frac{\overline{t_E}}{\overline{t_C}} = \exp\left(-\frac{\Delta G}{k_{\mathrm{B}}T}\right) \tag{10.33}$$

at best. Note that $\overline{t_C} = \frac{1}{k_{-1}^{C}}, \overline{t_E} = \frac{1}{k_{-1}^{E}}$ represents the average coupling time of C and E to N. In other words, the difference in binding times is used to discriminate the correct or wrong molecular recognition. This result, however, clearly shows that a simple thermodynamic mechanism cannot achieve accurate molecular recognition. This is because, to achieve an error rate as low as 10^{-9}, as in the DNA replication reaction described earlier, a free energy difference of about $\Delta G = 20\,k_{\mathrm{B}}T$ is required, while a hydrogen bond mutation involved in base recognition can only produce $\Delta G = 2\,k_{\mathrm{B}}T$, resulting in an error rate of $\exp(-2) \sim 0.1$ (10%).

Now, let us add a new intermediate step ("proofreading step") to the simple molecular recognition model (Fig. 10.6). Here, for simplicity, NC_0 and NC_1 are assumed to have the same dissociation rate as the substrate and the production rate of the product. The binding rate is assumed to be the same due to the diffusion-limited reaction. The same assumption was made for NE_0 and NE_1, and if $k_{-1}^{E} > k_{-1}^{C} \gg k_2$ as described earlier, the molecular recognition steps are doubled. The error rate of molecular recognition in each step is $\frac{[NE_i]}{[NC_i]} \simeq P_0$ ($i = 0, 1$), so that the overall error rate can be lowered to $P_0^2 = \exp\left(-\frac{\Delta 2G}{k_{\mathrm{B}}T}\right)$. This means that although NC_1 is reached through NC_0, NC_0 is already selected with a higher probability than NE_0, and then NC_1 and NE_1 are also selected in the same way, resulting in double selection as a whole, and discrimination ability is increased. By expressing these results in a differential equation,

$$\frac{\mathrm{d}[NC_0]}{\mathrm{d}t} = k_1[N][C] - (k_{-1}^{C} + k_2)[NC_0],$$
$$\frac{\mathrm{d}[NC_1]}{\mathrm{d}t} = k_2[NC_0] - (k_{-1}^{C} + k_2)[NC_1] \tag{10.34}$$

hold for C, and in the steady state, $\frac{d[NC_0]}{dt} = 0$, $\frac{d[NC_1]}{dt} = 0$. Thus

$$[NC_0] = \frac{k_1[N][C]}{k^C_{-1} + k_2}, \quad [NC_1] = \frac{k_2[NC_0]}{k^C_{-1} + k_2} = \frac{k_1 k_2[N][C]}{(k^C_{-1} + k_2)^2} \tag{10.35}$$

are derived. The same procedure can be applied for E, leading to

$$[NE_0] = \frac{k_1[N][E]}{k^E_{-1} + k_2}, \quad [NE_1] = \frac{k_2[NE_0]}{k^E_{-1} + k_2} = \frac{k_1 k_2[N][E]}{(k^E_{-1} + k_2)^2}. \tag{10.36}$$

If $[C] = [E]$, then the error rate is expressed as

$$P_1 = \frac{k_2[NE]}{k_2[NC]} = \frac{\frac{k_1 k_2[N][E]}{(k^E_{-1}+k_2)^2}}{\frac{k_1 k_2[N][C]}{(k^C_{-1}+k_2)^2}} = \left(\frac{k^C_{-1} + k_2}{k^E_{-1} + k_2}\right)^2 = P_0^2, \tag{10.37}$$

and if $k^E_{-1} > k^C_{-1} \gg k_2$,

$$P_0^2 \simeq \left(\frac{k^C_{-1}}{k^E_{-1}}\right)^2 = \exp\left(-\frac{2\Delta G}{k_B T}\right) \tag{10.38}$$

is confirmed. (In principle, further enhancement of discrimination ability is possible by increasing the number of intermediate states.) From the kinetic interpretation of this result, it is important to note that this mechanism causes a time delay in the synthesis of the final product. Since the production rate of final product is time-dependent that rises from time $t = 0$, the molecular recognition of the left side of the model is generated after the accumulation of the product. Hence with the time delay, the sorting ability is enhanced. Since this delayed process cannot be produced in thermal equilibrium, it is necessary to make the system nonequilibrium for this purpose (see Maxwell's demon in Section 10.7). This is how the kinetic proofreading works. In fact, in the translation reaction from mRNA, a two-step selection mechanism to identify the correct codon exists when the ribosome binds to the tRNA-EF-Tu-GTP complex. It provides an actual example of kinetic proofreading (the second step involves the hydrolysis of GTP). Also, this kinetic proofreading explains the significance of receptor phosphorylations and multistep reactions in the signal transduction system, whereas it will also provide a mechanism for the antibody recognition in the immune system [11].

10.6 Entropy of Statistical Mechanics and Information

Various kinds of entropy are known, and the relationship between them has been discussed. In this section, we will discuss the relationship between Boltzmann entropy in statistical mechanics and Shannon entropy in information theory. Boltzmann entropy (S_B) can be formulated by expressing the maximization of the number of states on the isoenergetic plane of the system by Shannon entropy (H). It is also consistent with thermodynamic entropy (S). We will confirm this in the following discussion.

Suppose there is an isolated system consisting of many N_t particles in thermal equilibrium at temperature T [K], where the total energy E_t [J] is constant while there is an exchange of energy between the particles (microcanonical ensemble). There are W ways to distribute E_t among N_t particles, which involves a large number expressed as a factorial of E_t or N_t, where every microstate is realized with equal probability (the principle of equipartition). If we focus on N_1 elements ($N_1 \ll N_t$) in this ensemble, the energy of the subsystem fluctuates around the average energy $\langle E \rangle = E_t$ and takes various values (canonical ensemble). Therefore, if the total number of states taken by the subsystem is W_1 and the number of energy states is N ($W_1 \geq N$, including the degenerate energy), then the probability of existence of the microstate P_j in the energy state E_j is $\sum_{j=1}^{W_1} P_j = 1$, and $\sum_{j=1}^{W_1} E_j P_j = \langle E \rangle$, and the average information of each state can be defined as $H = -\sum_{j=1}^{W_1} P_j \log_2(P_j)$ in terms of Shannon entropy. The most easily realized state corresponds to the state in which the number of possible microstates W_1 is maximized, that is, the state in which Shannon entropy H is maximized. Therefore, the probability distribution that maximizes Shannon entropy under the conditions of probability conservation and energy conservation can be expressed as

$$\frac{\partial}{\partial P_i}\left(-\sum_{j=1}^{W_1} P_j \log_2(P_j) - \lambda_1 \sum_{j=1}^{W_1} E_j P_j - \lambda_2 \sum_{j=1}^{W_1} P_j\right) = 0 \tag{10.39}$$

by using the Lagrange multiplier method. By solving this equation, $-\log_2(P_i) - 1 - \lambda_1 E_i - \lambda_2 = 0$, that is, $\log_2(P_i) = -\lambda_1(E_i + \frac{\lambda_2+1}{\lambda_1})$ is derived. If we convert the constants as $\beta = \lambda_1 \log_e(2)$, $\gamma = -\frac{\lambda_2+1}{\lambda_1}$, P_i is expressed as

$$P_i = \exp(-\beta(E_i - \gamma)) = \frac{\exp(-\beta E_i)}{\sum_{j=1}^{W_1} \exp(-\beta E_j)}. \tag{10.40}$$

Here, we used $\exp(-\beta\gamma) = \sum_{j=1}^{W_1} \exp(-\beta E_j)$, which is derived from the probability normalization condition, $\sum_{j=1}^{W_1} P_j = \sum_{j=1}^{W_1} \exp(-\beta(E_j - \gamma)) = 1$. If $\beta = 1/(k_B T)$ holds, the shoulder βE_i of the exponential function of the probability P_i is also a dimensionless quantity and $\gamma = F$ (free energy), and P_i becomes a Boltzmann distribution (canonical distribution). Then it is denoted as

$$H = -\sum_{j=1}^{W_1} P_j \log_2(P_j) = -\sum_{j=1}^{W_1} \exp(-\beta(E_j - \gamma)) \log_2(\exp(-\beta(E_j - \gamma)))$$

$$= \frac{\beta}{\log_e(2)} \left(\sum_{j=1}^{W_1} E_j \exp(-\beta(E_j - \gamma)) - \gamma \sum_{j=1}^{W_1} \exp(-\beta(E_j - \gamma)) \right)$$

$$= \frac{\beta}{\log_e(2)}(\langle E \rangle - \gamma). \tag{10.41}$$

If $S = (k_B \log_e(2))H$, the Shannon entropy H obtained from a microstate is consistent with the macroscopic thermodynamic entropy S, recalling that the thermodynamic entropy $S = (E - F)/T$. This $(k_B \log_e(2))H$ is the Boltzmann entropy S_B. Returning to the microcanonical ensemble, the total number of microstates is W and the probability of taking each microstate is equal to $P = 1/W$, so that the entropy is maximum and the famous expression of

$$S_B = -k_B \log_e(2) \sum_{j=1}^{W} \left(\frac{1}{W} \right) \log_2 \left(\frac{1}{W} \right)$$

$$= k_B \log_e(2) \cdot \log_2(W) \sum_{j=1}^{W} \left(\frac{1}{W} \right)$$

$$= k_B \log_e(W) \tag{10.42}$$

is obtained. Of course, this is not a derivation of statistical mechanics. It does not demonstrate how the state of an equal weight ratio is imposed from physics. What is shown here is the consistency with thermodynamics when the preceding conditions are imposed.

Exercise 10.6.1 In Section 9.4, we considered polymers with the sequences of certain N monomers. We assumed that only a certain sequence has a function. We then investigated if such a sequence can be preserved by selection, and stated that an error catastrophe occurs when the mutation rate is high. In this example, one particular type of sequence is selected from among the possible 2^N sequences, and this sequence is considered to have some information. Find the Shannon entropy in this case, and explore the relationship between it and the selection pressure required to maintain it (ratio of fitness, $s = a_m/a_\ell$). Rewrite the error catastrophe condition in relation to the ratio s and Shannon entropy. In addition, discuss the relationship between the error catastrophe and the thermodynamic phase transition by mapping the degree of fitness to the energy and the mutation rate to the temperature, based on the correspondence between Shannon entropy and thermodynamic entropy. Here recall that the phase transition in thermodynamics can be described by considering the minimization of the free energy $E - TS$ (where E is the internal energy, T is the temperature, and S is the entropy) as a change from a low-energy state to a high-entropy state with the increase of temperature T.

10.7 Demon and Information

In the previous section, we discussed the relation between statistical entropy and information. Since statistical entropy and thermodynamic entropy are connected, it should be possible to discuss the connection between information and thermodynamic entropy. Historically, the connection between thermodynamic entropy and information was pointed out by Szilard, prior to Shannon. This is a groundbreaking argument that links Maxwell's demon with information. Now, we first discuss this demon.

The second law of thermodynamics is a fundamental law of physics that expresses the irreversibility of macroscopic phenomena. It is expressed as the impossibility of the second kind of perpetual motion, that is, the impossibility of converting the heat into work from a single heat bath with one temperature without leaving any other change. It was formulated in the form that entropy does not decrease in isolated systems.

Maxwell's demon was proposed, which might seemingly violate the second law paradoxically, by considering certain microscopic operations. Consider, for example, a partition that divides a box into left and right compartments. Initially, suppose that the whole box is in equilibrium and the temperature of the left and right sides is equal. Suppose that a small door was set on the partition and the demon located beside the door, makes operations to let only the fast-moving particles pass to the right, and slow-moving ones to the left. Then, the temperature on the right side of the door would gradually increase, and as a result, a temperature difference between the two sides would be generated, and the entropy would decrease. Also, one can extract work by using the induced temperature difference. At first glance, this produces a paradox because the entropy would decrease in an isolated system.

Of course, this is a false paradox because the increase in the entropy of the demon is not considered. Here, the demon obtains information by measuring the motion of a molecule, which gives the information and the demon makes a choice of operations based on it. In other words, the demon performs the process of obtaining information to choose the action between opening and closing a door. In this process of (observation)-(information)-(operation), the demon generates heat and so entropy increases, which neutralizes the decrease in the entropy generated by the temperature difference.

Szilard was the first to clarify this point and noted the relationship between information and entropy by introducing a different version of Maxwell's demon. He considered an idealized situation in which a box (volume V_0) in a heat bath at temperature T contains an ideal gas of one particle. Suppose a partition is infinitely light and thin. This partition is inserted at a certain stage. Then, from the side with a particle, pressure P is applied to the partition, so that the partition is moved to the

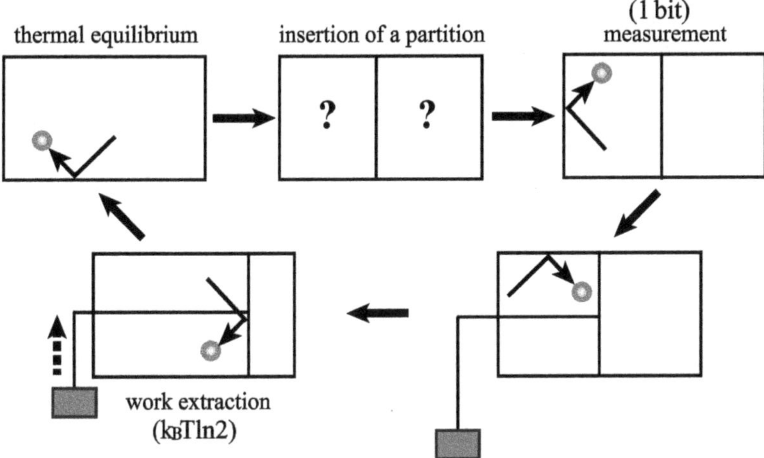

Figure 10.7 Schematic diagram of Szilard engine.

edge of the box. Depending on whether the particle is on the left or the right side, a weight is attached to the partition on the side of the particle. Then the work by the particle that drives the partition to the edge is given as

$$\int_{V_0/2}^{V_0} PdV = k_B T \int_{V_0/2}^{V_0} \frac{dV}{V} = k_B T \log_e(2). \tag{10.43}$$

Since the partition mass and thickness are infinitely small, the box returns to its original state before the partition was inserted. This operation can be repeated. Therefore, it appears that the system returns to its original state after getting the work from a heat bath of one temperature T. This is Szilard's engine (Fig. 10.7). If one can get heat from a heat bath of one temperature T and convert that heat to work without changing anything else, this would break the second law of thermodynamics.

Of course, this is not actually the case. The demon puts in a partition and puts in a weight according to the position of the particle. Here, the demon gains the information of the particle's position and rewrites its own state based on the information to perform the operation. Then, the demon is not returned to its original state. To return, the demon needs to erase the information. The second law of thermodynamics says that the entropy must increase at some point in the cycle where the demon returns to its original state after completing the series of (observation)-(state change)-(operation)-(return), and that the entropy is generated at that point. Where the entropy generation occurs depends on the specific way in which this demon operates. However, what is important is that this sequence of actions always produces more heat than the work that can be extracted.

Exercise 10.7.1 Consider a specific setting for the demon in the examples of Maxwell and Szilard so that the demon returns to the original state after the manipulation, and show that entropy is actually generated. Compare various theories on this.

Here, we can consider that this demon obtains information about whether a particle is on the left or the right, converts it into free energy, and extracts the work. Then, since whether the particles exist on the left or right is equally probable, this demon obtains $\log_2(2) = 1$ [bit] information and makes use of it. On the other hand, if the heat generated in this process is divided by T, an entropy of at least $k_B \log_e(2)$ is generated. In other words, the demon operates by making a choice among a bundle of possible actions, but this choice is made according to the possible states of the system. The demon creates "correlations," which did not exist in the thermal equilibrium, by manipulating the system depending on the state of the system. In the Maxwell version, it is the correlation between the speed of the particle and the opening or closing of the door, and in the Szilard version it is the position of the particle and the position of the weight. In doing so, directional change is generated that cannot exist in equilibrium. This is the result of selective operation by the demon. This requires interaction between the system and the demon, and as a result, heat is generated when the demon returns to its original state.

Szilard showed the relationship between information and thermodynamic entropy [21]. Later, when Shannon formulated information entropy, this relationship was revealed. In other words, Szilard showed the possibility of extending the second law by including the demon's gain of the information derived from the thermodynamic entropy. At this time, the demon creates a "correlation" between the state of the demon and its surroundings by operation, so that the correlation can be expressed in terms of mutual information. Szilard's proposal is now formulated elegantly by using recent advances in so-called stochastic thermodynamics [16]. Szilard's engine was recently implemented experimentally in a one-electron experiment [8].

An organism receives information about the state of the external world and acts accordingly. For example, in chemotaxis, cells perceive the concentration of an external chemical component and moves in a specific direction. In cell proliferation and differentiation, information received by membrane sensors is transmitted through the signal transduction system, and gene expression is changed accordingly. In this way, the (observation)-(information)-(operation) linkage is fundamental to the response of an organism. In this sense, the Maxwell demons, in particular, their analysis in terms of the information, will provide important perspectives for considering biological phenomena.

Let us compare the machine with the biological system. For instance, there is a marked difference between the operating energy of a man-made computer and that of a biological cell [23]. In a computer, the transistor, which is the functional element, is designed to operate accurately by injecting a large amount of power energy (order of $10^8\,k_{\mathrm{B}}T$) for 1 [bit] of information processing. Under such energy cost, the computer operates with high precision even though it generates high heat, but it is vulnerable to noise and environmental fluctuations. A computer is, so to speak, a system to achieve a definite goal quickly in a well-prepared environment, while consuming a large amount of energy.

On the other hand, in a cellular system, protein molecules, which are the functional elements of a cellular system, fluctuate stochastically at an energy level of the order of $10\,k_{\mathrm{B}}T$. In this respect, there is similarity to the picture of a demon working stochastically in a heat bath at temperature T. As collective action of these molecules, the cellular system operates with low precision but generating only low heat, and is robust to noise and environmental changes. The cellular system is, so to speak, a system that can operate in a fluctuating environment while saving operating energy, whereas its purpose is not clearly prescribed. This thermodynamic difference in information processing underlines the distinction between artificial and biological systems [20], and for the latter to function, the viewpoint on Maxwell's demon will be important.

For example, as mentioned in Section 5.2, molecular motors such as myosin, kinesin, and RNA polymerase perform directional motion using about $20\,k_{\mathrm{B}}T$ of free energy for the hydrolysis of ATP and GTP under physiological conditions, and the process operates with biased Brownian motion (Brownian ratchet; a ratchet is a claw wheel device that rotates only in one direction). In this respect, it is thought that the unidirectional motion is created by making a correlation between the state of the molecular motor and its motion within the heat bath. Hence, one may consider Maxwell's demon as being implemented through evolution.

At the beginning of this chapter, we mentioned bioinformatics. As mentioned therein, the current bioinformatics has a focus on statistical analysis of DNA sequences. However, for a DNA sequence to act as information, it must be read through processes controlled by chemical reactions. Furthermore, to act as genetic information, the DNA sequence must be replicated. As discussed in Section 9.9, only molecules minority in number carry genetic information in this situation, whereas many other reactions are controlled by such minority molecules. It is possible that such molecules may act as demons in this situation. The kinetic proofreading in Section 10.5 is also considered from this point of view, because demons operate by using energy to move away from the equilibrium distribution to make correlation. Thus, it will be important to consider how demons work and are replicated under thermal fluctuations in biological information.

10.8 Dynamics and Information

A perspective that is not adequately addressed in the discussion of information in this chapter is its relationship to dynamics, which is a major theme throughout this book. The recent studies that discuss the relationship between information and entropy with the use of stochastic thermodynamics assumes interaction with the equilibrium heat bath. The obtained relationship is a statistical result after averaging out the processes in time. In the future, discussion that incorporates dynamics will be necessary. For example, Wiener, the founder of cybernetics, stated that the transient state-change of an enzyme acts like Maxwell's demon, and mentioned that even though the second law of thermodynamics cannot be violated on average over a long period of time, still, for the transient time, apparent violation might exist. If it takes a long time for the enzyme itself to fall into equilibrium, the generation of heat in the "demon" can be postponed for a while [26]. The possibility of performing information processing by the emergence of slow changes will be an important issue to be discussed in the future.

Indeed, the dynamic information processing will be important in the signal information processing discussed in Chapter 3 and in the problems of cell–cell interaction and cell differentiation discussed in Chapter 7. On the other hand, the adaptive process by attractor selection mentioned in Chapter 6 can also be considered as a demon in the stochastic dynamical system, since the state selection in one direction is made through noise.

It will then be necessary to consider a series of operations of demons, such as acquisition of information, information processing, and state selection by the information processing, in terms of the dynamical systems. In general, in the nonequilibrium process in a cell, correlations between elements and processes are created and mutual information is generated.

In cells, the time scales of each element or process are diverse. Then, slow processes or elements can sample fast changes and respond accordingly, which in turn influences the fast processes. In other words, these slow elementary processes can be considered to make correlations and control other processes by using them. In that sense, such slow processes can be considered to work as a demon. Of course, in this case, the fast changes are not thermal noise. The degree to which the fast change can be eliminated to be regarded as noise needs to be examined. In order to solve this problem, it is necessary to take into account the interference between the processes with multiple time scales, and to study how elements with slower dynamics control dynamical systems of fast elements with many degrees of freedom. If this can be formulated, it will provide a fresh view to understand life as information dynamics, which goes beyond the framework to just put the system into an equilibrium heat bath.

References

[1] S. V. Buldyrev *et al.* Long-range correlation properties of coding and noncoding DNA sequences: GenBank analysis. *Physical Review E,* **51**(5): 5084–5091, 1995.

[2] J. W. Drake. A constant rate of spontaneous mutation in DNA-based microbes. *PNAS,* **88**(16): 7160–7164, 1991.

[3] The FANTOM Consortium, RIKEN Genome Exploration Research Group and Genome Science Group. The transcriptional landscape of the mammalian genome. *Science,* **309**(5740): 1559–1563, 2005.

[4] P. Hogeweg. The roots of bioinformatics in theoretical biology. *PLOS Computational Biology.* **7**(3): e1002021, 2011.

[5] J. J. Hopfield. Kinetic proofreading: A new mechanism for reducing errors in biosynthetic processes requiring high specificity. *PNAS,* **71**(10): 4135–4139, 1974.

[6] International Human Genome Sequencing Consortium. Initial sequencing and analysis of the human genome. *Nature,* **409**: 860–921, 2001.

[7] K. Kaneko and I. Tsuda. *Complex Systems: Chaos and Beyond: A Constructive Approach with Applications in Life Sciences.* Berlin, Springer, 2000.

[8] J. V. Koski *et al.* Experimental realization of a Szilard engine with a single electron. *PNAS,* **111**(38): 13786–13789, 2014.

[9] W. Li and K. Kaneko. Long-range correlation and partial $1/f^\alpha$ spectrum in a noncoding DNA sequence. *Europhysics Letters,* **17**: 655–660, 1992.

[10] W. Li, T. G. Marra and K. Kaneko. Understanding long-range correlations in DNA sequences. *Physica* D **75**: 392–416, 1994.

[11] T. W. McKeithan. Kinetic proofreading in T-cell receptor signal transduction. *PNAS,* **92**(11): 5042–5046, 1995.

[12] J. Ninio. Kinetic amplification of enzyme discrimination. *Biochimie,* **57**: 587–595, 1975.

[13] S. Ohno. *Evolution by Gene Duplication.* New York, Springer, 1970.

[14] F. Oosawa. Spontaneous activity of living cells. *BioSystems,* **88**: 191–201, 2007.

[15] C.-K. Peng *et al.* Long-range corrclations in nucleotide sequences. *Nature,* **356**: 168–170, 1992; Mosaic organization of DNA nucleotides. *Physical Review Journals E,* **49**(2): 1685–1689, 1994.

[16] T. Sagawa. Thermodynamics of information processing in small systems. *Progress of Theoretical Physics,* **127**(1): 1–56, 2012.

[17] C. E. Shannon. *Prediction and Entropy of Printed English.* Bell system technical journal, Wiley Online Library, 1951. DOI: https://doi.org/10.1002/j.1538-7305.1951.tb01366.x

[18] R. Shaw. *The Dripping Faucet as A Model Chaotic System.* Santa Cruz, Aerial Press, 1984.

[19] J. M. Smith. The concept of information in biology. *Philosophy of Science,* **67**: 177–194, 2000.

[20] T. Yanagida *et al.* Brownian motion, fluctuation and life. *BioSystems,* **88**: 228–242, 2007.

Books for general reference:

[21] L. Brillouin. *Science and Information Theory.* Mineola, Dover Publications, 2004.

[22] T. M. Cover and J. A. Thomas. *Elements of Information Theory.* Wiley Series in Telecommunications and Signal Processing, 2006

[23] R. P. Feynman and A. Hey. *Feynman Lectures on Computation.* Boulder, Westview Press, 2000.

[24] H. Leff and A. F. Rex. *Maxwell's Demon 2 Entropy, Classical and Quantum Information, Computing*, Boca Raton, CRC Press, 2002.

[25] C. E. Shannon and W. Weaver. *Mathematical Theory of Communication*. Champaign, University of Illinois Press, 1963.

[26] N. Wiener. *Cybernetics*: *Or Control and Communication in the Animal and the Machine*. New Orleans, Quid Pro LLC, 2013.

Index

For EU product safety concerns, contact us at Calle de José Abascal, 56–1°,
28003 Madrid, Spain or eugpsr@cambridge.org.